Advanced Materials for the Conservation of Stone

Majid Hosseini • Ioannis Karapanagiotis
Editors

Advanced Materials for the Conservation of Stone

 Springer

Editors
Majid Hosseini
Department of Manufacturing and Industrial
Engineering
The University of Texas Rio Grande Valley
Edinburg, TX, USA

Ioannis Karapanagiotis
Department of Management and
Conservation of Ecclesiastical Cultural
Heritage Objects
University Ecclesiastical Academy
of Thessaloniki
Thessaloniki, Greece

ISBN 978-3-319-72259-7   ISBN 978-3-319-72260-3   (eBook)
https://doi.org/10.1007/978-3-319-72260-3

Library of Congress Control Number: 2017963362

Printed on acid-free paper

This Springer imprint is published by Springer Nature
The registered company is Springer International Publishing AG
The registered company address is: Gewerbestrasse 11, 6330 Cham, Switzerland

Preface

This book identifies novel advanced materials that can be utilized as protective agents for the preservation of stone. Biomimetic, superhydrophobic/superoleophobic, water-/oil-repellent coatings, antifungal coatings, anti-graffiti coatings, and photocatalytic, self-cleaning coatings, as well as advanced consolidants and cleaning agents, are some examples of the advanced nanostructured materials, which provide promising avenues for conservation purposes. These new, innovative methodologies are compatible with the treated substrates' physicochemical characteristics while also minimizing the potential for stone alteration. Therefore, the use of these compatible advanced materials is expected to mitigate undesirable effects while granting long-term stability and reducing the cost of restoration interventions. These solutions to stone conservation issues result in an increase in sustainability, a reduction in environmental impact, and may provide several social, ecological, and economic benefits. This book provides an overview of the recent trends and progress in advanced materials applied to stone protection, explores the scientific principles behind these advanced materials, and discusses their applications to different types of stone preservation efforts. Essential information, as well as knowledge on the availability and applicability of advanced nanostructured materials, is provided, focusing on the practical aspects of stone protection. The book highlights the interdisciplinary efforts regarding novel applications of nanostructured materials in advancement of stone protection, with particular emphasis on expected developments in the field. Currently, adherence to traditional methods and conventional materials in the conservation of cultural heritage items is often incompatible with the original works of art and lacks both durability and performance. New, advanced nanostructured materials are being designed and developed with the aim of being chemically, physically, and mechanically compatible with stone. Their physicochemical, morphological properties, eco-toxicity, mechanisms of degradation, and functionality are considered in this book. The authors present a thorough overview of cutting-edge discoveries and recent technological developments, which depict breakthroughs in novel nanomaterials and utilization strategies for applications in cultural heritage. They also address the current status and future outlook of the topic regarding a wide range of global issues.

Chapter 1 discusses the applications of superhydrophobic and water-repellent coatings for the protection of monuments and other stone objects of cultural heritage. The chapter presents an overview of the methods to induce enhanced hydrophobicity and water repellency to natural stone. The fundamental concepts on the wettability of solid surfaces are briefly described in this chapter. A case study of siloxane-nanoparticle dispersions sprayed on sandstone and marble specimens is presented. This study demonstrates the superhydrophobic and water-repellent properties of the deposited polysiloxane-nanoparticle composite coatings. How to achieve superhydrophobicity and water-repellent characteristics by using exclusively aqueous products and inherent hydrophilic materials, accompanied by superoleophobicity and oil repellency, is also discussed. Finally, this chapter provides an evaluation of the coatings' effects on the color, vapor permeability, and water absorption by capillarity of the treated sandstone and marble.

Chapter 2 presents a detailed design of an advanced fracture mechanics-based join repair using adhesives, pins, and a protective adhesive to enhance the join's mechanical performance. A pinning design is laid out, focusing on material selection, study of the stresses in the substrate and pin, and failure modes. Such a design is meant to preserve the integrity and longevity of Adam.

Chapter 3 gives an overview of nanostructured protective treatments based on the use of SiO_2, TiO_2, ZnO, and Ag nanoparticles to confer superhydrophobic, self-cleaning, and antifouling properties to marble surfaces. Particular attention is focused on the development of photocatalytic nanoTiO$_2$-based treatments and its advantages, drawbacks, and critical issues. The most recent advances using modified innovative TiO_2 nanoparticles in dispersion and as nanocomposites are also described in this chapter. Finally, a comparison of the results obtained in controlled lab conditions and on real deteriorated surfaces is presented.

Chapter 4 presents a hybrid consolidant with high affinity to carbonaceous substrates by adding synthesized nano-hydroxyapatite into TEOS sol, including amylamine ($CH_3(CH_2)_4NH_2$) as a surfactant. The use of such a hybrid consolidant provides an efficient means of protecting gels from cracking by reducing the capillary pressure. The chapter also examines the role of hydroxyapatite and amylamine in the silica structure by comparing the synthesized consolidant with other nanocomposites containing TEOS, TEOS and hydroxyapatite, and TEOS and amylamine as basic reagents. The synthesized products have been characterized and evaluated for their effectiveness as strengthening agents on limestone. The chapter concludes that nanocomposite-treated stone showed an improvement in hygric properties, drilling resistance, and tensile strength due to the crack-free structure of the nanocomposite.

Chapter 5 discusses an integrated methodological approach for the selection of restoration mortar for optimal compatibility and performance in conservation/restoration interventions. Three case studies are described to illustrate the methodological approach: the Byzantine Monastery of Kaisariani in Athens, Greece; the traditional bridge of Plaka in Epirus, Greece; and the Holy Aedicule in Jerusalem.

Chapter 6 highlights the use of inorganic nanoparticles for the consolidation and antifungal protection of stone heritage. A brief overview on the main synthesis methods and the most common analytical techniques employed for the physico-chemical characterization of nanoparticles are presented. The chapter discusses the factors influencing the success of the nanomaterials as protective agents. Finally, the loss of stone cohesion and biodeterioration as two of the most common issues that affect stone substrates are highlighted in this chapter.

Chapter 7 addresses the main achievements in the field of nanomaterials applied to stone consolidation, discussing the principles that underpin materials development and application to artifacts. Consolidation systems comprise both inorganic (e.g., dispersions of alkaline earth hydroxide nanoparticles) and hybrid nanomaterials (e.g., organic-inorganic silica gels) to account for the preservation of carbonate and sandstone.

Chapter 8 discusses laboratory verification tests for determining the efficiency of various consolidation treatments on stone. This chapter introduces a portable, ultrasonic, double-hole probe for measuring material properties along a depth profile and assessing penetration depth in the near-surface material layer between two drilled holes. The information in this chapter provides restorers with a unique capability to follow the consolidation progress during repetitive impregnation and/or agent maturing.

Chapter 9 describes drawbacks and challenges of carbonate stone consolidation. There is a focus on important phenomena relating to the carbonate media particularities (possible disturbance of sol-gel routes and lack of strong chemical bond with calcite), issues related to extended hydrolysis reactions, and susceptibility of most common alkoxysilane-based products to crack. An in-depth discussion is also provided on novel alkoxysilane-based products that have shown potential to treat carbonate stones. The chapter further explains limitations and constraints, as well as new proposals to consolidate porous carbonate stones with new alkoxysilane-based formulations that are able to provide consolidation actions at different degrees and at major or minor depths.

Chapter 10 discusses various forms of stucco degradation processes and their significance, external factors of degradation, methods, laboratory techniques used for the analysis of stucco and its degradation products, and diagnostic techniques for stucco damage. The chapter gives an overview of treatment methods with nanomaterials (hydroxyapatite and derivatives, calcium and magnesium hydroxides) that are able to offer stucco consolidation. Examples of stucco models prepared and studied in the laboratory and treated with nanoparticles are also provided in this chapter.

Chapter 11 provides an overview of nanoproducts for treatments against biological growths. In this chapter, results of tests with nanoproducts on natural stone and other porous inorganic materials that have data on the actual effect on biocoatings are discussed while considering the impacts on the materials. The environmental impact of nanoparticles and their effects are also explored in this chapter.

Chapter 12 presents the development of an innovative sol-gel route for preserving cultural heritage stonework. Specifically, the chapter discusses a surfactant-assisted sol-gel synthesis to produce, in situ on the stonework, crack-free nanomaterials to be used as long-term consolidants. Additionally, hydrophobic, water-repellent, self-cleaning, and biocidal properties that can be incorporated into the product by innovative chemical modifications of the proposed synthesis route are presented.

Chapter 13 discusses applications of nanoparticle-based materials in the conservation of cultural heritage for their consolidating and self-cleaning abilities. The chapter provides an extended literature survey on testing procedures and the antimicrobial properties and effectiveness of nanomaterials and major types of nanoparticles (TiO_2, Ag, ZnO, CuO), as well as their limitations and advantages for applications in built cultural heritage. Recommendations for new research directions to further investigate the antimicrobial effectiveness of nanomaterials are provided in this chapter.

Chapter 14 highlights the configuration of assessment criteria, methodology compilation, and the strategic planning of cleaning interventions applied on architectural surfaces of monuments. The chapter also presents an integrated decision-making system to assess cleaning interventions on stone architectural surfaces that show the characteristic decay pattern of black crusts. The developed integrated decision-making system utilizes a fuzzy logic model that is incorporated into GIS thematic maps which depict decay patterns and applied pilot cleaning interventions. The authors have demonstrated their developed methodology in practice.

Finally, a number of people have helped make this book possible. We hereby acknowledge Ms. Anita Lekhwani, our senior editor; Mr. Brian Halm, our project coordinator; Ms. Faith Pilacik, our editorial assistant, at Springer Science+Business Media, respectively; Mr. Murugesan Tamilselvan, our project coordinator from SPi Global; Ms. Megan Rohm; and all contributors and reviewers, without whose contributions and support this book would not have been written. We thank you all for the excellent work and assistance that has been provided in moving this book project forward.

Edinburg, TX, USA   Majid Hosseini
Thessaloniki, Greece   Ioannis Karapanagiotis
Fall 2017

Contents

About the Editors

Majid Hosseini has earned both his Ph.D. and M.S. degrees in chemical engineering from *The University of Akron* in Ohio, USA. He has also completed an MSE degree in manufacturing engineering at UTRGV in Texas, USA, and a bachelor's degree in chemical engineering at *Sharif University of Technology* in Tehran, Iran. He has edited high-caliber books and book chapters, authored multiple research articles, and coinvented patent application technologies. He has served as a key speaker at national and international conferences and has been actively engaged in technology development. Dr. Hosseini's research interests, expertise, and experiences are diverse, ranging from smart bio-/nanomaterials, smart polymers and coatings, nanoparticles, and bio-/nanotechnology to bioprocess engineering and development, biomanufacturing, biofuels and bioenergy, and sustainability. Dr. Hosseini works at *The University of Texas Rio Grande Valley* in Edinburg, Texas, USA.

Department of Manufacturing and Industrial Engineering, The University of Texas Rio Grande Valley, Edinburg, TX, USA

Ioannis Karapanagiotis has obtained his Ph.D. in materials science and engineering from the *University of Minnesota*, USA, and his diploma in chemical engineering from the *Aristotle University of Thessaloniki*, Greece. He serves as a member in editorial boards and reviewer in several journals (more than 70), and he has published multiple research papers (more than 130) in peer-reviewed journals, books, and conference proceedings. Dr. Karapanagiotis specializes in interfacial engineering and its applications on the protection and conservation of cultural heritage and in the physicochemical characterization and analysis of cultural heritage materials which are found in historic monuments, paintings, icons, textiles, and manuscripts. Dr. Karapanagiotis is an associate professor and head of the *Department of Management and Conservation of Ecclesiastical Cultural Heritage Objects, University Ecclesiastical Academy of Thessaloniki*, Greece.

Department of Management and Conservation of Ecclesiastical Cultural Heritage Objects, University Ecclesiastical Academy of Thessaloniki, Thessaloniki, Greece

Chapter 1
Superhydrophobic Coatings for the Protection of Natural Stone

Ioannis Karapanagiotis and Majid Hosseini

1.1 Introduction

Conservation and protection of stone and historical buildings from extreme environmental conditions that affect the stones are important. By utilizing polymer coatings capable of repelling water on the stone surfaces, their degradation may be prevented [1]. Although polymer coatings have been widely utilized for stone protection, their effectiveness mainly relies on the substrate's characteristics (e.g., porosity and roughness) [2]. In order to effectively protect and conserve monuments of cultural heritage, it is important to develop new methodologies that can help to improve the polymer coatings' hydrophobicity [2]. In general, hydrophobicity is governed by the interfacial tension between the solid and the liquid and the geometrical characteristics of the surface. Specific examples of hydrophobic coatings utilized as protective agents for stone and monuments' preservation include fluorinated materials, organosilicon compounds, vinyl polymers, and acrylic polymers [3] as well as acrylic resins [4], fluoropolymers including perfluoro-polyethers [5], and partially fluorinated acrylic copolymers [6] to name a few.

Moreover, hybrid coatings have also been utilized as a means of protection. In this regard, nanoparticles (NPs) (e.g., SiO_2, Al_2O_3, SnO_2, and TiO_2) have widely been applied to create nanocomposites with better properties for stone protection. These coatings will help to decrease the porous materials' wettability and potentially reduce water penetration into stone while minimizing the decomposition rate

I. Karapanagiotis (✉)
Department of Management and Conservation of Ecclesiastical Cultural Heritage Objects,
University Ecclesiastical Academy of Thessaloniki, Thessaloniki, Greece
e-mail: y.karapanagiotis@aeath.gr

M. Hosseini (✉)
Department of Manufacturing and Industrial Engineering, The University
of Texas Rio Grande Valley, Edinburg, TX, USA
e-mail: majid.hosseini01@utrgv.edu

© Springer International Publishing AG 2018
M. Hosseini, I. Karapanagiotis (eds.), *Advanced Materials for the Conservation of Stone*, https://doi.org/10.1007/978-3-319-72260-3_1

and increasing the surface roughness, which result in the creation of a superhydrophobic and water-repellent surface characteristic [7].

Although a mixture of roughening the surface and reducing the surface free energy has been used in the development of superhydrophobic materials, the reported methods in literature typically involve the utilization of multiple processes that have hindered their large-scale adoption for protection of stone. Using superhydrophobic treatments will help to preserve stone by preventing water penetration and agents of decay from entering into stone's pore structure. However, synthesizing superhydrophobic materials to be used as protective agents and consolidants is considered a challenging task. The applicability and long-term durability of the newly developed nanostructured materials as a fundamental requirement for stone protection are typically explored using advanced analytical techniques in order to optimize these developed technologies. The ongoing progress in advanced materials may help to maximize the effectiveness of their application in stone protection, easing the adoption of such technology to the end user. The urgency in finding creative, efficient, and sustainable solutions for environmentally conscious stone protection is considered in this chapter and briefly reviewed.

This chapter discusses superhydrophobic coatings which can be used for the surface protection of natural stone, used as building materials in monuments and other outdoor objects of the cultural heritage. These protective coatings possess furthermore the ability to repel water drops and to remain completely dry after rainfall (i.e., the net interaction between a superhydrophobic surface and a water drop is repulsive). Consequently, these superhydrophobic and water-repellent coatings can, in principle, offer good protection to natural stone against the degradation effects of rainwater.

In particular, the chapter begins (Sect. 1.2) with some fundamental concepts and equations related to the wettability of solid surfaces along with describing the key findings of some important reported studies. Previously published methods, which were devised to induce enhanced hydrophobicity and water repellency to natural stone, are reviewed in the next section (Sect. 1.3). In Sect. 1.4, a case study is described: polysiloxane-nanoparticle coatings are deposited on sandstone. Using NPs in appropriate concentration, superhydrophobicity and water repellency are achieved. The discussion is extended to marble surfaces. Finally, the conclusions are summarized in Sect. 1.5.

1.2   Superhydrophobicity: A Brief Review

The interaction between a solid surface and a water drop has been studied systematically first by T. Young [8]. Consider a drop of water resting on a horizontal, passive, and atomically smooth surface (Fig. 1.1a). Neglecting the effect of gravity, the exact shape of the liquid-water drop is specified by the Young equation which is a force balance equation:

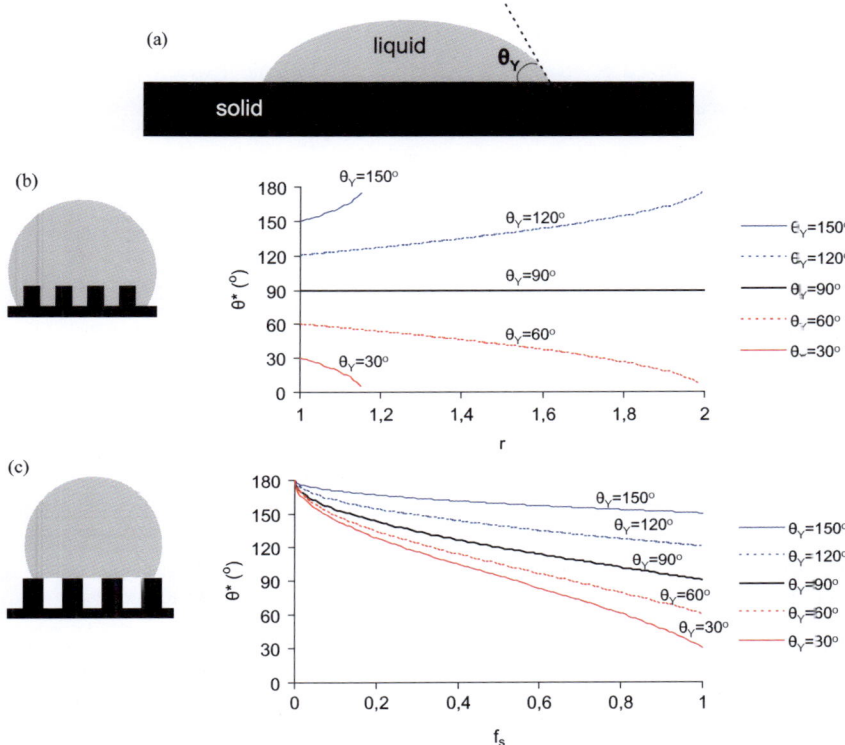

Fig. 1.1 (a) Definition of the Young contact angle (θ_Y) on a smooth surface. (b) The Wenzel scenario where a water drop follows the roughness of the underlying substrate—homogeneous wetting regime. The plot shows the apparent contact angle θ^* (Eq. 1.2) as a function of the roughness factor r for five materials with $\theta_Y = 30°$, $\theta_Y = 60°$ (inherent hydrophilic materials), $\theta_Y = 120°$, $\theta_Y = 150°$ (inherent hydrophobic materials), and $\theta_Y = 90°$. The atomically smooth surface corresponds to $r = 1$. (c) The Cassie-Baxter scenario where a water drop sits on a "composite" surface of air and solid—heterogeneous wetting regime. The plot shows the apparent contact angle θ^* (Eq. 1.3) as a function of the solid fraction f_s for five materials with $\theta_Y = 30°$, $\theta_Y = 60°$ (inherent hydrophilic materials), $\theta_Y = 120°$, $\theta_Y = 150°$ (inherent hydrophobic materials), and $\theta_Y = 90°$. The atomically smooth surface corresponds to $f_s = 1$

$$\gamma_{sv} = \gamma_{lv} \cos\theta_Y + \gamma_{sl} \tag{1.1}$$

where γ_{sv}, γ_{lv}, and γ_{sl} are the solid-vapor (sv), liquid-vapor (lv), and solid-liquid (sl) interfacial surface tensions, respectively, and θ_Y the equilibrium-Young contact angle, defined in Fig. 1.1a. Inherent hydrophilic surfaces correspond to $\theta_Y < 90°$, and inherent hydrophobic surfaces are defined by larger contact angles ($\theta_Y > 90°$). The Young contact angle (θ_Y) is influenced exclusively by the intermolecular interactions; the effect of surface roughness in the wettability of the solid surface is not taken into account. The role of surface roughness was first introduced, in a quantitative way, by R.N. Wenzel [9].

Wenzel recognized that there is a distinction between the total surface of a drop-solid interface and the geometric surface, which is the surface as measured in the plane of the interface [9] (Fig. 1.1b). Hence, Wenzel defined a roughness factor (r) as the ratio of the actual versus the geometric surface. This dimensionless factor, r, is a number larger than unity and equals unity only for ideally flat surfaces. According to Wenzel, the apparent contact angle of a water drop on a rough surface (θ^*) is related with θ_Y through the roughness factor (r) as follows:

$$\cos\theta^* = r\cos\theta_Y \tag{1.2}$$

Some years after the pioneering work of R.N. Wenzel, A.B.D. Cassie and S. Baxter studied the wettability of porous surfaces [10]. The term "porous surfaces" was intentionally used by the two researchers to clearly emphasize that, according to their concept, water cannot fully penetrate and wet the whole surface, as it was the case in the Wenzel scenario. In the Cassie-Baxter scenario, air is trapped between the drop and the surface pores of the substrate (Fig. 1.1c). Therefore, the drop only contacts the solid through the top of the asperities, on a fraction that is usually denoted as f_s. This dimensionless open porosity factor f_s is a number smaller than unity and equals unity only for ideally flat surfaces. According to Cassie and Baxter, the apparent contact angle of a water drop on a porous surface (θ^*) is related with θ_Y through the factor (f_s) as follows:

$$\cos\theta^* = -1 + f_s\left(\cos\theta_Y + 1\right) \tag{1.3}$$

It is noteworthy that from a thermodynamic point of view, only the Wenzel state is a stable state, while the Cassie-Baxter scenario describes a metastable state. Hence, a transition from the Cassie-Baxter state to the stable Wenzel state is often observed. Both Eqs. 1.2 and 1.3 can predict large contact angles θ^*, even larger than 150°, which is usually taken as the threshold for superhydrophobicity. In the Wenzel model, however, θ^* can become larger than 150° only for rough inherent hydrophobic materials ($\theta_Y > 90°$), as shown in the plot of Fig. 1.1b. The Wenzel model suggests that roughness increases the hydrophobic or hydrophilic characteristics of a surface. Considering that $r > 1$, Eq. 1.2 suggests that $\theta^* < \theta_Y$ for inherent hydrophilic ($\theta_Y < 90°$) materials, while $\theta^* > \theta_Y$ for inherent hydrophobic ($\theta_Y > 90°$) materials, as demonstrated in the plot of Fig. 1.1b. In the Cassie-Baxter model, θ^* of an either inherent hydrophilic or hydrophobic material increases with roughness (i.e., surface porosity) and can therefore become $\theta^* > 150°$ for any material, as shown in the plot of Fig. 1.1c. Consequently, the Cassie-Baxter model suggests that roughness enhances the hydrophobic character of an either inherent hydrophilic or hydrophobic surface. Considering that $f_s < 1$, Eq. 1.3 suggests that $\theta^* > \theta_Y$ for either inherent hydrophilic ($\theta_Y < 90°$) or hydrophobic ($\theta_Y > 90°$) materials, as demonstrated in the plot of Fig. 1.1c.

Fig. 1.2 Advancing θ_A and receding θ_R contact angles of a drop on a tilting surface

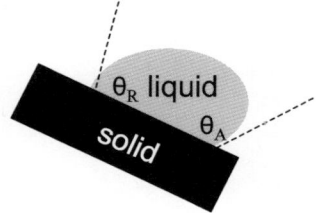

The static contact angle (θ^*), which describes the contact of a drop on a horizontal surface, does not describe to an acceptable extent the wettability of a surface. The significant role of the contact angle hysteresis, referred to as the difference of the advancing (θ_A) and receding (θ_R) contact angle (Fig. 1.2), has been recognized several years ago [11, 12], and it was clearly emphasized in 2000 by Öner and McCarthy, who described the following scenario [13]. Consider that there are two surfaces, A and B, with static contact angles $\theta(A)^* > \theta(B)^*$. If we focus exclusively on the static situations, then we reach the conclusion that surface A is more hydrophobic than surface B, because surface A corresponds to a larger static contact angle. What if when the surface B is tilted 1° from the horizontal, the drop slides off, and when the surface A is tilted to any angle, the drop remains pinned to the surface [13]? Then we must recognize that surface B is more water repellent than surface A. Consequently, in a static situation, clearly surface A is more hydrophobic, but if the goal is to obtain a water-repellent surface, the static contact angle is irrelevant; in such a case, the drop should move with very little applied force [13]. The critical line force per unit length (F) needed to start a drop moving over a solid surface is given by Eq. 1.4 as follows [14]:

$$F = \gamma_{lv} \left(\cos\theta_R - \cos\theta_A \right) \tag{1.4}$$

where F is the critical line force per unit length and θ_R and θ_A are the receding and advancing contact angles, respectively. According to Eq. 1.4, as the contact angle hysteresis ($\theta_A - \theta_R$) decreases, the critical force F needed to make the drop to move decreases. Consequently, surfaces with enhanced water repellency must correspond to small $\theta_A - \theta_R$.

In 2008, it was clearly shown that the aforementioned scenario describing the wettability of surface A is not just semantics [15]. A drop on the surface of the rose petal corresponds to $\theta^* = 152°$, implying that the natural surface is very hydrophobic, and even more it can be considered as a superhydrophobic surface ($\theta^* > 150°$). However, "a drop cannot roll off even when the surface is turned upside down, thus implying high drop adhesion" [15]. This wetting behavior has been described as the "rose petal effect" [15] and since 2008 has been recognized and emphasized by some researchers [16–19]. Clearly, the rose petal effect is a totally different wetting behavior than that observed on the surface of the lotus leaf [20] where superhydrophobicity ($\theta^* > 150°$) is accompanied by water repellency, corresponding to $\theta_A - \theta_R \sim 4°$ [21], and therefore, a drop can effortlessly roll off. This is the famous

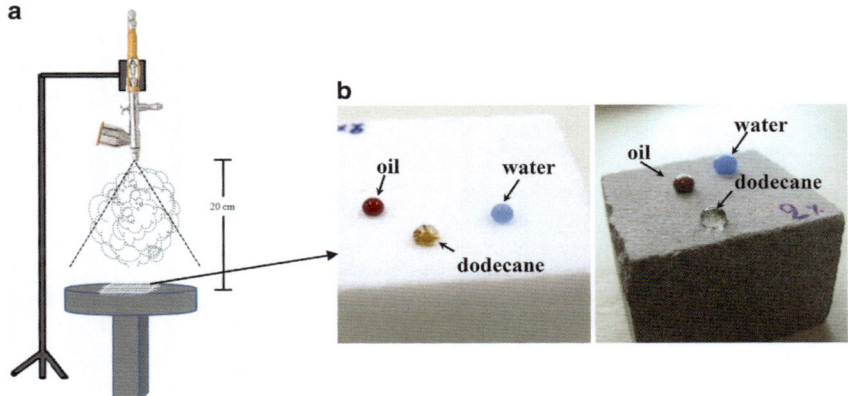

Fig. 1.3 (**a**) Schematic presentation of the method which was devised to deposit polysiloxane-nanoparticle coatings on stone (and other substrates) and to induce special wetting properties: oxide nanoparticles are dispersed in a silane/siloxane solution at appropriate concentration, and the dispersion is sprayed on the target substrate. Reprinted with permission from reference number [25], Copyright (2016), Elsevier. (**b**) Coated marble and sandstone specimens which show super-hydrophobicity and superoleophobicity. Even the contact angles of dodecane, which corresponds to very low surface tension, with the treated surfaces are very large. Reprinted with permission from reference number [26], Copyright (2016), Elsevier

"lotus leaf effect" and since 1997 [20] has been commonly reported in the literature.

1.3 Coatings of Enhanced Hydrophobicity and Water Repellency for the Protection of Natural Stone

Superhydrophobic and water-repellent coatings can, in principle, offer better protection to natural stone against the degradation effects of rainwater. The first method, devised to induce superhydrophobicity and water repellency to natural stone for the protection of monuments and other stone objects of the cultural heritage, was published in 2007 [2]. Since then, the method has been improved and applied on various natural stones offering some very promising results [1, 7, 22–26]. The method is schematically presented in Fig. 1.3a [25]. In brief, oxide nanoparticles are dispersed in an alkoxy-silane/siloxane solution, usually in an organic solvent. The concentration of the NPs plays a key role in the resulting wetting properties, as discussed in the next section. The dispersion is then applied by spray (or brush) on the target stone substrate or any other substrate (e.g., textile, metal, and other). The alkoxy-silane/siloxane undergoes spontaneously a sol-gel process under the effect of air humidity resulting in a continuous poly(alkyl siloxane) material of very high viscosity [27], enriched with NPs. The surface of the resulting polysiloxane-nanoparticle composite coating exhibits superhydrophobic and water-repellent properties, provided that an appropriate nanoparticle concentration was selected for the preparation of the dispersion. Using this method, superhydrophobic coatings were produced and deposited on marble and sandstone, as demonstrated in Fig. 1.3b [26]. The latter shows, moreover, that superhydrophobicity can be accompanied by superoleophobicity.

The aforementioned method has some important advantages which promote its potential application for the protection of historic monuments: (1) It can be used to treat large surfaces under ambient conditions, as the coatings are deposited simply by spray or brush. (2) Alkoxysilanes are extensively used in stone consolidation [27]. Consequently, materials which are already used in stone conservation are mostly included. The silanes/siloxanes are mixed with NPs which, however, are added in small amounts just to affect the morphology of the surface of the coating and therefore to adjust its wetting properties. (3) Apart from the alkoxy-silane/siloxane materials, other macromolecular organosilicon products or even solutions of organic polymers can be used to achieve superhydrophobicity and water repellency, following the method of Fig. 1.3 [2, 22, 24]. Consequently, the method is very flexible with respect to the type of the selected polymer. (4) Various oxide NPs, such as SiO_2, Al_2O_3, SnO_2, and ZnO, ranging from 7 to 150 nm [1, 2, 7, 22–26] can be used to obtain superhydrophobicity. Consequently, the method is very flexible with respect to the type and size of the selected nanoparticles.

Another interesting method was developed by Mosquera et al. [28] who have synthesized, via the "co-condensation of tetraethoxysilane (TEOS) and hydroxyl-terminated polydimethylsiloxane (PDMS)", an organically modified silicate (ormosil) when a nonionic surfactant (n-octylamine) was present. During drying, improvements in the material's flexibility and cracking prevention can be attributed to the incorporation of PDMS to the TEOS [28]. The authors reported that the nonionic surfactant had an important role in TEOS and PDMS' co-condensation, as well as during the synthesis of a homogeneous organic-inorganic (O-I) hybrid xerogel for the application as a hydrophobic treatment and consolidant [28]. The synthesized nanomaterials were reported to be effective in stone treatments and increased the stone's mechanical resistance without any adverse effect on the treated stone [28].

De Ferri et al. [29] investigated the water repellency of a thin protective coating applied to granite samples, sandstone, and limestone which consisted of alkoxysilanes and modified silica NPs from commercially available products. After being exposed to atmosphere for 4 months, a 150° static contact angle was reported for NPs at suitable concentration for each of the stone sample [29]. However, the water capillary absorption analysis showed that the coatings had a different behavior [29]. Due to their hydrophobic effect, these treatments were reported to be effective especially for granite, although their effectiveness on other stone species requires additional studies [29], particularly when the contact with water is lengthy.

A one-step method was reported for the synthesis of superhydrophobic surface coatings without using any volatile organic components for applications on stones and other building materials through spraying or brushing [30]. The devolved synthesis procedure was reported to have the potential for large-scale production due its simple operational requirement which eliminated additional processing such as heating [30]. It was reported that incorporating silica NPs to an O-I silica oligomer mixture when a surfactant was present effectively imparted superhydrophobicity to the material [30]. The synthesized NPs were also capable of producing a densely packed coating that could trap air and prevent the penetration of water into the coating, resulting in a minimized contact area between the surface and the droplet [30]. Moreover, the materials' surface free energy was reduced, yielding a large static contact angle ($\sim$150°) and a low hysteresis value ($\sim$7°) which showed effective water repellency [30]. The newly developed nanocomposite, when used as a protective agent, could make a homogeneous coating with a quantifiable hydrophobic and repellency property [30].

Functionalized polyolefins have also been synthesized for the protection of stone [31]. In this study [31], there were two copolymerizations of ethylene (10-undecen-1-ol (UDO) and 7-methyl-1, 6-octadiene (MOD)) which were catalyzed by the Ziegler-Natta metallocene/methylaluminoxane (MAO) catalyst. Evaluation of the color change, photooxidative stability, and the water repellency of the treated stone showed an effective performance compared to that of Paraloid B-72 (an acrylic resin used as consolidant) [31]. The ethylene/UDO copolymer with a crystalline character and a low solubility was reported to keep its protective efficiency due to its stability to photooxidative degradation [31]. The authors also reported that the functionalization through hydro-esterification enhanced the compatibility and adhesion of the film to the stone surface, as well as stability to weathering conditions [31]. The employment of hydrogen as a chain transfer agent in the course of the polymerization allowed to greatly decrease the molecular weight, thus increasing solubility [31]. This material was highly stable to photooxidative aging and maintained its water repellency properties even after 1000 h of exposure to solar radiation [31].

By using the additive PDMS-OH with the TEOS-containing catalyst di-n-butyltin dilaurate (DBTL), an O-I hybrid crack-free xerogel was synthesized by Dan Li et al. [32] via the solgel method. In this work, the catalyst (i.e., DBTL) was reported to effectively decrease the gelling time [32]. An increase in viscosity of the sol and an improvement of the xerogel's cracking were also observed when PDMS-OH was included [32]. When the PDMS/TEOS mole ratio was higher than 0.153, the xerogels were opaque [32]. Moreover, when the sol had 0.1% (w/v) of silica NPs added to it, both the surface roughness and hydrophobicity of the xerogel increased without any changes in the xerogels' color [32]. When treated with a hybrid TEOS-SiO$_2$-PDMS-OH sol, the limestone's surface had a contact angle up to 97° [32].

Fermo et al. [33] have utilized Alpha®SI30 (a commercial siloxane resin) to improve the hydrophobicity features of Angera stone, Botticino limestone, and Carrara marble. The authors reported that the obtained coatings were hydrophobic based upon the contact angle measurements [33]. SEM and XPS analyses revealed crack-free and homogeneous surfaces along with the resin's homogeneous distribution across the stones [33]. From the characterization of the surface topography by

AFM, it was noted that the coating lacked the thickness required to hide the roughness of the material and that a thin film was formed by resin which followed the sample's features [33].

Another hybrid coating with superhydrophobic characteristics was developed by Corcione et al. [34] that when applied on glass substrate and photopolymerized by a UV lamp resulted in a significant improvement of surface properties. The authors have also tested their developed hybrid mixture on pietra leccese (PL), typical of the Apulia Region (Italy), which resulted in an increase in all the physical and surface proprieties of the hybrid film while the coating was transparent [34]. SEM analysis of the cured film confirmed that the silica NPs were interconnected with the methacrylic-siloxane matrix [34]. The hybrid film had a high Tg which limited organic phase's mobility in the silica [34]. The developed hybrid coating was also hydrophobic in nature and capable of preserving the stone from water penetration [34].

Luo et al. [35] utilized composites of nano-hydroxyapatite (n-HAp) and TEOS in order to improve sandstone consolidation and protection. Along with a neutral catalyst, n-Hap and hydroxyl-terminated PDMS were combined with TEOS to yield a coarser network hydrophobic property and vapor transport simultaneously when applied to sandstones [35]. The developed TEOS/PDMS/HAp enhanced sandstone consolidation and protection [35]. Moreover, n-HAp was reported to effectively improve mechanical properties and was resistant to aging but did not impart any changes to the sandstones' color [35]. A reduction in the formation of cracks during the drying process was observed in TEOS-based gels that had PDMS and n-HAp added [35]. Hydrophobicity was attributed to PDMS by reducing the surface tension and an increase in surface roughness by n-Hap [35]. Aging analyses revealed that n-HAp was critical in improving weathering resistance [35].

By using sol-gel route, a TEOS stone consolidant formulation with additives such as "PDMS-OH and cetyl trimethyl ammonium bromide (CTAB)" surfactant was created by Y. and J. Liu [36]. Data indicated that PDMS-OH was effective in improving of the sandstones' hydrophobicity [36]. By adding CTAB to the siloxane composition, micro- and nanostructural changes to the surface morphology and roughness prevented cracking when the coating was dried [36]. Additionally, it was reported that the Chinese Chongqing Dazu sandstone sculptures showed improved resistance against acid and salt crystallization weathering and exhibited hydrophobicity and a crack-free surface after being treated with the developed film [36]. It was concluded that TEOS/PDMS-OH/CTAB composite coatings imparted hydrophobic properties and were effective consolidants for the damaged sandstone [36].

Due to their unique physicochemical characteristics, inorganic NPs (e.g., ZnO, TiO_2, SiO_2, etc.) are expected to display better performance in comparison with typical chemical compounds. Specifically, TiO_2 and SiO_2 NPs can potentially increase water repellency through an increase in surface roughness. Besides, the addition of these nanocomposites grants superhydrophobic and self-cleaning properties. Moreover, the photocatalytic properties of nano-TiO_2 impart self-cleaning properties while being cost-effective as well as easy to apply and already have been used in restoring stone heritage.

In this regard, by utilizing a sol-gel method, where the precursor TEOS, under acid catalysis (oxalic acid), was mixed with PDMS and TTIP (titanium tetraisopropoxide), Kapridaki and Maravelaki-Kalaitzaki [37] designed a crack-free and transparent hydrophobic SiO_2-TiO_2 hybrid coating (with a crystallite size of 5 nm). The gel was protected from cracking during the drying process by the controlling properties of the oxalic acid as well as by the TiO_2 and PDMS incorporated into the silica matrix [37]. Moreover, the presence of TiO_2 was reported to be effective against bacterial activity and the absorption of pollutants [37]. The coating's self-cleaning properties were confirmed on marble samples and the subsequent removal of biofilms and methylene blue dye [37]. A larger contact angle and a smaller water capillary coefficient were noted after the coating was applied [37]. The marble's hydrophobicity was improved by PDMS' methyl moieties, with changes to the water capillary coefficient and contact angles confirming the results [37]. The synthesis of a clear SiO_2-TiO_2-based hydrophobic nanocomposite is novel as it is self-cleaning, and practically it does not change the marble's color or vapor permeability [37].

SiO_2 and TiO_2 NPs used to modify consolidant properties have also been analyzed in a work conducted by D'Amato et al. [38] for their application on marble samples. The authors prepared different solutions of Paraloid B-72 with dispersed SiO_2 and TiO_2 NPs [38]. The results showed that nanocomposites of Paraloid B-72 modified with SiO_2 and TiO_2 NPs improved the consolidant's effectiveness and enhanced the conservative performances on the marble specimens [38]. When SiO_2 with a concentration of 1% (w/v) was added, the resulting coating outperformed other sample in terms of hydrophobicity and water repellency [38]. Also, for SiO_2 and TiO_2 mixtures, the ideal SiO_2 concentration was found to be 1% (w/v) [38].

Ugur [39] modified "porous intraclastic limestone" (IL) and "welded tuff" (WT) surfaces with a waterborne FPS (fluorinated polysiloxane) and further made them impermeable to water by dip-coating process [39]. Contact angle, attenuated total reflectance Fourier transform infrared spectroscopy (ATR-FTIR), and capillary water absorption analyses indicated that hydrophobicity was a result of a surface-fluorinated carbon chain [39]. Furthermore, stone treated with FPS showed a highly hydrophobic surface without significant color variation between the control and treated samples [39]. It was concluded that the hydrophobicity and the stones' initial color were maintained up to a certain degree even at high temperatures [39]. Moreover, the film effectively imparted superhydrophobicity to IL and WT as well [39].

In a reported study by D'Orazio and Grippo [40], a polymeric coating based on "a water dispersed TiO_2/poly (carbonate urethane) nanocomposite by means of cold mixing of single components via sonication" was developed and characterized [40]. Several analytical methods were utilized to determine the thermal and viscoelastic properties as well as the nanocomposite's morphology and structural characteristics. Moreover, a photocatalytic analysis carried out on nanocomposite samples indicated that the 1% (w/w) content of TiO_2 NPs within the coating yielded self-cleaning ability for pollution [40].

Cappelletti et al. [41] coated three different stone surfaces (i.e., Angera stone, Carrara, and Botticino marbles) by a hybrid NPs-based treatment ("Alpha®SI30 + Ti sols") as protection agent and compared them against a commercially available polysiloxane (Si-based resin, Alpha®SI30) treatment [41]. The treatments were studied in an attempt to enhance the surfaces' hydrophobic properties [41]. Moreover, the hybrid coatings were applied to three different stone surfaces so as to obtain superhydrophobicity [41]. The reported synthetic method produced transparent hybrid layers [41]. The results showed that only the Carrara marble's final contact angle achieved superhydrophobicity ($\theta > 150°$). Decent hydrophobicity ($138° < \theta < 141°$) was also obtained for the Angera and Botticino samples [41]. The hybrid coatings were also reported to reduce the formation of salts with respect to the pure resin [41].

Centering upon total immersion with a fluorosurfactant (Capstone FS-63), a hydrophobization process was proposed by Kronlund et al. [42] for efficient stone protection. By utilizing total immersion treatments, "mechanical grinding and capillary absorption measurements" confirmed that the fluorosurfactant diffused and reacted within the interior-most pores, indicating gains in successful functionalization depth from the micro- to millimeter scale [42]. In order for this to occur, the fluorosurfactant solution utilized less polar cosolvents (e.g., ethylene glycol, ethanol) in lieu of water alongside higher reaction temperatures which drove the "vesicle-surfactant equilibrium toward free surfactants" and was confirmed via dynamic light scattering (DLS) measurements [42]. Protection efficiency was found to be significantly influenced by solvent polarity and the temperature of reaction, with performance improvements observed as increases to the cosolvent to water ratio, and temperature were made [42]. Upon removing the exterior of the stones' surface, capillary absorption measurements indicated functionality of the hydrophobic surfactant molecules at up to 2 mm in depth [42]. Optimal treatments presented $0.2 \pm 0.3\%$ capillary absorption, when compared to that of grinded standards, underscoring the significance of internally functionalized pore surfaces [42]. DLS analysis revealed that if the temperature increased, solvent polarity decreased (e.g., addition of cosolvent), or a combination of the two, the vesicles either became smaller or degraded into free surfactants [42]. This action allows for rapid surfactant diffusion into interior pores resulting in enhanced stone performance [42]. Correlations can be made between increased functionalization efficiency along with the mixture's measured particle size distributions and diffusion values as a function of temperature [42].

Gherardi et al. [43] have reported a nano-TiO$_2$ (NA_TiO$_2$) treatment, "based on a dispersion of solar-light activated TiO$_2$ nanocrystals," for applications in stone protection and maintenance. This photocatalytic treatment was developed by a simple and relatively cheap method that can be applied by spraying or brushing (substrate dependent), allowing the anatase NPs to be activated by both UV rays and solar irradiation [43]. Stable dispersions (i.e., with water or ethylene glycol) can be achieved with NA_TiO$_2$, circumventing precipitation issues and produces homogeneous nano-TiO$_2$ treatments with greater aesthetic compatibility and photocatalytic activity on Noto stone and Carrara marble over commercial TiO$_2$ (i.e., P25_TiO$_2$)

treatments [43]. The treatment's and the stone's morphological properties contribute to this phenomena as well as benzyl alcohol molecules attached to the surface of the anatase NP [43]. As such, photo-efficiency increased due in part to improvements to solar light absorption, which in turn improved charge trapping by supplying additional surface electron traps (Ti^{3+} centers) [43]. Moreover, the NA_TiO_2 treatments did not significantly affect the stone's capillary water absorption, only slightly increasing its wettability [43]. For marble, the NA_TiO_2 treatments were also highly homogeneous, able to preserve not only the active surface area of the NPs but also the stone substrate's optical quality [43]. However, NA_TiO_2 treatments on Noto stones were unable to form continuous films and led to NP aggregation within the pores, causing the stone's photoactivity to diminish [43]. The obtained data in this study indicated that combining the properties of a greater amount of surface electron traps, solar light absorption, high dispersion, and reduced recombination effects in NPs all have an important role in creating effective photocatalytic treatments for stone restoration and conservation [43].

In another reported study, La Russa et al. [44] evaluated the efficiency of three nano-TiO_2 coatings applied on Noto calcarenite and Carrara marble. The authors have evaluated the photocatalytic properties of nano-TiO_2 combined with the self-cleaning and hydrophobic properties of acrylic and fluorinated polymers along with their appropriateness to the stone surfaces [44]. These properties were best exhibited by an acrylic water suspension (Fosbuild) combined with TiO_2 [44]. Further analysis revealed that good hydrophobic and photodegradation properties were achieved with the fluorinated polymer-TiO_2 mixture (AKP-TiO_2), even after accelerated UV aging [44]. Additionally, the high stability of the C–F bond altered the surface morphology insignificantly, in spite of its poor penetration depth. On the other hand, Paraloid B-72 was found to be a poor binder when combined with TiO_2 NPs, causing intense superficial alteration of each of the stone samples without any discernible, desirable properties [44]. For conservation intervention, the suggested amounts by the authors were 1600 g/m^2 for calcarenite and 100 g/m^2 for marble [44]. This study indicated that coating performance was related to stone penetration depth, as securing the polymer on the pore structure ensured both durability and hydrophobicity [44].

1.4 Case Study: Superhydrophobic and Water-Repellent Polysiloxane-Nanoparticle Composite Coatings for the Protection of Sandstone and Marble

Using the method of Fig. 1.3, coatings of siloxanes mixed with silica NPs with a 7 nm mean diameter were deposited on sandstone and marble surfaces. Attention is first focused on sandstone (Sect. 1.4.1); results on treated marble are presented and discussed later in Sect. 1.4.2. In the studies of both Sects. 1.4.1 and 1.4.2, silica NPs are added in Rhodorsil 224 (Rhodia Silicones) which is a (alkyl siloxane) dissolved in white spirit (7 wt %).

The versatility of the method of Fig. 1.3 is demonstrated in Sects. 1.4.3, 1.4.4, and 1.4.5 as it is shown that superhydrophobicity and water repellency can be (1) achieved using exclusively aqueous products, (2) achieved using inherent hydrophilic materials, and (3) accompanied by superoleophobicity and oil repellency.

In addition to the superhydrophobicity and water repellency, several other factors should also be considered when selecting a proper product for stone protection. For example, "a good protective material must (1) not affect the color of the stone substrate, (2) assure a good permeability for water vapor and (3) decrease the amount of water absorbed by capillarity" [7]. The efficacy of polysiloxane-nanoparticle coatings to fulfil the aforementioned properties is discussed in Sect. 1.4.6.

1.4.1 Superhydrophobicity and Water Repellency on Sandstone

Figure 1.4a shows the variations of the static contact angle (θ^*) and the contact angle hysteresis ($\theta_A - \theta_R$) with nanoparticle/siloxane mass ratio for water drops placed on Rhodorsil-silica coatings which were deposited on Opuka specimens [7]. Opuka is a sandstone which has been used for the restoration of the castle of Prague [7]. According to the results of Fig. 1.4a, θ^* initially increases with particle concentration from 141°, for pure Rhodorsil without NPs, to 159° for coatings which were prepared using 0.18 nanoparticle/Rhodorsil mass ratios. The nanoparticle concentration does not have practically any effect on θ^* for coatings which correspond to nanoparticle/siloxane mass ratio >0.1: in this regime, θ^* is (1) stable, corresponding to the plateau of the curve in Fig. 1.4a, and (2) is extremely high (>150°) corresponding to superhydrophobicity.

The variation of θ^* with the concentration of NPs which are added in a polymer matrix to affect the surface morphology and therefore the wetting properties of the composite coating has been reported in previously published studies [1, 2, 7, 22–26, 45, 46]. NPs form microscale clusters which enhance the roughness of the coating surface at the micrometer/nanometer scale. Initially, the clusters are separated by smooth areas of continuous polysiloxane coating, as shown in the left SEM image of Fig. 1.4a, captured for a coating with 0.05 nanoparticle/siloxane mass ratio. Larger (coalesced) clusters are formed at elevated particle concentrations, resulting in a continuous rough surface according to the right SEM image, which corresponds to 0.27 nanoparticle/siloxane mass ratio. This dense rough surface structure is responsible for the observed superhydrophobicity, that is, $\theta^* > 150°$.

The hysteresis ($\theta_A - \theta_R$) varies with the nanoparticle/siloxane mass ratio as follows (Fig. 1.4a): the hysteresis initially increases, it reaches a maximum value, and then it decreases. At the mass ratio of around 0.1, the hysteresis is minimized (~5°) and becomes constant. This, rather complicated, type of variation of the hysteresis with the nanoparticle concentration is interpreted by the SEM images included in Fig. 1.4a. A water drop may fill the large, smooth interspaces that exist, among the clusters, in coatings prepared using low particle concentration. Therefore, the clusters may act as obstacles against any motion of the drop, thus inducing an increase

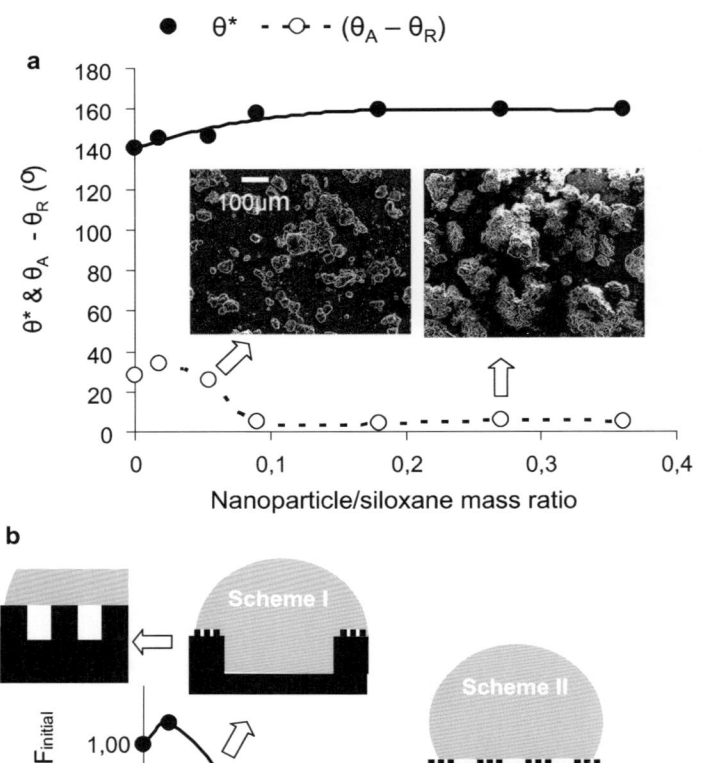

Fig. 1.4 (**a**) Static contact angle (θ^*) and contact angle hysteresis ($\theta_A - \theta_R$) versus the nanoparticle/Rhodorsil mass ratio. The composite coatings were deposited on Opuka sandstone. SEM images revealing the evolution of surface structure with nanoparticle concentration are included. Modified with permission from reference number [7], Copyright 2009, Springer. (**b**) Force per unit length (F) normalized to its initial value ($F_{initial}$) versus the nanoparticle/Rhodorsil mass ratio. Schemes I and II show the wetting scenarios for coatings prepared using low and high nanoparticle concentration, respectively. In scheme I water fills the large, smooth areas that exist among the clusters. However, water does not penetrate the nanocrevices that exist on the surface of the protruding clusters, as shown in the inset of the expanded scale (left). In any case, scheme I cannot be rationalized by the Cassie-Baxter model which can be applied to interpret the scenario of scheme II, where a dense rough structure exists inducing nonsticking properties

of the hysteresis. As the particle concentration increases, the smooth interspaces become extremely small and a continuous rough structure is formed. Consequently, the hysteresis drops and reaches a very small value, as evidenced by the data of Fig. 1.4a [7]. The surface structure of Rhodorsil-silica coatings prepared using nanoparticle/siloxane mass ratio >0.1 becomes saturated, and therefore both ϑ^* and $\theta_A - \theta_R$ become constant, corresponding to very high and low values, respectively. Consequently, both superhydrophobicity and water repellency occur at the surfaces of the coatings with nanoparticle/siloxane mass ratio >0.1.

According to the SEM images of Fig. 1.4a, as the concentration of the additives-NPs increases, the surface roughness of the deposited coatings increases. Consequently, it can be argued that the horizontal axis of the plot in Fig. 1.4a corresponds qualitatively to surface roughness. Based on this argument, it is concluded that the variation of the hysteresis in Fig. 1.4a is in agreement with the results reported for water drops on rough wax [12], PDMS [47], and poly(tetrafluoroethylene) [48] surfaces. Moreover, the hysteresis variation reported in Fig. 1.4a for composite coatings deposited on Opuka is in agreement with the results reported for the same coatings (Rhodorsil-silica) deposited on glass [22].

The critical line force, F, described by Eq. 1.4 is plotted in Fig. 1.4b versus the nanoparticle/siloxane mass ratio. In particular, in the perpendicular axis of the plot, the force F normalized to its initial value ($F_{initial}$), which corresponds to the polysiloxane coating without NPs, is included. As it was expected, the curve in Fig. 1.4b follows the behavior of $\theta_A - \theta_R$ reported in Fig. 1.4a. Two schemes illustrating water drops on rough surfaces are provided. In scheme I (low particle concentration), a water drop fills the large, smooth areas that exist among the clusters, as it was previously discussed, but it probably does not penetrate the nanocrevices that exist on the surface of the protruding clusters. Apparently, the Cassie-Baxter model [10] cannot describe the wetting scenario of scheme I, but it can be applied to interpret the scenario of scheme II (high particle concentration) where a dense rough structure exists. The surface of scheme II corresponds to the nonsticking state, as evidenced by the small F values reported in Fig. 1.4b. In particular, for nanoparticle/siloxane mass ratio >0.1, $F/F_{initial}$ is <0.1, implying that it takes less than ten times as less force to move a drop on Rhodorsil-silica than on a pure Rhodorsil coating. On the contrary, a higher F than $F_{initial}$ must be applied when the scenario of scheme I is realized for nanoparticle/siloxane mass ratio <0.1.

The origin of the $\theta_A - \theta_R$ variation reported in Fig. 1.4a is elucidated in Fig. 1.5. Advancing (θ_A) and receding (θ_R) contact angles are plotted as a function of the nanoparticle/siloxane mass ratio. The results in Fig. 1.5 show that θ_A increases as the nanoparticle/siloxane mass ratio increases from 0 to around 0.1. In the same nanoparticle/siloxane mass ratio (0–0.1), a very slight increase of θ_R is initially reported, followed by a very rapid increase of θ_R. Fig. 1.5 suggests, therefore, that as the nanoparticle/siloxane mass ratio increases from 0 to 0.1, the difference $\theta_A - \theta_R$ (or the force F) should first increase and then decrease, as it was reported in Fig. 1.4a (or Fig. 1.4b). At elevated particle concentrations (nanoparticle/siloxane mass ratio >0.1), the hysteresis nearly vanishes as both θ_A and θ_R become large and comparable (i.e., the difference $\theta_A - \theta_R$ becomes small).

Fig. 1.5 Advancing (θ_A) and receding (θ_R) contact angles versus the nanoparticle/Rhodorsil mass ratio. The composite coatings were deposited on Opuka sandstone

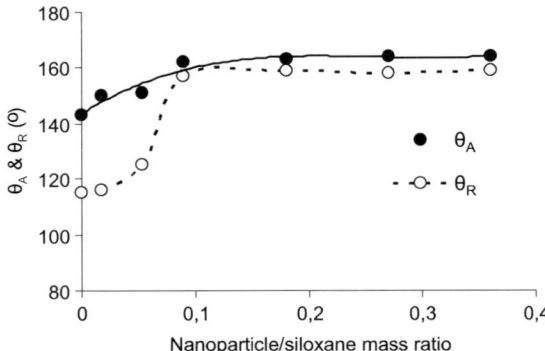

Fig. 1.6 Static contact angle (θ^*) versus the nanoparticle/Rhodorsil mass ratio. The composite coatings were deposited on Thassos marble and Opuka sandstone. The data of Thassos marble were adapted with permission from reference number [1], Copyright 2009, Elsevier

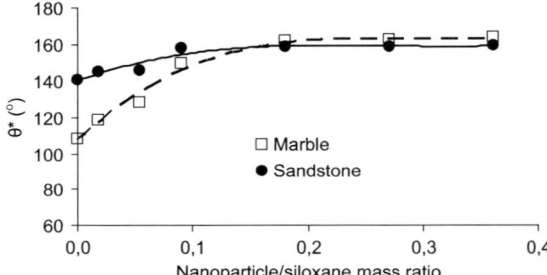

The θ_A and θ_R measurements presented in the plot of Fig. 1.5 were carried out on Rhodorsil-silica coatings which were deposited on Opuka sandstone specimens [7]. It is reported that the trend of the θ_A and θ_R variations of Fig. 1.5 is in excellent agreement with results reported for the same coatings (Rhodorsil-silica) deposited on glass [22]. Furthermore, considering that the horizontal axis of the plot of Fig. 1.5 corresponds qualitatively to surface roughness, the curves of Fig. 1.5 appear to be qualitatively similar with corresponding measurements which were performed on rough wax [12], PDMS [47], and poly(tetrafluoroethylene) [48] surfaces.

1.4.2   Superhydrophobicity and Water Repellency on Marble

Figure 1.6 shows the variation of θ^* with nanoparticle/siloxane mass ratio, for water drops placed on Rhodorsil-silica coatings which were deposited on marble specimens [1]. For comparison, the results of θ^* on coated sandstone are reproduced from Fig. 1.4a and are included in the plot of Fig. 1.6. The two substrates, Thassos marble and Opuka sandstone, exhibit different surface morphologies, as marble is smoother than sandstone. For example, the open porosity of the Thassos marble is only 0.2% [1], whereas the corresponding value for the Opuka stone is 7.7% [7]. This

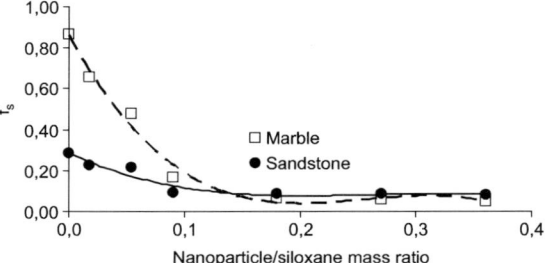

Fig. 1.7 Open porosity factor (f_s) versus the nanoparticle/Rhodorsil mass ratio. The f_s values were calculated using Eq. 1.3. The composite coatings were deposited on Thassos marble and Opuka sandstone. Only the results obtained for coatings with nanoparticle/siloxane mass ratio >0.1, which are coatings corresponding to the Cassie-Baxter nonsticking state, are useful for the discussion

difference, however, does not affect the maximum θ^* reported in Fig. 1.6 for the two sets of data. The maximum θ^* is achieved for water drops on coatings with nanoparticle/Rhodorsil mass ratio of roughly >0.1, and it is practically unaffected by the underlying substrate. However, according to Fig. 1.6, the role of the substrate roughness becomes important for coatings prepared using low particle concentration that are coatings corresponding to nanoparticle/Rhodorsil mass ratio <0.1. The θ^* on a smooth polysiloxane coating (no NPs) on marble is only 108°. As marble is quite smooth, the deposited polysiloxane coating is also relatively smooth. However, the polysiloxane coating on sandstone exhibits an augmented roughness which originates from the roughness of the underlying substrate. Hence, the θ^* measured on Rhodorsil (no particles) deposited on sandstone is 141°, which is substantially higher than the 108° value measured on coated marble. Embedding NPs in the polysiloxane matrix results in an increase of θ^*, which is more dramatic when the coating is deposited on marble than on sandstone (Fig. 1.6). At the nanoparticle/Rhodorsil mass ratio of around 0.1, the coating morphologies become saturated, and the effect of the substrate roughness on θ^* practically vanishes. The same conclusion was drawn for the effect of the substrate on $\theta_A - \theta_R$. The same hysteresis of ~5° was measured for water drops on composite coatings with nanoparticle/Rhodorsil mass ratio >0.18 which were deposited on either Thassos marble or Opuka sandstone [1, 7].

Using Eq. 1.3, the open porosity factors (f_s), which are predicted by the Cassie-Baxter model, can be calculated for the polysiloxane-nanoparticle coatings of Fig. 1.6. The contact angle of a water drop on extremely smooth Rhodorsil surface, which is produced by spin coating onto clean glass slides, is 102° [22]. This can be considered as the Young contact angle (θ_Y) and was used in the Cassie-Baxter equation along with the θ^* values of Fig. 1.6, to calculate the corresponding f_s values. The results are plotted in Fig. 1.7 versus the nanoparticle/siloxane mass ratio. It is stressed that the calculations may be elusive for coatings with nanoparticle/siloxane mass ratio <0.1, as the Cassie-Baxter model can be rationalized only for coatings prepared using high particle concentration. For the rough composite coatings which correspond to the Cassie-Baxter state (nanoparticle/siloxane mass ratio >0.1), the results of Eq. 1.3 are useful and show that the f_s is independent of (1) the nanoparticle concentration (i.e., the coating surfaces become saturated) and (2) the underlying substrate, and (3) moreover f_s receives a very small value of <0.1 (Fig. 1.7). Typically, the f_s of a truly superhydrophobic, nonsticking, water-repellent surface where the Cassie-Baxter scenario is realized is indeed <0.1, as reported by the literature [24, 45, 46].

1.4.3 Waterborne Superhydrophobic and Water-Repellent Coatings

The self-/easy cleaning process that takes place on a water-repellent surface is extremely useful for the conservation of monuments and outdoor objects of the cultural heritage. The self-/easy cleaning process is demonstrated in Fig. 1.8 for a block of white marble, coated by Silres BS 4004, which is a mixture of silanes and siloxanes in water, and silica NPs [23]. The coated marble surface was intentionally contaminated with a blue pigment which was easily removed by water drops, as demonstrated in the successive snapshots of Fig. 1.8. No organic solvent was used for the preparation and deposition of the water-repellent coating of Fig. 1.8. Consequently, waterborne siloxanes can be easily used to induce water repellency to stone and may therefore replace the harmful solvent-based alkoxysilanes which are extensively used currently in the conservation of stone.

1.4.4 Superhydrophobic and Water-Repellent Coatings from Inherent Hydrophilic Materials

In Sect. 1.3, it was described that the method of Fig. 1.3 can be applied to produce superhydrophobic and water-repellent coatings using various silane/siloxane materials and organic polymers. Two examples for silanes/siloxanes were discussed in detail so far: Rhodorsil 224 (Figs. 1.4, 1.5, 1.6, and 1.7) and Silres BS 4004 (Fig. 1.8) which were both mixed with silica NPs. An example of an organic polymer is discussed herein, that is, poly(methyl methacrylate) (PMMA). PMMA is an acrylic polymer which is used in several applications, and it is commonly used as a reference material for case studies. Acrylic polymers, such as, for instance, the Paraloid products, are extensively used in conservation science as adhesives and protective coatings. However, their use in stone consolidation and conservation is limited by their big molecular sizes and high viscosities which result to low depths of penetration through the porous network of the stone. Consequently, in principle, consolidants composed of small silane and siloxane molecules of low viscosity are

Fig. 1.8 The self-/easy cleaning property of a water-repellent coating deposited on marble is demonstrated in these successive snapshots. The coating was prepared using a waterborne silane/siloxane product (Silres BS 4004) and silica nanoparticles; the nanoparticle/siloxane mass ratio was 0.2. The coated surface of the marble segment was intentionally contaminated with a blue pigment. Water drops that were left to roll off removed the pigment particles. A sequence number (1–3) is provided in the lower left corner of each photograph. Reprinted with permission from reference number [23], Copyright (2013), John Wiley & Sons, Inc

advantageous, as these have the ability to penetrate deep into the stone network, before they become highly viscous gels through the sol-gel process [27].

PMMA is an inherent hydrophilic material. The contact angle of a water drop on extremely smooth PMMA surface, which was produced by spin coating onto clean glass slides, was measured and found 72° [22]. This value can be taken as the Young contact angle (θ_Y) and therefore $\theta_Y < 90°$. Following the method of Fig. 1.3, silica NPs were dispersed in a PMMA solution in toluene, and the dispersion was sprayed on calcium carbonate (CaCO$_3$) surfaces which were prepared in the laboratory and used as model, marble-like surfaces [2]. The static contact angle (θ^*) of water drops on the deposited composite coating was 157°, implying that superhydrophobicity was achieved using exclusively inherent hydrophilic materials such as PMMA and silica. Likewise, high θ^* (>150°) was achieved for water drops on the PMMA-nanoparticle coatings, which were sprayed on glass slides [22, 24]. The contact angle hysteresis was only 5°, thus suggesting that the composite coatings exhibit water repellency [22, 24]. The Cassie-Baxter model (but not the Wenzel model) can support superhydrophobicity on inherent hydrophilic materials, according to the data and the relative discussion of Fig. 1.1.

1.4.5 Superoleophobic and Oil-Repellent Coatings

The production of oil-repellent surfaces is a difficult task considering the low surface tension of oil (32 mN/m) compared, for instance, to that of water (72 mN/m). As shown in Fig. 1.3b, with careful selection of an appropriate siloxane product, the resulting polysiloxane-nanoparticle coating can show both superhydrophobicity and superoleophobicity [26]. This can be an important advantage for coating materials designed for the protection of outdoor surfaces of monuments and other objects of the cultural heritage, which are often exposed to urban air pollution.

The results of Fig. 1.9 were achieved using Silres BS29A which, according to the manufacturer (Wacker), is an aqueous silicone emulsion composed of alkoxysilanes and a fluoropolymer. Silica NPs were added in the emulsion, and the dispersion was sprayed on marble blocks [26]. The nanoparticle/siloxane mass ratio of the composite coatings included in Fig. 1.9 was 0.28. For comparison, coatings of pure Silres BS29A (without NPs) were deposited on marble. Drops of water, glycerol, ethylene glycol, and olive oil were placed on the treated marbles, and contact angles (θ^*) were measured. The surface tensions of the tested liquids follow the sequence water → glycerol → ethylene glycol → olive oil, where water and oil are the liquids with the highest and lowest surface tensions corresponding to 72 and 32 mN/m, respectively. The surface tensions of glycerol and ethylene glycol fall within the water-oil range, corresponding to 60 and 47.3 mN/m, respectively. The results of Fig. 1.9 show that embedding NPs in the Silres BS29A matrix results in the production of a surface that exhibits superhydrophobic and superoleophobic properties, as $\theta^* > 150°$ for both water and oil drops. Likewise, high θ^* (>150°) were measured for glycerol and ethylene glycol drops placed on the composite coatings. When the

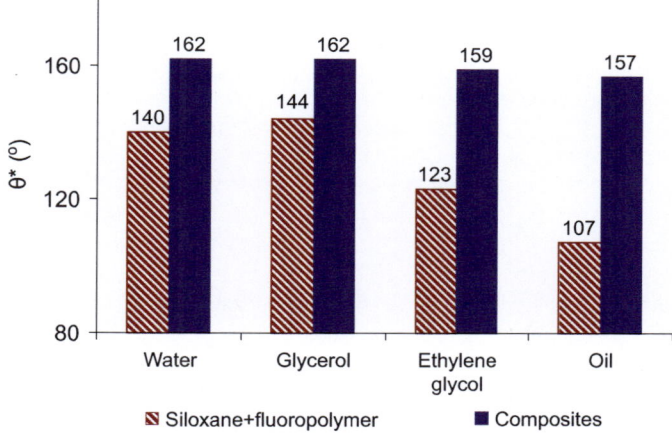

Fig. 1.9 Contact angles (θ^*) of water, glycerol, ethylene glycol, and oil drops on coatings of Silres BS29A (siloxane + fluoropolymer) and Silres BS29A enriched with silica nanoparticles (composites). The latter were prepared using a nanoparticle/siloxane mass ratio of 0.28. Coatings were sprayed on marble surfaces. Drops on composite coatings correspond to $\theta^* > 150°$ [26]

marble specimens coated with the composite coatings were slightly tilted from the horizontal (<7°), water and oil drops slided off implying water and oil repellency [26].

1.4.6   Other Properties of Superhydrophobic and Water-Repellent Polysiloxane-Nanoparticle Coatings

In principle, superhydrophobic and water-repellent coatings are advantageous over the traditional hydrophobic coatings, suggested in the past for the protection of stone. However, coatings devised for the protection of monuments and other stone objects of the cultural heritage should fulfil some other requirements. For instance, a good protective material must (1) not affect the color of the stone substrate, (2) assure a good permeability for water vapor, and (3) decrease the amount of water absorbed by capillarity. These are discussed in the following, summarizing the results obtained for polysiloxane-nanoparticle coatings deposited on Opuka sandstone and Thassos marble specimens [1, 7]. The coatings were produced following the standard method of Figure 1.3 and using Rhodorsil 224 and silica NPs with a nanoparticle/Rhodorsil mass ratio of 0.36. According to Sects. 1.4.1 (for sandstone) and 1.4.2 (for marble), these coatings fall in the superhydrophobic and water-repellent regime, as evidenced by the results of Figs. 1.4 and 1.6. For comparison, hydrophobic coatings made of pure Rhodorsil 224 (without NPs) are included in the following studies.

Table 1.1 Results of colorimetric, capillary water absorption, and vapor permeability measurements which were carried out on sandstone and marble specimens, covered by hydrophobic polysiloxane coatings (Rhodorsil) and superhydrophobic polysiloxane-nanoparticle coatings (Rhodorsil-silica). The latter were prepared using a nanoparticle/siloxane mass ratio of 0.36. Measurements were carried out using Eqs. 1.5, 1.6, and 1.7 [1, 7]

	Polysiloxane coating			Polysiloxane-nanoparticle coating		
Substrate	ΔE^*	RC%	RVP%	ΔE^*	RC%	RVP%
Opuka sandstone	1.2	88.5	19.8	4.3	94.0	14.4
Thassos marble	1.5	69.4	10.0	3.8	89.5	35.0

Colorimetric measurements were carried out, on a comparative basis, on coated and uncoated sandstone and marble specimens [1, 7]. The global color differences (ΔE^*) imposed by the coating applications were calculated according to Eq. 1.5:

$$\Delta E^* = \sqrt{\Delta L^{*2} + \Delta a^{*2} + \Delta b^{*2}}$$ (1.5)

where L^*, a^*, and b^* are the standard coordinates of the CIE 1976 scale.

Capillary water absorption measurements were performed by the gravimetric sorption technique, as described in detail elsewhere [1, 7]. The % reduction of water absorption by capillarity (RC%) was calculated according to Eq. 1.6:

$$RC\% = \left(\frac{m_{uw} - m_{tw}}{m_{uw}} \right) \times 100$$ (1.6)

where m_{uw} and m_{tw} are the masses of water absorbed by capillarity by the uncoated and coated specimens, respectively.

Finally, for the vapor permeability tests, the % reduction of vapor permeability (RVP%) was calculated according to Eq. 1.7 [1, 7]:

$$RVP\% = \left(\frac{m_{uv} - m_{tv}}{m_{uv}} \right) \times 100$$ (1.7)

where m_{uv} and m_{tv} are the masses of water vapor penetrating the uncoated and coated specimens, respectively.

An ideal coating should not affect the color ($\Delta E^* = 0$) and the vapor permeability (RVP% = 100) of the treated substrate and should eliminate the amount of water absorbed by capillarity (RC% = 0). The results for the Opuka sandstone and Thassos marble specimens are summarized in Table 1.1. The latter shows that the use of NPs in the protective coatings has a negative impact on the aesthetic appearances of the specimens, as higher ΔE^* values were measured for the specimens covered by polysiloxane-nanoparticle coatings compared to the specimens treated with pure polysiloxane. To reduce the color change induced by the NPs, composite coatings with smaller nanoparticle concentration can be used, maintaining, however, the superhydrophobic and water-repellent character of the coatings. The composite coatings included in Table 1.1 were produced using a nanoparticle/siloxane mass ratio of 0.36. As shown in Figs. 1.4 and 1.6, superhydrophobicity is maintained for coatings with smaller nanoparticle/siloxane mass ratios.

The use of NPs promotes hydrophobicity, inducing superhydrophobicity. As a result, higher RC% values were recorded for the stones treated with superhydrophobic composites, compared to the corresponding RC% values measured on stones which were treated with pure polysiloxane (Table 1.1). Finally, the use of NPs increased the RVP% measured on marble specimens, but it did not have any considerable effect on the water vapor permeability tests, which were carried out on the sandstone specimens. Interestingly, a slight decrease of the RVP% from 19.8 to 14.4 is reported in Table 1.1 for the Opuka specimens treated with pure polysiloxane and polysiloxane-nanoparticle coatings, respectively. This effect may be attributed to the augmented porosity of the Opuka sandstone, as it was previously argued [7].

1.5 Conclusion

The use of NPs as additives in siloxane-based coatings, for the protection of sandstone and marble, provides some very positive results. NPs induce superhydrophobicity and water repellency to the surface of the composite coatings, as evidenced by the high static contact angle and low contact angle hysteresis as well as the enhanced reduction of water by capillarity. Interestingly, superhydrophobic and water-repellent coatings can be produced using (1) exclusively aqueous products, (2) inherent hydrophilic materials, and moreover (3) can have superoleophobic and oil repellent behavior. Apparently all these properties are very attractive, as they enhance the resistance of the coated stone to the negative effects of rainwater and pollutants, because of the self-cleaning (lotus) effect. On the other hand, NPs affect the aesthetic appearance of the stone, and they can have a negative impact on vapor permeability, as it was reported for marble specimens. Reduction of the negative effects of the NPs can be achieved by tuning crucial parameters such as the nanoparticle concentration. The potential use of polysiloxane-nanoparticle composite films for monument protection is a very promising idea, which, however, must be further examined. For instance, before making the new products available to conservators, the durability of the good properties of the composite materials must be investigated through accelerating aging and in situ pilot-scale experiments. Moreover, the long-term side effects of the protective coatings should be evaluated.

References

1. Manoudis P, et al. Fabrication of super-hydrophobic surfaces for enhanced stone protection. Surf Coat Technol. 2009;203(10):1322–8.
2. Manoudis P, et al. Polymer-silica nanoparticles composite films as protective coatings for stone-based monuments. J Phys Conf Ser. 2007;61:1361. IOP Publishing
3. Cappelletti G, Fermo P. 15—Hydrophobic and superhydrophobic coatings for limestone and marble conservation A2. In: Montemor MF, editor. Smart composite coatings and membranes. Cambridge: Woodhead Publishing; 2016. p. 421–52.
4. Chiantore O, Lazzari M. Photo-oxidative stability of paraloid acrylic protective polymers. Polymer. 2001;42(1):17–27.
5. Doherty B, et al. Efficiency and resistance of the artificial oxalate protection treatment on marble against chemical weathering. Appl Surf Sci. 2007;253(10):4477–84.
6. Poli T, Toniolo L, Chiantore O. The protection of different Italian marbles with two partially fluorinated acrylic copolymers. Appl Phys A Mater Sci Process. 2004;79(2):347–51.
7. Manoudis P, et al. Superhydrophobic films for the protection of outdoor cultural heritage assets. Appl Phys A Mater Sci Process. 2009;97(2):351–60.
8. Young T. An essay on the cohesion of fluids. Philos Trans R Soc Lond. 1805;95:65–87.
9. Wenzel RN. Resistance of solid surfaces to wetting by water. Ind Eng Chem Res. 1936;28(8):988–94.
10. Cassie A, Baxter S. Wettability of porous surfaces. Trans Faraday Soc. 1944;40:546–51.
11. Good RJ. A thermodynamic derivation of Wenzel's modification of Young's equation for contact angles; together with a theory of hysteresis1. J Am Chem Soc. 1952;74(20):5041–2.
12. Johnson RE Jr, Dettre RH. Contact angle hysteresis. III. Study of an idealized heterogeneous surface. J Phys Chem. 1964;68(7):1744–50.
13. Öner D, McCarthy TJ. Ultrahydrophobic surfaces. Effects of topography length scales on wettability. Langmuir. 2000;16(20):7777–82.
14. Chen W, et al. Ultrahydrophobic and ultralyophobic surfaces: some comments and examples. Langmuir. 1999;15(10):3395–9.
15. Feng L, et al. Petal effect: a superhydrophobic state with high adhesive force. Langmuir. 2008;24(8):4114–9.
16. Bhushan B, Her EK. Fabrication of superhydrophobic surfaces with high and low adhesion inspired from rose petal. Langmuir. 2010;26(11):8207–17.
17. Teisala H, Tuominen M, Kuusipalo J. Adhesion mechanism of water droplets on hierarchically rough superhydrophobic rose petal surface. J Nanomater. 2011;2011:33.
18. Manoudis PN, Gemenetzis D, Karapanagiotis I. A comparative study of the wetting properties of a superhydrophobic siloxane material and rose metal. Sci Cult. 2017;3(2):7–12.
19. Karapanagiotis I, Aifantis KE, Konstantinidis A. Capturing the evaporation process of water drops on sticky superhydrophobic polymer-nanoparticle surfaces. Mater Lett. 2016;164(Supplement C):117–9.
20. Barthlott W, Neinhuis C. Purity of the sacred lotus, or escape from contamination in biological surfaces. Planta. 1997;202(1):1–8.
21. Zorba V, et al. Biomimetic artificial surfaces quantitatively reproduce the water repellency of a lotus leaf. Adv Mater. 2008;20(21):4049–54.
22. Manoudis PN, et al. Superhydrophobic composite films produced on various substrates. Langmuir. 2008;24(19):11225–32.
23. Chatzigrigoriou A, Manoudis PN, Karapanagiotis I. Fabrication of water repellent coatings using waterborne resins for the protection of the cultural heritage. Macromol Symp. 2013;331–332(1):158–65.
24. Manoudis PN, Karapanagiotis I. Modification of the wettability of polymer surfaces using nanoparticles. Prog Org Coat. 2014;77(2):331–8.
25. Aslanidou D, Karapanagiotis I, Panayiotou C. Superhydrophobic, superoleophobic coatings for the protection of silk textiles. Prog Org Coat. 2016;97:44–52.

26. Aslanidou D, Karapanagiotis I, Panayiotou C. Tuning the wetting properties of siloxane-nanoparticle coatings to induce superhydrophobicity and superoleophobicity for stone protection. Mater Des. 2016;108:736–44.
27. Wheeler G. Alkoxysilanes and the consolidation of stone. Los Angeles, CA: Getty Publications; 2005.
28. Mosquera MJ, de los Santos DM, Rivas T. Surfactant-synthesized ormosils with application to stone restoration. Langmuir. 2010;26(9):6737–45.
29. de Ferri L, et al. Study of silica nanoparticles–polysiloxane hydrophobic treatments for stone-based monument protection. J Cult Herit. 2011;12(4):356–63.
30. Facio DS, Mosquera MJ. Simple strategy for producing superhydrophobic nanocomposite coatings in situ on a building substrate. ACS Appl Mater Interfaces. 2013;5(15):7517–26.
31. Pedna A, et al. Synthesis of functionalized polyolefins with novel applications as protective coatings for stone cultural heritage. Prog Org Coat. 2013;76(11):1600–7.
32. Li D, et al. The effect of adding PDMS-OH and silica nanoparticles on sol–gel properties and effectiveness in stone protection. Appl Surf Sci. 2013;266:368–74.
33. Fermo P, et al. Hydrophobizing coatings for cultural heritage. A detailed study of resin/stone surface interaction. Appl Phys A. 2014;116(1):341–8.
34. Esposito Corcione C, Striani R, Frigione M. Hydrophobic photopolymerizable nanostructured hybrid materials: an effective solution for the protection of porous stones. Polym Compos. 2015;36(6):1039–47.
35. Luo Y, Xiao L, Zhang X. Characterization of TEOS/PDMS/HA nanocomposites for application as consolidant/hydrophobic products on sandstones. J Cult Herit. 2015;16(4):470–8.
36. Liu Y, Liu J. Synthesis of TEOS/PDMS-OH/CTAB composite coating material as a new stone consolidant formulation. Constr Build Mater. 2016;122:90–4.
37. Kapridaki C, Maravelaki-Kalaitzaki P. TiO₂–SiO₂–PDMS nano-composite hydrophobic coating with self-cleaning properties for marble protection. Prog Org Coat. 2013;76(2):400–10.
38. D'Amato R, et al. Development of nanocomposites for conservation of artistic stones. Proc Inst Mech Eng N J Nanoeng Nanosyst. 2014;228(1):19–26.
39. Ugur I. Surface characterization of some porous natural stones modified with a water-borne fluorinated polysiloxane agent under physical weathering conditions. J Coat Technol Res. 2014;11(4):639–49.
40. D'Orazio L, Grippo A. A water dispersed titanium dioxide/poly (carbonate urethane) nanocomposite for protecting cultural heritage: preparation and properties. Prog Org Coat. 2015;79:1–7.
41. Cappelletti G, Fermo P, Camiloni M. Smart hybrid coatings for natural stones conservation. Prog Org Coat. 2015;78:511–6.
42. Kronlund D, et al. Hydrophobization of marble pore surfaces using a total immersion treatment method–product selection and optimization of concentration and treatment time. Prog Org Coat. 2015;85:159–67.
43. Gherardi F, et al. Efficient self-cleaning treatments for built heritage based on highly photoactive and well-dispersible TiO₂ nanocrystals. Microchem J. 2016;126:54–62.
44. La Russa MF, et al. Nano-TiO₂ coatings for cultural heritage protection: the role of the binder on hydrophobic and self-cleaning efficacy. Prog Org Coat. 2016;91:1–8.
45. Tiwari MK, et al. Highly liquid-repellent, large-area, nanostructured poly (vinylidene fluoride)/poly (ethyl 2-cyanoacrylate) composite coatings: particle filler effects. ACS Appl Mater Interfaces. 2010;2(4):1114–9.
46. Basu BJ, Dinesh Kumar V. Fabrication of superhydrophobic nanocomposite coatings using polytetrafluoroethylene and silica nanoparticles. ISRN Nanotechnol. 2011;2011:803910.
47. Tserepi A, Vlachopoulou M, Gogolides E. Nanotexturing of poly (dimethylsiloxane) in plasmas for creating robust super-hydrophobic surfaces. Nanotechnology. 2006;17(15):3977.
48. Morra M, Occhiello E, Garbassi F. Contact angle hysteresis in oxygen plasma treated poly (tetrafluoroethylene). Langmuir. 1989;5(3):872–6.

Chapter 2
Advanced Conservation Methods for Historical Monuments

Jessica Rosewitz and Nima Rahbar

2.1 Introduction

On the evening of October 6, 2002, a fifteenth-century Carrara marble statue of *Adam*, by the Venetian sculptor Tullio Lombardo, fell and broke after the pedestal supporting the statue collapsed in the Velez Blanco Patio of the Metropolitan Museum of Art (MMA) in New York [1]. The accident caused the sculpture to fracture into 28 large pieces and hundreds of smaller ones. The statue was repaired by conservators and restored in 2014. The history of the statue and the subsequent research and treatment from the art conservator's point of view can be found in reported literature [2–4].

In general, the conservation of sculptures follows several principles. The original materials should remain intact, and the repairs should not cause long-term damage. The restoration should be completely reversible, if possible. In the case of using a pinning repair to reconnect a break, the pinning site should not cause crack initiation in the sculpture. Furthermore, the application of adhesives should be completely reversible. Specifically, within the bounds of the conservation of *Adam*, the goal was to maintain a tight join and a thin bond line with a reversible adhesive and a pinned repair. The breaks were fresh and the mating surfaces still fit tightly together, different than other historic sculptures with centuries old breaks. Reports have been published on the selection of adhesives that made the thinnest bond line, as this was critical to the recreation of *Adam* [5, 6]. A second but equally important consideration was to design a repair system with the least number of pins and drilled holes. The pin material was a critical choice [7]. From considerable study of the history and past applications [4, 8, 9], conservators needed a pin that would not create stresses above what Carrara marble can endure. The restored Adam statue is on

J. Rosewitz • N. Rahbar (✉)
Worcester Polytechnic Institute, Worcester, MA, USA
e-mail: jarosewitz@wpi.edu; nrahbar@wpi.edu

© Springer International Publishing AG 2018
M. Hosseini, I. Karapanagiotis (eds.), *Advanced Materials for the Conservation of Stone*, https://doi.org/10.1007/978-3-319-72260-3_2

Fig. 2.1 The restored
Adam statue

display at the MMA, Fig. 2.1. This chapter highlights the implications of engineering mechanics and numerical modeling in monument conservation.

One of the key components of the restoration process is the selection of the adhesives that can preserve historical pieces for centuries. Adhesives are widely used in monument conservation. However, conservators' basic understanding of the robustness of marble/adhesive interfaces is still limited. Furthermore, in most cases, robustness is assessed in terms of interfacial strengths that do not capture the combined effects of stress state, flaw size, and geometry. Fracture mechanics relates the crack length, the inherent resistance to crack growth, and the stress at which the crack propagates. Hence, fracture mechanics can address the prescribed parameters by estimating the interfacial fracture toughness and analyzing the interfaces between a range of adhesives and marble.

Prior work on the strength of marble/adhesive join has mainly focused on the fracture properties of marble/mortar compounds [10]. This has initiated considerable research into the use of adhesives in the preservation and restoration of historical structures [5]. The range of adhesives that can be used has been described by Horie [11]. Koob [12] has also studied the use of acrylic adhesives, while Jorjani [13] has extended the use of interfacial fracture mechanics to the analysis of joins in the restoration of historical monuments. Thermoplastic adhesives are generally weaker but more reversible than thermosetting adhesives, but both are used in the restoration of monuments. Therefore, there is a need to develop interfacial fracture mechanics techniques for the analysis of robustness in joins between marble structures and thermoplastic/thermosetting polymers. This criterion is addressed in this chapter.

However, under static loading conditions such as the self-weight of statues, it is possible for subcritical crack growth to occur in ceramics or the interfaces between adhesives and ceramics [5]. Since such interfacial crack growth can give rise to fracture under loading conditions that are typically considered to be safe for the long term, there is a need to consider the mechanisms of subcritical crack growth. Furthermore, since the adhesives can creep at room temperature, a likely mechanism of subcritical fracture is creep crack growth. The study of subcritical crack growth in adhesive/marble interfaces will be further discussed in the chapter.

For the selection of the optimum pin materials, experimental and numerical studies of adhesives and structural pin materials were conducted. FEA were also used to study the failure mechanisms in the repaired system. By understanding the possible failure mechanisms in the join repair, this study identified the adhesive and pin material most suitable to repair this statue. These experiments provided insight into the overall behavior of a pinned join under loading and guidance on the best pin material to use in this conservation effort. Further discussion regarding the study of pinning materials is provided in this chapter.

2.2 Study of Mixed Mode Fracture of Marble/Adhesive Interfaces

The section presents the results of a combined experimental and theoretical study of the interfacial fracture between a range of adhesives and Carrara marble using Brazil nut specimens [14, 15]. The interfacial fracture toughness is predicted and measured for joins produced with thermoplastic and thermosetting resins, as well as combinations applied in sequence. The results are discussed with regard to robustness and the potential to intentionally undo joins.

2.2.1   Materials

2.2.1.1   Acrylic Resins

Acrylate and methacrylate acrylic resins have been common in art conservation since the 1950s, when Lucite 44, a polybutyl methacrylate, was used as a varnish for oil paintings [16], and still fulfill criteria for present-day art conservation [17]. Copolymers with a desired glass transition temperature (Tg) can be made by varying the percentage of each monomer [16]. Two such acrylic copolymers, Paraloid B-72 and Paraloid B-48N (Rohm & Haas, Philadelphia, PA) as well as a 3:1 mix of B-72/B-48 N, were tested. Paraloid B-72 is a copolymer of ethyl methacrylate/ methyl acrylate with a Tg of 45 °C [12]. In art conservation, this resin is used as a coating, consolidant, and adhesive. For this study, a 40% by weight solution of B-72 was prepared in an acetone/ethanol solvent solution. Paraloid B-48N is a copolymer of methyl methacrylate and butyl acrylate with a Tg of 69 °C and often used as a protective film for canvas paintings and metal. A 40% by weight solution of B-48N was prepared in an acetone/ethanol solvent solution. A mixture of Paraloid B-72 and B-48N at the ratio of 3:1 (by weight) with a Tg of 46 °C was produced by Jorjani and reported in the literature [18]. Further details of the production of this acrylic resin blend can be found in the literature [13].

2.2.1.2   Polyvinyl Acetate Resins

Polyvinyl acetate resins are made by a reaction between polyvinyl alcohol and an aldehyde. Mowital B60HH (Kuraray America, Inc., Houston, TX), a polyvinyl butyral (PVB), was used in this study. It has a Tg of 65 °C, placing it between the relatively low Tg of Paraloid B-72 and the high Tg of PMMA, at 105 °C [19]. It was made as a 40% by weight solution with ethanol. No reference to the use of PVB as a stone adhesive was found, although it has been used as an adhesive for glass and wood and as a consolidant for wood, textiles, fossils, and paper [16].

2.2.1.3   Polyester Resins

The use of polyesters in art conservation began in the 1940s, and they have been used as consolidants and fillers for wood, as well as adhesives and consolidants for stone. While polyesters cannot be dissolved in organic solvents, they can be removed after swelling [16]. The polyester tested was Marmorkitt 1000 (Akemi North America, Holbrook, NY), a commercially available thermosetting adhesive that has a Tg of ~70 °C [20]. Polyester resins are mixed in a two-step process, in which an initiator is mixed with a resin prepolymer containing a reactive monomer, normally styrene. The polymerization first results in a gel and then a hard solid [16]. Adjusting the amount of initiator used at room temperature can regulate the working time to 12–20 min. Some amount of shrinkage is involved in the polymerization process.

2.2.1.4   Epoxy Resins

Epoxy resins consist of an epoxide component that reacts with a hardener. Diglycidyl ether of bisphenol-A (DGEBA) is the epoxide component, while the hardeners are aliphatic amines and amides. Used widely in the conservation of glass, stone, and wood, epoxy resins are thermosetting adhesives that are known for strength and creep resistance [21], but some are susceptible to yellowing when exposed to UV [22]. Epotek 301-2 (Epoxy Technology, Billerica, MA), HXTAL NYL-1 (Restorer Supplies Inc., Naples, FL) and Akepox 2000 (Akemi North America, Holbrook, NY) were used in this study. All three are clear, colorless, and optical grade adhesives, and specific manufacturer information is supplied here. Epotek 301-2 has a Tg ~80 °C and is specified for resistance to impact and vibrations. HXTAL NYL-1 has a Tg of ~50 °C and is not susceptible to discoloration under UV exposure. Akepox 2000 is reported to be thermally stable up to 60–70 °C, and its low viscosity makes very thin joins possible, but it is susceptible to discoloration.

2.2.2   Experimental Procedures

The adhesives described were used to attach semicircular half disks of Brazil nut specimens cut from Carrara marble (Fig. 2.2) [15, 16]. Interfacial fracture toughness was measured over a wide range of mode mixities by varying the angle between the direction of the loading axis and the long axis of the notch. Full details of the Brazil nut specimen preparation can be found in literature [5]. The adhesives used are summarized in Table 2.1.

2.2.3   Interfacial Fracture Testing

Interfacial fracture tests were conducted in an 8872 Instron servo-hydraulic dual-column mechanical analyzer-controlled testing machine (Instron, Canton, MA). Testing was performed under displacement control at a ramp rate of 5 μm/s, and load/displacement data were recorded. The critical load, P_C, at which the specimen failed, was recorded for each loading angle, θ. The relation between the critical load, P_C, and interfacial fracture toughness, G_C, is described in full detail in published literature [5]. The corresponding total energy release rate, G, is given by the sum of the energy release from modes I and II:

Fig. 2.2 Schematic of
Brazil nut specimen [6].
Reprinted with permission
from Materials Science
and Engineering: A,
528/10-11, Ting Tan, Nima
Rahbar, Andrea Buono,
George Wheeler, Wole
Soboyejo. Sub-critical
crack growth in adhesive/
marble interfaces,
3697–3704, Copyright
(2011), Elsevier

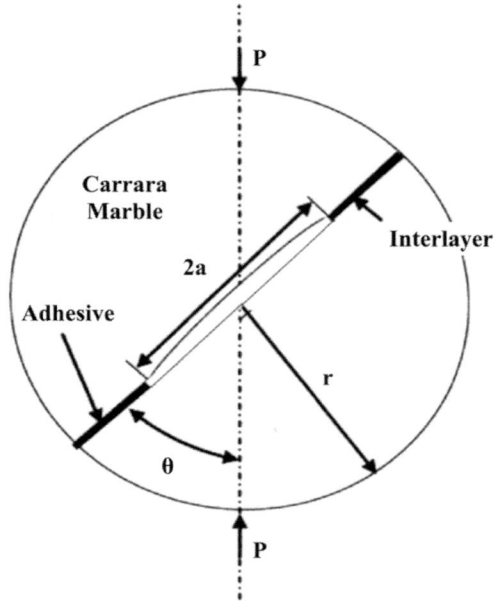

Table 2.1 Mechanical properties of adhesive materials

Material	E (GPa)	Poisson's ratio	α	ε	ω
Epotek 301-2	3.66	0.36	0.88	−0.0614	−11.9
Mowital B60HH	2.37	0.3–0.4	0.92	−0.0685	−12.1
HXTAL NYL-1	2.52	0.3–0.4	0.92	−0.0681	−12.1
B-48	3.0–3.5	0.3–0.4	0.92	−0.0662	−12.0
B-72	3.0–3.5	0.3–0.4	0.92	−0.0662	−12.0
B-72/B-48	3.0–3.5	0.3–0.4	0.92	−0.0662	−12.0
Marmorkitt 1000	3.0–3.5	0.3–0.4	0.92	−0.0662	−12.0
Akepox 2000	3.0–3.5	0.3–0.4	0.92	−0.0662	−12.0

$$G = G_\mathrm{I} + G_\mathrm{II} = \frac{K_\mathrm{I}^2 + K_\mathrm{II}^2}{E'} \tag{2.1}$$

where K_I and K_II are the mode I and II stress intensity factors, respectively, and E' is
the plane strain Young's modulus of the substrate. The corresponding mode mixity,
ψ, controlled by the loading angle, θ [23], is given by

$$\psi = \tan^{-1}\left(\frac{f_\mathrm{II}}{f_\mathrm{I}}\right) + \omega + \varepsilon \ln\left(\frac{L}{h}\right) \tag{2.2}$$

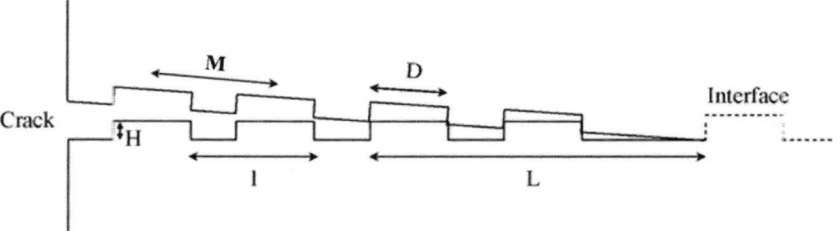

Fig. 2.3 Schematic of the zone model [5]. Reprinted with permission from Materials Science and Engineering: A, 527/18-19, N. Rahbar, M. Jorjani, C. Riccardelli, G. Wheeler, I. Yakub, Ting Tan, W.O. Soboyejo. Mixed mode fracture of marble/adhesive interfaces, 4939–4946, Copyright (2010), Elsevier

where ω is a shift parameter accounting for dissimilar Lame's constants between the substrate and adhesive and ε is a Dundurs parameter, both tabulated in the literature [24]. The dimensions L and h are approximately equal to the thickness of the adhesive layer.

Equations (2.1) and (2.2) were used to calculate the energy release rate and mode mixities for the Brazil nut specimens tested over a range of mode mixities by varying the loading angle, θ, between 0° and 25° (Fig. 2.2). The fracture toughness of Carrara marble was also measured over the same range of mode mixities. The crack profiles and fracture modes were examined in a scanning electron microscope (SEM). The crack profiles were then incorporated into zone shielding models that were used for the estimation of the mode mixity dependence of the interfacial fracture toughness.

2.2.4 Modeling

Evans and Hutchinson have previously used asperity contact models to study the dependence of interfacial fracture toughness on mode mixity, called the zone model. Full details on the derivation of the interfacial fracture toughness, G_C, based on the zone model for these experiments have already been reported [5]. It is valid for the simplified contact conditions depicted in Fig. 2.3 in which the shear stresses and displacements are elastic and analogous to those associated with a linear array of microcracks. As illustrated in Fig. 2.3, L is the zone length, H is the height of the interface step, D is the facet length, and M is the spacing between facet (microcracks) centers. For all adhesives considered in this study, the interfacial fracture toughness, G_C, is a function of the local energy release rate, G_0, and the mode mixity, ψ, which is given by

$$G_C = G_0 \left(1 + \tan^2 \psi \right) \tag{2.3}$$

The interfacial fracture toughness shown in Eq. (2.3) was used to model the effects of mode mixity on the interfacial fracture toughness.

2.2.5   Results and Discussion

2.2.5.1   Bond Line Width

The smooth-surface adhesive bonds measured from 26 to 46 μm, smaller than the range stated by Podany et al.; the largest bond width for the smooth surface, at 46 μm, was observed with the Akepox 2000. The largest bond width for the fractured surface was observed in the fractured B-72/Epotek sandwich, at 58 μm. Seven of nine adhesive layers were thicker in the fractured samples than in the smooth samples. Dry joins of both pre-fractured and smooth Brazil nut specimens had an average resulting bond width of 23 and 21 μm, respectively. The difference of ~2 μm between the two types does not explain the discrepancy between the widths of the two surfaces after the adhesive is applied, reaching a difference of almost 30 μm in the B-72/Epotek composite system.

2.2.5.2   Fracture Toughness

The mode mixity dependence of the fracture toughness of Carrara marble is presented in Fig. 2.4a. Similarly, the measured interfacial fracture toughness values for different classes of adhesives on pre-fractured and smooth surfaces are presented as functions of mode mixity in Fig. 2.4b–j. This also shows that the interfacial toughness is higher for the pre-fractured samples compared to the smooth samples for all adhesives. As for many materials [14, 15], the fracture toughness increases with increasing mode mixity. However, the interfacial toughness values for PVB are essentially independent of mode mixity (Fig. 2.4d). This increase of fracture toughness with mode mixity is attributed largely to the effects of contact-induced crack tip shielding that will be discussed in the next section. Over the range of phase angles measured, all adhesives exhibit similar interfacial fracture toughness values.

2.2.5.3   Crack/Microstructure Interactions

Crack/microstructure interactions are important in interface robustness. Typical crack/microstructure profiles occurred at the interfaces (Fig. 2.5a–c) and were schematically sketched (Fig. 2.5d–f). Type I failure was observed when an interface crack initiated and remained at the interface with kinks (Fig. 2.5a), corresponding to mode I failure in which a crack opens due to normal stresses orthogonal to the crack plane. The interfacial fracture toughness was computed with these failure loads.

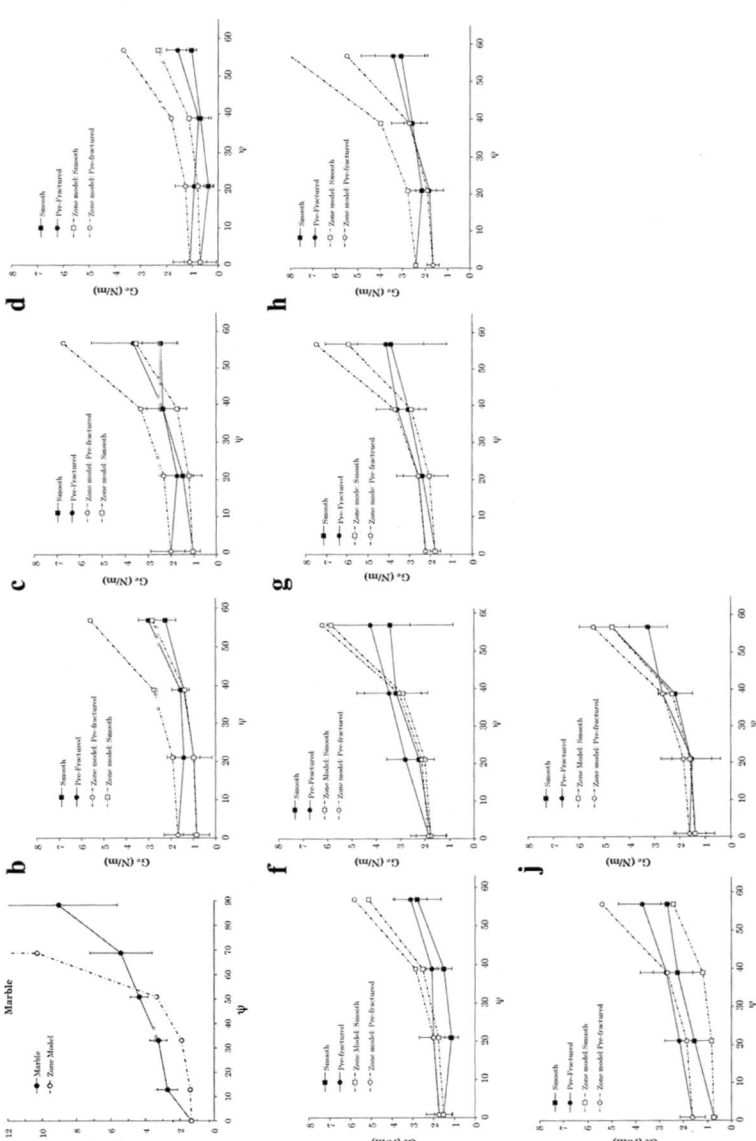

Fig. 2.4 (**a**) Fracture toughness of bulk marble as a function of mode mixity. (**b–d**) Interfacial fracture toughness of marble/thermoplastic adhesive acrylic resin: (**b**) B-48; (**c**) B-72 and polyvinyl acetate, and (**d**) Mowital B60HH. (**e–h**) Interfacial fracture toughness of marble/thermosetting adhesive polyester resin: (**e**) Marmorkitt 1000; (**f**) Akepox 2000; (**g**) Epotek 301-2; and (**h**) HXTAL NYL-1. (**i, j**) Interfacial fracture toughness of marble/mixture: (**i**) B-72/B-48 and acrylic resin and epoxy resin applied in sequence; and (**j**) B-72 Epotek [5]. Reprinted with permission from Materials Science and Engineering: A, 527/18-19, N. Rahbar, M. Jorjani, C. Riccardelli, G. Wheeler, I. Yakub, Ting Tan, W.O. Soboyejo. Mixed mode fracture of marble/adhesive interfaces, 4939–4946, Copyright (2010), Elsevier

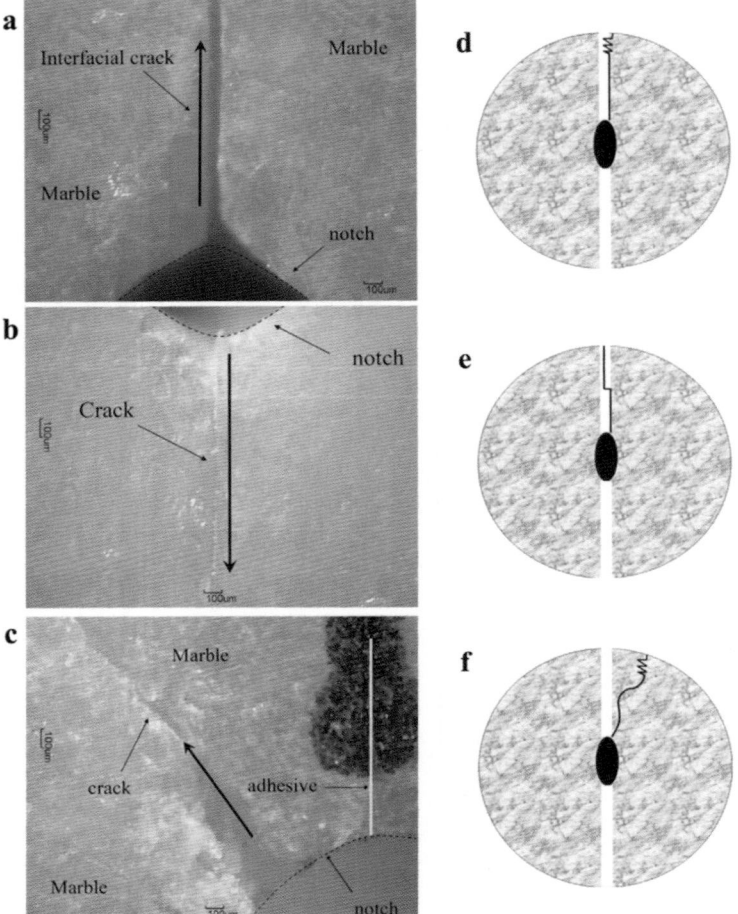

Fig. 2.5 (**a–c**) Crack/microstructure interaction for a crack advancing in the marble for (**a**) an interfacial crack, (**b**) a crack advancing in the adhesive, and (**c**) a crack advancing in the marble. (**d–f**) Schematic illustration of typical fracture patterns observed: (**d**) type I fracture pattern, (**e**) type II fracture pattern, and (**f**) type III fracture pattern [5]. Reprinted with permission from Materials Science and Engineering: A, 527/18-19, N. Rahbar, M. Jorjani, C. Riccardelli, G. Wheeler, I. Yakub, Ting Tan, W.O. Soboyejo. Mixed mode fracture of marble/adhesive interfaces, 4939–4946, Copyright (2010), Elsevier

Type II failure in the adhesive/marble interfaces occurred predominantly in the polyvinyl acetal resin smooth samples, with fracture starting at the interface and kinking into the adhesive (Fig. 2.5b), corresponding to mode II failure in which crack surface slides across one another, due to in-plane shear. Type III failure occurred by the kinking of the cracks into the marble (Fig. 2.5f), corresponding to mode III failure in which the crack tears open out of plane, due to out-of-plane shear. Generally, higher mode failures occurred for interfaces where the interface toughness is higher than marble toughness.

2.2.5.4 Modeling

The predicted fracture toughness values obtained from the zone model are presented in Fig. 2.4b–j and compared to the experimental data. The overall predictions from the zone model using Eq. (2.3) for Carrara marble and the three types of adhesives are greater than the experimental fracture toughness values. The fracture toughness of pre-fractured samples was greater than that of the smooth samples for all types of adhesives.

Fracture toughness of Carrara marble increases with the increase in mode mixity (Fig. 2.4a). For the case of Carrara marble, the fracture toughness predictions from the zone model as a function of mode mixity appear to be an upper bound to experimental values. Although far from ideal, it presents the fracture behavior of a brittle material. The increase in fracture toughness as a function of mode mixity is attributed mainly to the presence of surface forces that inhibit sliding of crack surfaces [25].

In the case of pre-fractured samples prepared with acrylic resins (Fig. 2.4b–d), the zone model predictions appear to be an upper bound for the experimental results, indicating the interfacial fracture toughness is mode II dependent [26]. In the case of smooth samples prepared with acrylic resins, the predictions from the zone model in acrylic thermoplastics are in good agreement with the experimentally measured interfacial fracture toughness values for a wide range of mode mixity. Marble/PVB adhesive interfaces exhibit straightforward fracture behavior, independent of mode mixity. Such behavior is consistent with the observed crack growth scenarios in these samples.

The mode mixity dependence of the interfacial fracture toughness values of the thermosetting adhesives is presented in Fig. 2.4e–h. Smooth samples prepared with epoxy resin adhesives exhibited similar behavior to the predictions from the zone model. The pre-fractured sample of epoxy resins also exhibits similar trends, except for the HXTAL NYL-1(Fig. 2.4h), for which the predictions are greater than the experimental measurements. For the polyester resin adhesive, the predictions for smooth and pre-fractured samples are upper bounds. This clearly presents the contribution of mode II fracture toughness, KII, in addition to KI, to the roughness induced shielding mechanism in the toughness of polyester resin/marble interfaces. Finally, in the case of the B-72/epoxy resin adhesive system, for both smooth and pre-fractured samples, the predictions from the zone model are in good agreement with the experimental values.

Therefore, the predictions from the zone model appear to be essentially upper-bound predictions for the interfacial fracture toughness data. This clearly indicates the presence of contact-induced shielding mechanisms as suggested by the observed crack/microstructure interactions that are presented in Fig. 2.5.

2.2.5.5   Implications

The current work provides the basis for the design and repair of historical monuments fabricated from Carrara marble. Critical stresses and crack lengths associated
with a range of mode mixities are estimated from measurements of fracture toughness. The interfacial fracture toughness of interfaces is similar, even though the
strengths of materials vary significantly. The critical crack lengths and stresses
should, therefore, be similar to the range of join materials examined in this study.
The mode mixity dependence of the interfacial fracture toughness values is also
well predicted by the zone contact shielding model. This provides a quantitative
basis for design and fabrication of robust monuments that will not fail under monotonic loads or the self-weight of such structures in stationary positions. This work
also provides the basis for the ranking of different adhesives that can be used for the
repair of such structures. However, it is important to note that subcritical crack
growth may still occur in such structures under conditions in which the crack driving forces are less than those required for crack growth under monotonic loading. In
such scenarios, slow crack growth may occur within the marble or the interfaces,
eventually giving rise to catastrophic failure at the interface between marble and
adhesive. Owing to the complicated nature of bonding fractured stone surfaces, it
would be impossible to prevent flaws at the adhesive/marble interfaces. Such flaws
could be initiation sites of these eventual catastrophic cracks. The following section
will detail the estimation of subcritical interfacial crack growth under such static
loading and provide life span predictions for selected adhesive/marble interfaces.

2.3   Study of Subcritical Crack Growth in Adhesive/Marble Interfaces

This section presents the results of an experimental study of interfacial creep crack
growth between Carrara marble and adhesives relevant to the restoration of *Adam*.
Subcritical crack growth is studied using Brazil nut fracture mechanics experiments.
The interfacial crack growth rates are compared to Carrara marble and thermoset-
ting/thermoplastic adhesives. The implications of the results are discussed for the
long-term preservation of historical monuments.

2.3.1   Materials

The materials used in this study include Carrara marble and various adhesives supplied by the MMA in New York. Their mechanical properties are summarized in
Table 2.1, as measured in reported literature [6]. The acrylic resin adhesives and the
epoxy resin adhesives used in this study were previously described.

2.3.2 Experimental Procedure

Sample preparation used in this study was described previously. Both the fractured surface and smooth surface half disks relied on adhesives in between to bond them together. This process along with curing of adhesives was described in previous sections.

2.3.3 Interfacial Creep Crack Growth

Interfacial creep crack growth was studied using Carrara marble Brazil nut specimens (Fig. 2.2). The range of the loading angle, θ, between the loading direction and the initial notch simulates the full range of mode mixities between pure modes I and II. Full details of the experimental procedure to monitor crack growth can be found in literature [6].

2.3.4 Modeling

2.3.4.1 Interfacial Fracture Mechanics

The total energy release rate for the Brazil nut specimens with thin adhesive layers is dependent on the mode I and II stress intensity factors K_I and K_{II}, the plane strain Young's modulus E', the applied load P, and half the crack length a. The stress intensity factors are given in the literature [14, 27]:

$$K_I = f_I P \sqrt{a} \tag{2.4}$$

$$K_{II} = f_{II} P \sqrt{a} \tag{2.5}$$

The total energy release rate G is given as the sum of the energy release from modes I and II:

$$G = G_I + G_{II} = \frac{K_I^2 + K_{II}^2}{E'} = \frac{\left(f_I^2 + f_{II}^2\right)P^2 a}{E'} \tag{2.6}$$

With this approximation, the observed crack growth rates were compared to the total energy release rate for this study.

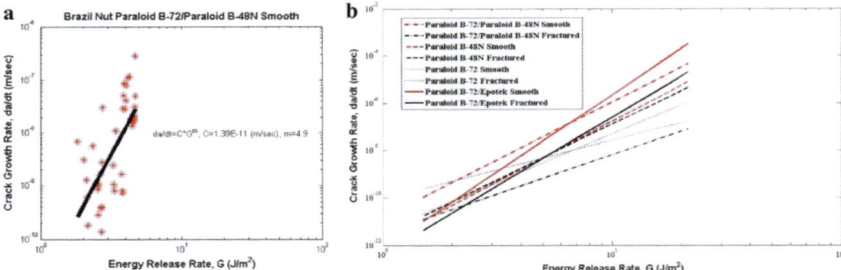

Fig. 2.6 (**a**) Interfacial creep crack growth rate data for Paraloid B-72/B-48 adhesive on smooth samples; (**b**) Best fit models of crack growth rate versus the energy release rate data for different adhesives [6]. Reprinted with permission from Materials Science and Engineering: A, 528/10-11, Ting Tan, Nima Rahbar, Andrea Buono, George Wheeler, Wole Soboyejo. Sub-critical crack growth in adhesive/marble interfaces, 3697–3704, Copyright (2011), Elsevier

2.3.4.2 Interfacial Crack Growth

The prediction of crack extension as a function of time [28] relates the energy release rate to the crack growth rate with a power law expression, which can be integrated as follows to obtain predictions of crack growth lives. This model is dependent on the initial flaw size, a_0; the final flaw size, a_f; the material constants C and m from the power law expression; the initial stress, σ_0; and the plane strain Young's modulus, E'. Thus, lifetime of a specific marble/adhesive interface is given as

$$\int_{a_0}^{a_f} \frac{\mathrm{d}a}{a^m \left(N_{\mathrm{I}}^2 + N_{\mathrm{II}}^2 \right)^m} = \int_0^{t_f} C \left[\frac{\sigma_0^2}{E'} \right]^m \mathrm{d}t \tag{2.7}$$

Full details of the derivation of Eqs. (2.6) and (2.7) have been published in a previous study [6]. In this way, the potential interfacial creep crack growth lives were compared to the different interfaces examined in this study.

2.3.5 Results and Discussion

2.3.5.1 Interfacial Crack Growth Rates

The interfacial creep crack growth rate data for Paraloid B-72/48N on smooth Brazil nut samples is presented in Fig. 2.6a using Eq. (2.6). The creep crack growth rate data for various adhesives obtained from the current study (Fig. 2.6b) along with the Paris fits (Table 2.2) were generally good across the range of experimental data. Furthermore, the Paris constants C and m, obtained for BN72/48S and BN72/48F,

Table 2.2 C and m constants for various adhesive/marble interfaces

Notation	
BN 48 S	Paraloid B-48N resin on smooth Brazil nut specimens
BN 48 F	Paraloid B-48N resin on fractured Brazil nut specimen
BN 72 S	Paraloid B-72 resin on smooth Brazil nut specimen
BN 72 F	Paraloid B-72 resin on fractured Brazil nut specimen
BN 72 EP S	Paraloid B-72 resin with Epotek sandwich on smooth Brazil nut specimen
BN 72 EP F	Paraloid B-72 resin with Epotek sandwich on fractured Brazil nut specimen
BN 72/48 S	Paraloid B-72/B-48N resin combination on smooth Brazil nut specimen
BN 72/48 F	Paraloid B-72/B-48N resin combination on fractured Brazil nut specimen

Paris' power law constants			
Material	BN 48 S	BN 48 F	BN 72 S	BN 72 F
C (m/s)	1.33 E-12	2.53 E-12	3.91 E-12	8.43 E-11
m	5.1	4.7	4.1	2.5
Material	BN 72 EP S	BN 72 EP F	BN 72/48 S	BN 72/48 F
C (m/s)	6.99 E-13	3.98 E-13	1.39 E-11	3.20 E-12
m	6.5	5.8	4.9	3.3

were close to the values obtained for BN72S and BN72F, as well as BN48S and BN48F. The one order of magnitude lower C and m values obtained for BN72EPS and BN72EPF suggested slower crack growth rates for the combination thermosetting/thermoplastic resins.

2.3.5.2   Crack Growth Mechanisms

Environmental SEM images (FEI Quanta 200 E-SEM, FEI, Hillsboro, OR) of Carrara marble/adhesive interfaces on Brazil nut specimens (Fig. 2.7a–i) reveal evidence of interfacial crack bridging. There was also microvoid formation in the adhesive layers for the side profiles of each bi-material pair, suggesting a microvoid creep mechanism in the adhesive layers [5]. The extent of these mechanisms is much less in the Epotek-based bi-material interfaces (Fig. 2.7h, i). Furthermore, more tortuous crack paths were observed in the bi-material interfaces between the bonded fracture specimens (Fig. 2.7e–i). However, no clear trends were observed between the subcritical crack growth rates and the incidence of crack deflection (Fig. 2.7a–i).

The crack/microstructure interactions in the combinations of Paraloid B-72/48N specimens are shown in Fig. 2.7g. There is evidence of crack initiation along the Carrara marble/adhesive interface, and a deformed bridge is also observed beside the crack tip regime, an evidence of toughening/crack tip shielding [28, 29]. In order to locate the crack path after fracture, energy-dispersive x-ray spectroscopy (EDX) analysis was carried out (FEI Quanta 200 E-SEM, FEI, Hillsboro, OR). SEM images of pairwise fracture surfaces on the Paraloid B-48N specimens are presented (Fig. 2.7j, k). The results obtained from the EDX analysis of the same fracture

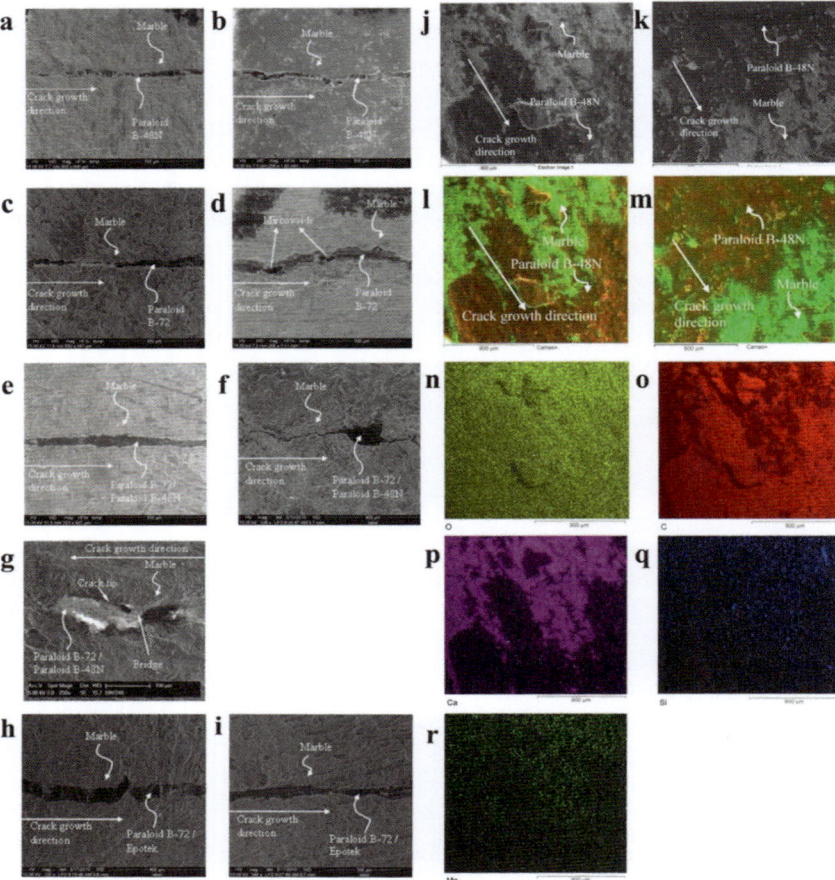

Fig. 2.7 (**a–d**) Interface between Carrara marble and different thermoplastic adhesives: (**a**) the Paraloid B-48N smooth surface, (**b**) Paraloid B-48N fracture surface, (**c**) Paraloid B-72 smooth surface, and (**d**) Paraloid B-72 fracture surface. (**e–g**) Interface between Carrara marble and thermoplastic adhesive combination: (**e**) Paraloid B-72/B-48N smooth surface, (**f**) Paraloid B-72/B-48N fracture surface, and (**g**) interaction between crack and microstructure of marble/adhesive interface. (**h, i**) Interface between Carrara marble and thermosetting adhesive: (**h**) Paraloid B-72/Epotek smooth surface and (**i**) Paraloid B-72/Epotek fracture surface. (**j, k**) SEM images of mating halves of the Paraloid B-48N Brazil nut specimens (**j**) left and (**k**) right. (**l**) EDX analysis on fracture surface of Brazil nut specimen with fractured surface and (**m**) integrated EDX analysis of mating Paraloid B-48N Brazil nut specimen halves; element maps of fracture surface (**l**) are in (**n–r**). (**n–r**) Element distribution of the fracture surface: (**n**) oxygen, (**o**) carbon, (**p**) calcium, (**q**) silicon, and (**r**) magnesium [6]. Reprinted with permission from Materials Science and Engineering: A, 528/10-11, Ting Tan, Nima Rahbar, Andrea Buono, George Wheeler, Wole Soboyejo. Subcritical crack growth in adhesive/marble interfaces, 3697–3704, Copyright (2011), Elsevier

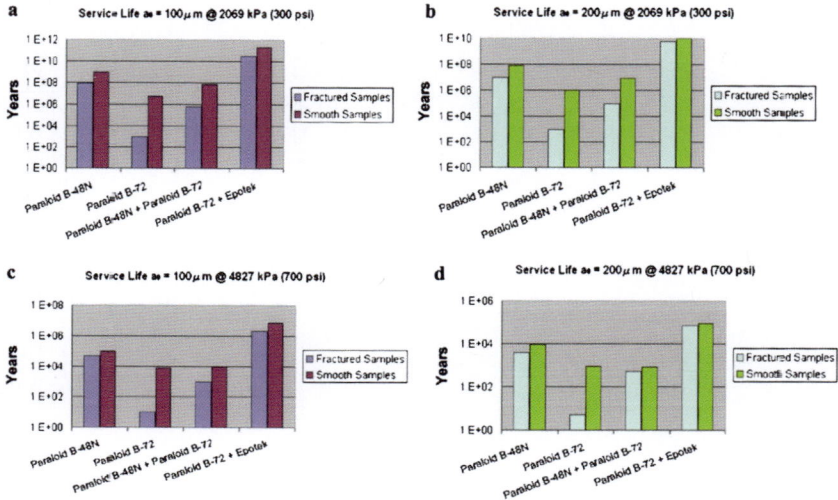

Fig. 2.8 Service life predictions for different marble/adhesive interfaces with different stress states: (**a**) 2069 kPa (300 psi) with initial flaw size a_0 = 100 μm, (**b**) 4827 kPa (700 psi) with initial flaw size a_0 = 100 μm, (**c**) 2069 kPa (300 psi) with initial flaw size a_0 = 200 μm, and (**d**) 4827 kPa (700 psi) with initial flaw size a_0 = 200 μm [6]. Reprinted with permission from Materials Science and Engineering: A, 528/10-11, Ting Tan, Nima Rahbar, Andrea Buono, George Wheeler, Wole Soboyejo. Sub-critical crack growth in adhesive/marble interfaces, 3697–3704, Copyright (2011), Elsevier

surfaces are presented in elemental maps (Fig. 2.7l–r). Combined, these suggest that interfacial crack growth occurred with some incidence of kinking in between the adhesive layers.

2.3.5.3   Crack Growth and Life Predictions

The predicted service lives obtained from Eq. (2.7) are summarized in Fig. 2.3. Two stress conditions, 2069 kPa (300 psi) and 4827 kPa (700 psi), simulated the real stress states using initial flaw size estimates (100 and 200 μm) in the marble structure [5]. For the 2069 kPa (300 psi) stress state, the service life predictions were greater than those from the 4827 kPa (700 psi) stress state. This is a reasonable prediction since higher remote stress facilitates crack growth. For the 4827 kPa (700 psi) condition with an initial flaw size of 200 μm, predicted lives obtained for Paraloid B-72 fractured and smooth were comparable to those predicted for Paraloid B-72/Paraloid B-48N fractured and smooth samples at several hundred years of the life. The predicted lives of the Paraloid B-72/Paraloid B-48N combinations are several thousand years, consistent with the life spans of several large-scale marble structures. However, the predicted lives obtained for Paraloid B-72/Paraloid B-48N fractured and smooth samples are over 10,000 years.

2.3.5.4 Implications

A protocol has been established for exploring subcritical crack growth in marble restorations which can occur at crack driving forces well below the critical conditions obtained from fracture toughness tests [1]. This can be used to rank crack growth resistance while providing vital inputs into the fracture mechanics estimation of structural lives. The mixed Paraloid B-72/Epotek samples emerged with the best combination of slow crack growth rates and predicted structural lives (Figs. 2.6b and 2.8). This was especially true for specimens that were bonded with combination thermoplastic/thermosetting systems. The slower crack growth rates in these systems are consistent with the observed high levels of crack tip shielding via crack bridging (Fig. 2.7g–i). Since actual restorations rely on the bonding of fractured pieces, these mixed adhesive results are most encouraging.

2.4 Study of Pinning Materials for Join Repair

2.4.1 Materials and Methods

The mechanical performances of six pinning materials as join reinforcement for the marble statue *Adam* were studied. The behavior of each material as a join reinforcement was experimentally and numerically investigated. Table 2.3 reports the pin materials and mechanical properties used in the experiments finite element simulations.

Table 2.3 Mechanical properties of materials used in finite element simulations

Material	Young's modulus (GPa)	Poisson's ratio
Carrara marble	49.0	0.19
Carbon fiber-reinforced polymer	28.2	0.30
Fiberglass, structural	14.8	0.30
Polycarbonate	3.90	0.37
Stainless steel, Type 316	200	0.27
Teflon® PTFE	1.80	0.46
Titanium, Grade 2	100	0.37
Akemi® Akepox® 2000 epoxy	3.3	0.30

Reprinted with permission from Materials and Design, 98, Jessica Rosewitz, Christina Muir, Carolyn Riccardelli, Nima Rahbar, George Wheeler. A multimodal study of pinning selection for restoration of a historic statue, 294–304, Copyright (2016), Elsevier

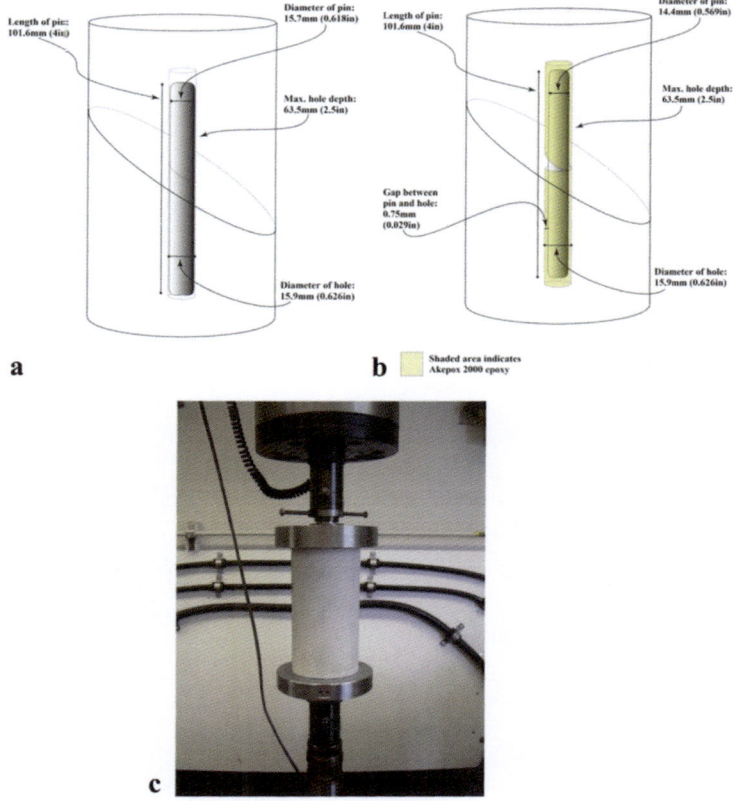

Fig. 2.9 (**a**) Schematic of the simulated dry join repair; (**b**) schematic of the simulated wet join repair with epoxy to bond the pin to the marble; and (**c**) experimental setup of a dry repair with fiberglass pin [7]. Reprinted with permission from Materials and Design, 98, Jessica Rosewitz, Christina Muir, Carolyn Riccardelli, Nima Rahbar, George Wheeler, A multimodal study of pinning selection for restoration of a historic statue, 294–304, Copyright (2016), Elsevier

2.4.1.1   Experimental Method

Mock join repair specimens were fabricated and tested to study the effects of stiffness and strength of the six pin materials on the mechanical performance of the repair (Fig. 2.9). Two sample sets of dry (without adhesive) and wet (with adhesive) join repairs were created. A wet join repair is commonly used in monument conservation, as opposed to dry join repair. Full details of the construction of the wet and dry join repairs can be found in literature [7]. The dry join is schematically represented in Fig. 2.9a and the wet join in Fig. 2.9b.

Uniaxial compression tests were performed on the dry and wet join repair specimens to simulate the overall load capacity (Fig. 2.9c) with a servo-hydraulic system until failure with an ultimate capacity 100 kN load cell (Instron 8501, Instron, Norwood, MA, USA). The experimental setup applied shear stresses in the pin and combined compression/tension and shear stresses in the marble substrate. The force and displacement were recorded during testing until failure occurred, or the maximum 95 kN load was reached.

2.4.1.2   Numerical Analysis of Join Repair

FEA was performed to further understand the mechanical performance of the join repair (Abaqus/CAE 6.12-2, Dassault Systèmes Simulia Corp., Johnston, RI, USA). Full details of the numerical analysis performed have already been reported [7]. Simulations followed the experimental setup shown in Fig. 2.9 with material properties presented in Table 2.3. The same tolerances were built into the FEA; therefore a slight amount of flexural deformation occurred before contact between the sides of the pin and the marble during simulated compressive testing. FEA provided the force-displacement curves and detailed stress distribution in the marble Carrara marble substrate around the hole for the six pin materials. A model was also created using a Carrara marble pin.

2.4.2   Experimental Results

2.4.2.1   Dry Join Repair Specimens

Testing was conducted until failure which was characterized as excessive horizontal join misalignment, visible failure of the marble substrate, pin shear, or the machine reached maximum load. There is inconsistent damage to the pin and marble at the interface in the dry repair specimens across the six pin materials. The force-displacement curves for dry repair specimens present initial soft behavior but then swiftly harden nonlinearly into a steep linear relationship indicative of hard contact with frictional behavior.

Due to their high stiffness and strength, the stainless steel, titanium, and carbon fiber pins crushed the marble. These specimens showed little horizontal join misalignment, and no fracture was visible on the marble surface. High compressive stresses fused the marble and metal pins, locking the crushed join together. Experiments with the stainless steel pin (Fig. 2.10a) exhibit a stiffer behavior and reached maximum force earlier than the experiments with the less stiff titanium pin (Fig. 2.10b).

The experimental results for the carbon fiber (Fig. 2.10c) and fiberglass pin (Fig. 2.10d) showed elastic to plastic behavior, explained by breakage of the pin's fibers, shearing, and pullout within the pin matrix material (Fig. 2.11a, b). The fiberglass pin shears along the repair plane and the marble fractures tangentially to the pin hole through the entire marble specimen (Fig. 2.11c). Two carbon fiber repair

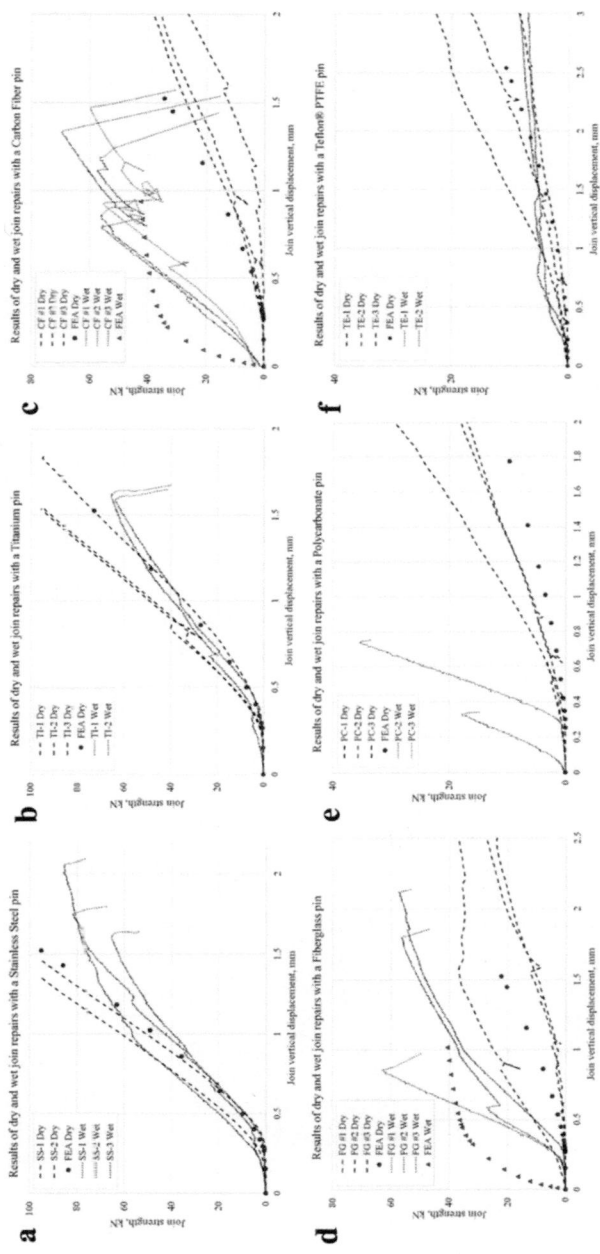

Fig. 2.10 (**a–f**) Force-displacement vs. FEA simulation results: (**a**) stainless steel pin, (**b**) titanium pin, (**c**) carbon fiber pin, (**d**) fiberglass pin, (**e**) polycarbonate pin, and (**f**) Teflon pin [7]. Reprinted with permission from Materials and Design, 98, Jessica Rosewitz, Christina Muir, Carolyn Riccardelli, Nima Rahbar, George Wheeler. A multimodal study of pinning selection for restoration of a historic statue, 294–304, Copyright (2016), Elsevier

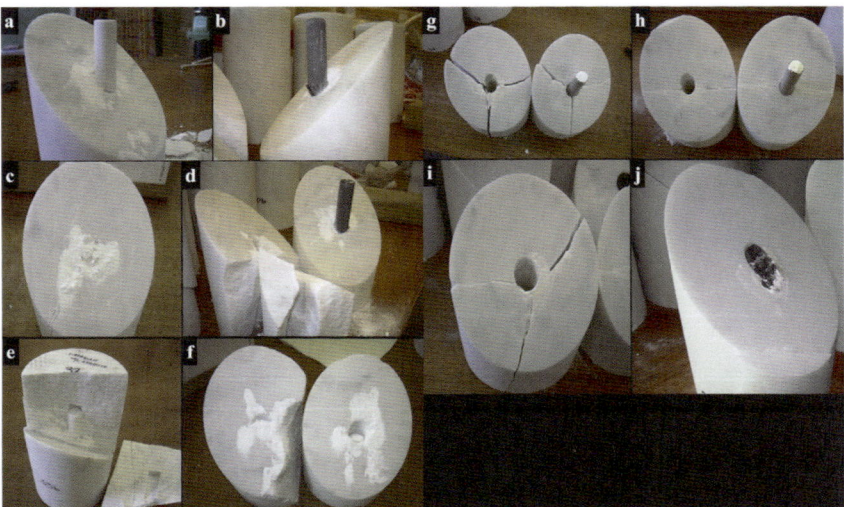

Fig. 2.11 (**a, b**) Fiber pin failure, dry repair: (**a**) Fiberglass pin fiber damage showing fiber shear; (**b**) Carbon fiber pin damage showing fiber pullout and kinking. (**c–f**) Marble fracture, dry repair: (**c**) fiberglass pin shear and marble fracture tangential to the pin hole; (**d**) carbon fiber pin kinking and marble fracture tangential and radial to pin hole; (**e**) crushing, powdering, and densification of marble around fiberglass pin; and (**f**) necking of Teflon® pin. (**g–j**) Marble fracture, wet repair: (**g**) Tri-axial radial marble fracture pattern on top and bottom cores with a stainless steel pin; (**h**) lateral radial marble fracture pattern with a stainless steel pin; (**i**) compressive ring in marble substrate around the pin hole with fracture with a titanium pin; and (**j**) compressive ring in marble substrate around the pin hole without fracture with a carbon fiber pin [7]. Reprinted with permission from Materials and Design, 98, Jessica Rosewitz, Christina Muir, Carolyn Riccardelli, Nima Rahbar, George Wheeler. A multimodal study of pinning selection for restoration of a historic statue, 294–304, Copyright (2016), Elsevier

specimens fused together. In the remaining carbon fiber specimen, the pin sheared, and one marble half fractured along a line tangential to the pin hole, similar to the fiberglass pin specimens (Fig. 2.11d). Of the fiberglass specimens, only one out of three fused together at conclusion of experiment. All fiberglass specimens caused marble fracture after pin shear, an important result signifying the fiberglass pin is sacrificial with respect to the marble.

The polycarbonate and Teflon pins failed in shear at low load and caused excessive horizontal displacement and sometimes fracture in the marble, with experiments stopped at the maximum 95 kN load. The experiments of the polycarbonate pin (Fig. 2.10e) and Teflon pin (Fig. 2.10f) displayed plastic behavior. The excessive horizontal displacement is indicative of early pin shear failure. The Teflon pin showed sufficient plasticity to deform severely and undergo necking before complete failure (Fig. 2.11f). Marble fractures in the dry repair with Teflon pin occurred tangential to the pin hole and radial across the pin hole. The excessive join misalignment under load suggests these two materials may not be suitable for sculpture repair.

2.4.2.2   Wet Join Repair Specimens

In the wet join repair specimens, the marble fractured at smaller displacement versus the dry join repair specimens. The stainless steel and titanium pins caused fracture around the pins, with cracking most often initiating outward from the pin in a trifold pattern (Fig. 2.11g, h) due to local mode I stresses. A distinct compressive ring in the marble formed around the pin hole at the join surface (Fig. 2.11i). The epoxy prevented the marble from crushing and pulverizing, as was observed in the dry join repair specimens. The wet join repairs with carbon fiber pins failed by pin shear failure, without marble fracture. The carbon fiber pin was sufficiently stiff enough to cause the same compression ring in the marble, similar to what was seen with the metal pins (Fig. 2.11j). The metal pins and the carbon fiber pin were precluded because of marble failure by fracture.

None of the specimens repaired with the fiberglass pin caused fracture in the marble, all failing by pin shear. The same compression ring phenomenon formed, but did not separate or lift off of the specimen before the fiberglass pin sheared. The polycarbonate pin went through shear failure at the join, but the marble did not fracture. The compression ring formed around the pin hole at the join surface, and some specimens caused that area to separate or lift off the marble before the polycarbonate pin sheared. The remaining differences between polycarbonate and fiberglass pins were minimal in the wet repair case, but fiberglass clearly outperformed all other pins in the experiments.

The epoxy failed to adhere to the Teflon before the experiment in two of the wet join repair specimens, due to the low surface roughness of Teflon and the low adhesion energy between Teflon with epoxy due to high resistance to van der Waals forces [30]. The join repairs failed by pin shear, and no marble fracture occurred. The same compression ring formed and did separate or lift away from the marble at the join surface, occurring prior to pin failure.

2.4.3   Simulation Results

2.4.3.1   Dry Join Repair Simulation

The finite element simulations can accurately predict the overall behavior of the systems (Fig. 2.10a–f). The failure of the marble can be described by two predominant modes: fracture of marble due to mode I cracks (Fig. 2.11d–f) and crushing of marble under compression which can be seen around the pin-to-hole interface (Fig. 2.11e). The highest stress concentrations are in the marble at the pin hole at the join surface in both compression and tension on opposite sides of the pin. These are the sites of crack propagations observed in the specimens. Figure 2.12a–e shows this propagation, and Table 2.4 shows the force and displacement when the maximum principal stress in marble exceeds the compression or tension limit stresses of 84.63 MPa and 6.9 MPa, respectively.

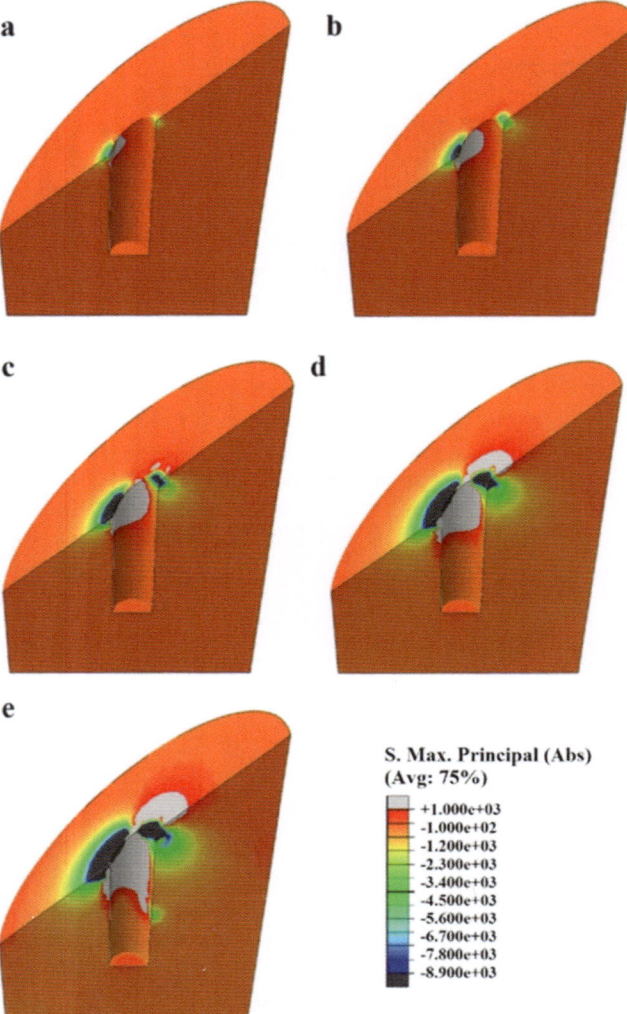

S. Max. Principal (Abs)
(Avg: 75%)

- +1.000e+03
- -1.000e+02
- -1.200e+03
- -2.300e+03
- -3.400e+03
- -4.500e+03
- -5.600e+03
- -6.700e+03
- -7.800e+03
- -8.900e+03

Fig. 2.12 (a–e) Principal stress propagation in the bottom half of the marble core, section cut (top half similar). Black zone exceeds compressive stress limit of 84.63 MPa; gray zone exceeds tensile stress limit of 6.9 MPa [7]. Reprinted with permission from Materials and Design, 98, Jessica Rosewitz, Christina Muir, Carolyn Riccardelli, Nima Rahbar, George Wheeler. A multimodal study of pinning selection for restoration of a historic statue, 294–304, Copyright (2016), Elsevier

Table 2.4 Force and vertical displacement at maximum principal compression and tension stresses in marble

Pin material	Tension failure in marble		Compression failure in marble	
	Force (kN)	Displacement (mm)	Force (kN)	Displacement (mm)
Stainless steel, Type 316	0.15	0.11	0.59	0.24
Titanium, Grade 2	0.13	0.16	0.59	0.28
Carrara marble	0.12	0.21	0.63	0.30
Carbon fiber-reinforced polymer	0.11	0.27	0.56	0.32
Structural fiberglass	0.11	0.28	0.53	0.35
Polycarbonate	0.06	0.30	0.67	0.49
Teflon® PTFE	0.06	0.30	0.76	0.65

Reprinted with permission from Materials and Design, 98, Jessica Rosewitz, Christina Muir, Carolyn Riccardelli, Nima Rahbar, George Wheeler. A multimodal study of pinning selection for restoration of a historic statue, 294–304, Copyright (2016), Elsevier

Figure 2.13a displays the corresponding vertical displacement when the maximum principal stress in marble exceeds the tension or compression limits from FEA of the dry join repair. Creating a join displacement ratio, T/C, for each pin material represents the overall efficiency of the system (Fig. 2.13b). The variable "T" is the join displacement at tension stress limit, and "C" is the compression stress limit in Carrara marble. A T/C ratio closer to 1.0 is desirable because it indicates the highest mechanical performance in the join. Fiberglass and carbon fiber pins nearly achieve that ratio, at 0.80 and 0.84, respectively. Most importantly, a join with these materials should prevent sudden or catastrophic join failure, a desirable trait for sculpture conservation and preservation.

The results in Table 2.4 show that the stress limits are reached at low force and displacement values. To prevent damage to marble, a pin material that holds the join together but is still less stiff than the substrate is the obvious choice. When taken in tandem with the experiments, fiberglass is the optimum pin material. The pin is less stiff than the marble, and the internal damage occurs at about 0.1 kN in tension and 0.5 kN in compression. Figure 2.10d shows that the entire join repair fails at about 20 kN. A reconstructed finite element model of *Adam* containing fractures was constructed without adhesive or pins [4]. A self-weight analysis provided stresses, from which the highest occur between the left calf and ankle fragments. The maximum compressive stress is 0.924 MPa at the front, and the tensile stress is 0.524 MPa at the back [4], quite low when compared to the tensile (6.9 MPa) and compressive (84.63 MPa) limits of Carrara marble. Therefore, fiberglass, the selected pin material, can also withstand the highest forces on the sculpture.

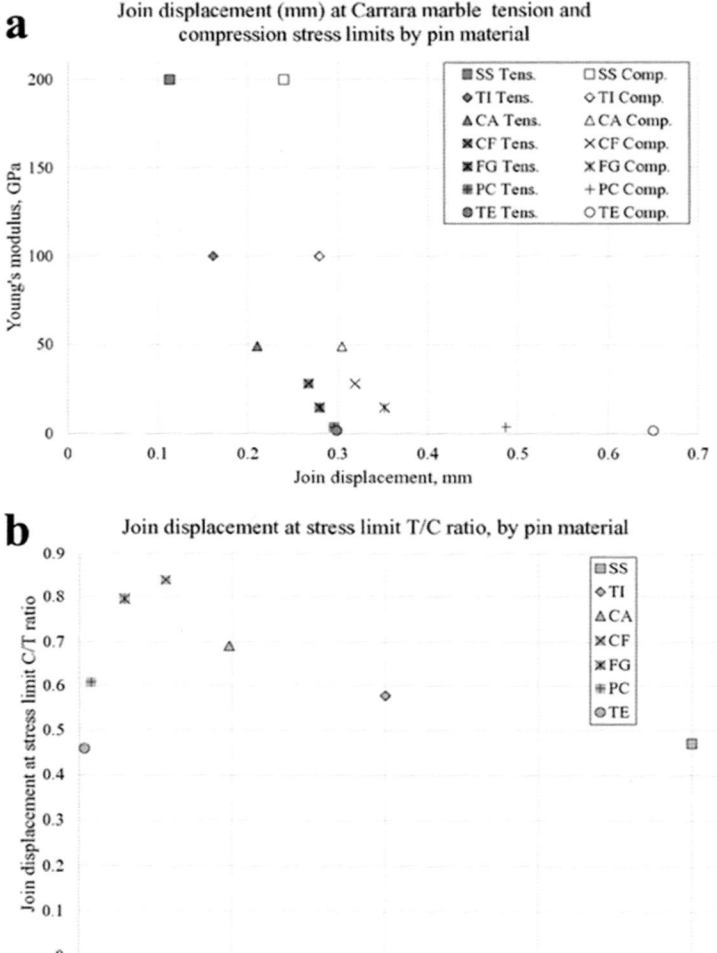

Fig. 2.13 (**a**) Join displacement (mm) characterized by pin material at tension and compression limits in Carrara marble. (**b**) Join displacement ratio *T/C* by pin material Young's modulus (*T* = join displacement at tension stress limit in Carrara marble, *C* = join displacement at compression stress limit in Carrara marble) [7]. Reprinted with permission from Materials and Design, 98, Jessica Rosewitz, Christina Muir, Carolyn Riccardelli, Nima Rahbar, George Wheeler. A multimodal study of pinning selection for restoration of a historic statue, 294–304, Copyright (2016), Elsevier

2.4.3.2 Wet Join Repair Simulation

The FEA simulations of the wet join repair modified the previous model by adding two sleeves of the epoxy in each hole surrounding half of the pin (Fig. 2.9b). Similar to the dry join repair, the six pin materials were studied. The bonding agent, Akemi Akepox 2000®, is specified for use as a superior adhesive for either natural or cast stone. The bond is stronger to silicate-bound stones than carbonate-bound stones, and the bonding agent easily bonds to damp stone. It is also specified to bond wood and ceramics and works well with glass fibers and polycarbonate.

For the tested wet join repairs with pin materials compatible with epoxy, the samples reached higher join strengths at lower join vertical displacements than the dry join repair specimens. However, for pin materials incompatible with epoxy, both sets of dry and wet repair results match. The results for the Teflon pin confirm observations that this pin material is not suitable.

Selected FEA results are presented for the wet join repair with the carbon fiber pin (Fig. 2.10c) and the fiberglass pin (Fig. 2.10d) to illustrate the disparity between the FEA and experimental results. The possible differences between the FEA and experimental results are hypothesized in the following. The application of epoxy with a syringe may have induced imperfections in the form of air pockets or bubbles. The method of setting the pin base using putty displaced a small volume of the epoxy. Finally, there may have been air gaps at interfaces due to temperature fluctuations induced by long curing times in multiple locations. For example, the coefficient of thermal expansion for type 304 stainless steel is 17 to 18×10^{-6} C^{-1}, while for Carrara marble, it is 2.4 to 6.7×10^{-6} C^{-1} [31]. These reasons may have reduced the effectiveness of the epoxy sleeve, creating locations where early failure might occur.

2.5 Conclusion

The mixed mode fracture toughness of Carrara marble and marble/adhesive interface were investigated in this chapter. The performance of the three types of adhesive systems (i.e., thermoplastic, thermosetting, and the combination of the two applied in sequence) for use on Carrara marble was evaluated. With the exception of the PVB adhesive, the fracture toughness values increased with increasing mode mixity. The results also showed that the interfacial toughness values were higher for the pre-fractured samples compared to the smooth samples. The trends in the mode mixity dependence of interfacial fracture toughness were consistent with predictions from a zone shielding model. However, the overall fracture toughness levels were similar to the different join materials that were used in this study. The current results provided the basis for the fracture mechanics-based design and fabrication of robust structures that can resist fracture under monotonic loading. Moreover, study of subcritical crack in adhesive/marble interfaces presented the results of Brazil nut studies of subcritical interfacial crack growth along interfaces between adhesives and Carrara marble. The measured crack growth rates were used to predict the life-time of these adhesive blends within a fracture mechanics framework. The results suggested that mixed thermosetting/thermoplastic polymer resins (Paraloid B-72/Epotek) provided the best resistance to subcritical crack growth in the restored fractured pieces. The predicted structural lives were also consistent with the expected lifetimes of marble structures. Although the results of combined experimental and numerical studies of the optimum pinning material to repair three fractured joins in *Adam* designed a matched pin and substrate pair, the methods reported herein may also be expanded to other conservation efforts with sculpture. The proposed repair should also be evaluated to preclude pinning, instead of using a reversible adhesive. While the experimental and FE simulations support the conservator's installing fiberglass pins in three locations, the most notable result from the treatment of *Adam* is that approximately 20 major joins were repaired with a completely reversible adhesive. This follows a common principal in conservation to leave materials intact wherever possible. By combining the experimental method and the FE simulations, it can be concluded that a join repair using a fiberglass pin is the most efficient join with regard to ultimate strength and should not cause long-term damage to the artwork.

References

1. Bohlen C. Met's 15th-century 'Adam' shatters as pedestal collapses. The New York Times; 2002.
2. Syson L, Cafa V. Adam by Tullio Lombardo. Metrop Mus J. 2010;49:8–31.
3. Cafa V. Ancient sources for Tullio Lombardo's Adam. Metrop Mus J. 2014;49:32–47.
4. Riccardelli C, et al. The treatment of Tullio Lombardo's Adam: a new approach to the conservation of monumental marble sculpture. Metrop Mus J. 2014;49:48–116.
5. Rahbar N, et al. Mixed mode fracture of marble/adhesive interfaces. Mater Sci Eng A. 2010;527:4939–46.
6. Tan T, et al. Sub-critical crack growth in adhesive/marble interfaces. Mater Sci Eng A. 2011;528:3697–704.
7. Rosewitz J, et al. A multimodal study of pinning selection for restoration of a historic statue. Mater Des. 2016;98:294–304.
8. Riccardelli C, et al. An examination of pinning materials for marble sculpture. AIC Objects Specialty Group Postprints. 2010;17:95–112.
9. Muir C. Evaluation of pinning materials for marble repair. In: Graduate school of architecture, planning and preservation. New York: Columbia University; 2008.
10. Tschegg EK, Jamek M, Schouenberg B. Fracture properties of marble-mortar compounds. Bull Eng Geol Environ. 2008;67(2):199–208.
11. Horie CV. Materials for conservation, organic consolidants, adhesives and coatings. London: Butterworths-Heinemann; 1990.
12. Koob SP. The use of Paraloid B-72 as an adhesive: its application for archeological ceramics and other materials. Stud Conserv. 1986;31(1):7–14.
13. Jorjani M. An evaluation of adhesives used for marble repair. In: Historic preservation program. New York: Columbia University; 2007.
14. Wang JS, Suo Z. Experimental determination of interfacial toughness curves using Brazil-nut sandwiches. Acta Metall Mater, 1990. 38(7): p. 1279–1290.

15. Rahbar N, Wang Y, Soboyejo WO. Mixed mode fracture of dental interfaces. Mater Sci Eng A. 2008;488:381–8.
16. Horie CV. Materials for conservation. Abingdon: Routledge; 2013.
17. Robinson MA. Early advances in the use of acrylic resins for the conservation of antiquities, in polymers in conservation. Bath (Great Britain): Bookcraft; 1992.
18. Jorjani M, et al. An evaluation of potential adhesives for marble repair. AIC Objects Specialty Group Postprints. 2008;15:95–107.
19. Andrews RD, Hammack TJ. The theoretical interpretation of dynamic mechanical loss spectra and transition temperatures. J Polym Sci B. 1965;3(8):655.
20. Kemp J. Fills for the repair of marble. J Archit Conserv. 2009;15(2):59–78.
21. Skeist I. Handbook of adhesives. New York: Reinhold Publishing Corporation; 1962.
22. Griswold J, Uricheck S. Loss compensation methods for stone. J Am Inst Conserv. 1998;37(4):89–110.
23. Atkinson C, Smelser RE, Sanchez J. Combined mode fracture via the cracked Brazilian disk test. Int J Fract. 1982;18:279.
24. Dundurs J. Edge-bonded dissimilar orthogonal elastic wedges under normal and shear loading. J Appl Mech. 1969;36:650–2.
25. He MY, Cao H, Evans AG. Mixed-mode fracture: the four-point shear specimen. Acta Metall Mater. 1990;38:839–46.
26. Kinloch AJ. Adhesion and adhesives. London: Chapman and Hall; 1987.
27. Hutchinson JW, Suo Z. Mixed mode cracking in layered materials. Adv Appl Mech. 1992;29:63–191.
28. Soboyejo WO. Mechanical properties of engineered materials. New York: Marcel Dekker; 2003.
29. Suo Z, Hutchinson JW. Sandwich test specimens for measuring interface crack toughness. Mater Sci Eng A. 1989;107:135–43.
30. Rahbar N, et al. Adhesion and interfacial fracture toughness between hard and soft materials. J Appl Phys. 2008;104(10):103533.
31. Siegesmund S, et al. Physical weathering of marbles caused by anisotropic thermal expansion. Int J Earth Sci. 2000;89:170–82.

Chapter 3
The Protection of Marble Surfaces: The Challenge to Develop Suitable Nanostructured Treatments

Lucia Toniolo and Francesca Gherardi

3.1 Introduction

Marbles have been extensively used in historical architecture, thanks to the specific mineralogical-petrographic, physical, and aesthetic characteristics and because of their easy workability, finishing, and polishing that have been highly appreciated by architects and sculptors of ancient and modern times. Marbles are rocks that have undergone metamorphism and contain >90% calcite or dolomite. Nevertheless, from the aesthetical point of view, marbles are characterized by the content and quality of accessory minerals of different natures, depending on the original rock type and on the grade of metamorphism (e.g., quartz, muscovite, feldspars, graphite, talc, pyrite, magnetite, hematite, etc.). Finely dispersed mineral phases impart specific colorations and tonalities, such as hematite creating red and chlorite and serpentine giving rise to green. For European historical architecture, the provenance of marbles is mainly from Greece or Italy, and this huge important patrimony constitutes an invaluable legacy to future generations.

The most important feature of marbles is the low-porosity values that characterize the fresh stone samples from quarries. Siegesmund [1] reported that plutonic and metamorphic stones have the lowest total porosity, while fresh marble porosity ranges between 0.05 and 1.00%. This porosity derives essentially from the cracking profile or fine intergranular detachments of the microstructure (Fig. 3.1a). After ageing and weathering, the porosity can even be doubled up to 2%, with a significant change in the pore size distribution (Fig. 3.1b) [1–3]. The phenomena and amplitude of changing depend on the type of marble, its mineralogical composition, as well as the average grain size, and it is more pronounced in the proximity of the surface.

L. Toniolo (✉) • F. Gherardi
Department of Chemistry, Materials and Chemical Engineering "Giulio Natta",
Politecnico di Milano, Milan, Italy
e-mail: lucia.toniolo@polimi.it

© Springer International Publishing AG 2018
M. Hosseini I. Karapanagiotis (eds.), *Advanced Materials for the Conservation of Stone*, https://doi.org/10.1007/978-3-319-72260-3_3

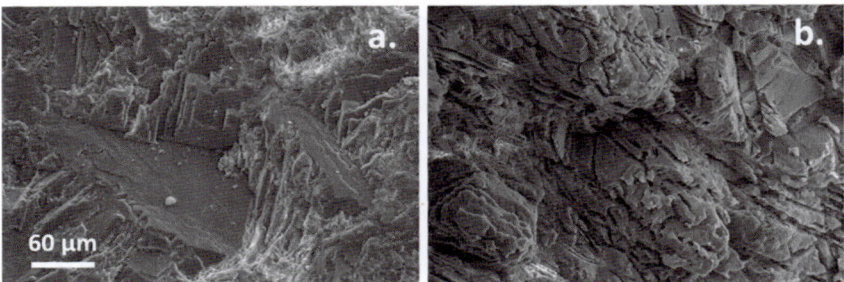

Fig. 3.1 ESEM images of fresh fractured, coarse grains Candoglia marble: (**a**) compact, sound crystalline microstructure; (**b**) naturally aged microstructure, showing inter- and intra-granular microfractures and corrosion patterns

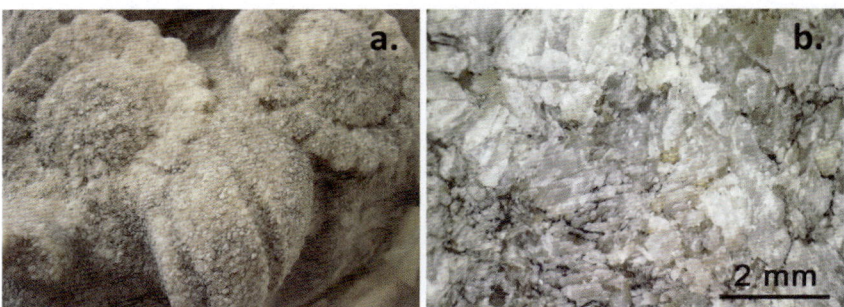

Fig. 3.2 State of conservation of Candoglia marble at the Cathedral of Milan. (**a**) Macrophoto of a decorative sculptured element with deep erosion of crystals; (**b**) microphotograph of the erosion and detachment of calcite grains

The weathering processes that take place at the interface material-environment are, in any case, determined by the presence of water, as far as marble deterioration is mainly due to a combination of physical and chemical phenomena whose driving force is the presence and ability of water retention: thermo-hygric expansion, frost damage, salt crystallization, and chemical corrosion [4]. Water, as it is well known, can easily penetrate the microstructure of the stone and fill the open porosity with different transportation mechanisms according to the pore size distribution [1, 4]. Even in low-porosity rocks like marbles, water is mainly responsible for severe deterioration phenomena (Fig. 3.2) such as loss of material, granular disintegration and pulverization, crust formation, and sulfation as well as chemical dissolution of calcite crystals [4, 5].

Marble is a critical substrate to protect [6–8], due to many intrinsic reasons, so far strictly linked to the mineralogical and physical properties reported above. Firstly, the very low open porosity does not allow the penetration of treatments in the bulk of the material, with many obvious disadvantages like the formation of a thin water-repellent layer (stone/treatment) and the accumulation of the coating on the stone's surface. The treatments have some difficulties to reach a certain depth inside the stone, and the application methodology is always critical [9].

The low intrinsic surface roughness caused by substrate compactness and usual surface smoothing results in rather low static contact angles of the treated surfaces. The contact angles measured on a treated marble tend to be near to that of the treatment itself, while on porous substrates, the surface hydrophobicity is enhanced by the natural roughness [10, 11]. Therefore, the surface water-repellent characteristic of a treated marble is, with most protective treatments, not completely satisfactory, while the reduction of water absorption by capillarity (relative capillary index, CIr—the ratio between the amount of water absorbed by capillarity by the treated stone specimen and that absorbed by the untreated one [12]) is usually around 40%, much less than for stones having 10% open porosity or higher. In absolute terms, the amount of water absorbed by capillarity (mg/cm^2 of liquid water absorbed into a 5×5 ×2 cm specimen) into the crystalline microstructure of both fresh (3–8 mg/cm^2 on Carrara marble) and thermally aged (25–35 mg/cm^2 on Carrara marble) marble is quite low. Actually, it is more difficult in the case of very low-porosity stones, that a hydrophobic coating can fully display its ability of modifying the properties of the internal microstructure of the substrate, as it happens in the case of rather porous stones. The ideal treatment should be able to penetrate into the pores, coating the internal walls without changing the pore size distribution and, therefore, altering the gas permeability of the porous material.

Finally, traditional treatments, mainly polymeric materials, tend to accumulate on the marble's surface and significantly change the stone's morphology and the aesthetic appearance. Besides that, they can constitute a barrier against the aggressive atmospheric agents but are more exposed to deterioration [8], and their durability is rather short. In the scientific literature, there are not any definitive evaluations of durability [13, 14]. Upon solar irradiation, thermal excursions, meteoric acid water, and gaseous pollutants, these materials generally lose their fundamental properties such as surface water repellency (significantly reduced after 1 year), adhesion, and elasticity, becoming brittle with diffused phenomena of surface cracking (Fig. 3.3).

Considering the great effort done by the European researchers to develop the common knowledge about the requirements of a protective material [7, 15, 16] and a shared standard protocol to assess the effectiveness of protective treatments [17], only a limited group or class of materials can be employed for marble protection.

In 2010, Dohene and Price [6] reviewed the scientific literature assessing that "water repellency has been provided largely by alkoxysilanes, silicones and fluoropolymers." According to the scientific community, a protective treatment is a product (liquid, solution, dispersion, suspension) that can be applied to the marble surface and does not alter the aesthetic properties of the stone, ensures a reduction of water absorption by capillarity, confers a water-repellent character to the surface, does not alter the water vapor permeability, does not produce aggressive secondary materials, or induces any damage to the substrate. For marble protection, other polymeric materials, such as acrylic resins [18], and inorganic hydrophilic low

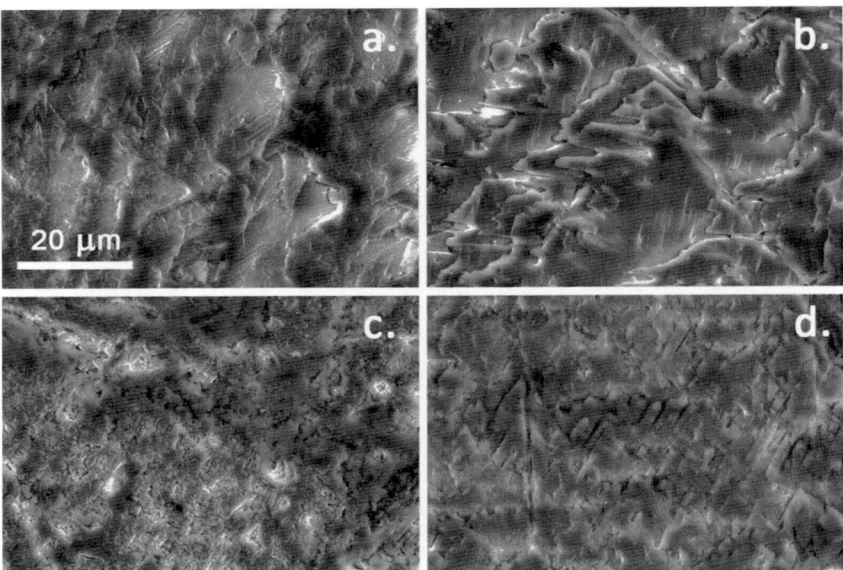

Fig. 3.3 ESEM images of Candoglia marble. (**a**) Untreated surface; (**b**) untreated surface after 9-month field exposure in Milan polluted environment; (**c**) surface treated with polysiloxane-based material; (**d**) surface treated with polysiloxane-based material after 9-month field exposure, showing subparallel micro-cracks in the protective layer

molecular weight compounds, such as ammonium oxalate to yield calcium oxalate [19, 20], quite insoluble and resistant to acidic environmental conditions, have been used. Fluoropolymers include perfluoropolyethers [19] and partially fluorinated acrylic copolymers [21, 22].

In recent years, the already interesting performances of silicone polymers or hybrid siloxanes [23] have been enhanced by the addition of inorganic oxide nanoparticles (NPs). Examples include silica (SiO_2), aluminum oxide (Al_2O_3), tin dioxide (SnO_2), and titanium dioxide (TiO_2), generating different nanocomposites that show good performances and properties according to the chemical nature and quality of both the polymeric material and the nanoparticle (chemical nature, size, and synthesis) [24]. In particular, the presence of nanostructured particles in the formulation of protectives grants a reduction of the free energy and an increase of surface roughness which can generate a superhydrophobic surface characteristic with high static contact angles ($>150°$) and very low values of contact angle hysteresis (generally $<10°$) [24, 25].

In the case of marble protection, the chemical and mineralogical characteristics, the size, and concentration of nanoparticles are key factors for the success and stability of the treatment. Even if superhydrophobicity can be achieved in certain conditions, the capillary absorption behavior of the treated marble with this kind of nanocomposites is always critical, with a reduction around 40–50% of the untreated stone. Karapanagiotis and co-workers [26] recently demonstrated that the concentration of very small SiO_2 nanoparticles in a silicone matrix is also a critical factor to maintain superhydrophobicity because it is necessary to have a homogeneous distribution of nanoparticles and specific continuous nanometric morphology on the surface. They also worked with purposely synthesized silica gels (ORMOSIL films) that show nanostructured morphology without the addition of inorganic NPs [27]. These materials exhibited good surface properties and aesthetical compatibility but have not been thoroughly tested as marble protection materials.

Cultural heritage in general, but specifically high-value historical buildings with marble sculptures and decorations in the European context, contributes to sustainable development by providing economic benefits through tourism and connected activities, as well as strengthening local identity, cultural values, and traditional practices. The impact of climate change on the conservation of this patrimony has been studied thoroughly by different points of view and disciplines, and a recent review accounts for the global consideration of this problem [28]. Unfortunately, the conclusion of this literature review is alarming because the authors assessed that current research is geographically limited (i.e., Europe, the UK, the USA) and confined in close disciplinary fields. However, there will be a deep necessity of a strong multidisciplinary approach in order to explore and profit from the opportunities that an innovative concept of preservation and adaptation could bring to the stakeholders and the entire social community [28]. Nevertheless, the impact of climate change and the increasing of atmospheric CO_2 levels has been studied [29] and modeled, taking into account carbonatic stone [30] and specifically Carrara marble [31]. In the future an increase of the recession rate of carbonatic surfaces is expected, corresponding to surface erosion damage for a wide range of low- and medium-porosity building stones caused by clean rainfall (karst effect, a clean rain effect due to rain at pH ~5.6 in equilibrium with atmospheric CO_2) [30].

The changing of environmental conditions, quality, and level of pollution in urban centers can determine a variation in the soiling and discoloration of building surfaces [32]. Urban aerosols are changing with climate change, containing different biological contaminants, while sulfur dioxide and smoke are being replaced by ozone, nitrogen oxides, and particles richer in organic compounds [32]. Particulate matter mainly derived by vehicular traffic and energy production shows an increasing organic carbon/elemental carbon ratio, making the soiling materials and deposits richer in organic compounds and brownish yellow in color. As reported by Brimblecombe and Grossi, there will be, over the course of the twenty-first century, a potential transition from blackening through yellowing to biological coloration processes [32].

In this research context and future scenario, it will be important that scientists will be able to diffuse and promote their results toward the stakeholders, building and cultural heritage managers, and decision-makers, so as to influence and support new investments and policies toward the practice of prevention and protection of marble surfaces and high-value outdoor stone heritage.

3.2 TiO$_2$ Nanoparticle-Based Materials for the Protection of Marble and Compact Limestones

The tentative application of nanoparticle dispersions and the development of formulations of protective treatments based on the combination of polymers or inorganic compounds with nanoparticles to obtain functionalized nanocomposites with enhanced properties lead in the last decade to the setup of improved superhydrophobic and superoleophobic coatings [25, 26]; photocatalytic, self-cleaning, and antifouling coatings; and coatings containing nano-TiO$_2$ [33], as well as nano-Ag, nano-ZnO, and nano-CuO, respectively [34, 35].

In the following sections, an overview about the properties of nano-TiO$_2$ and its application to marble surfaces for the preservation of historical buildings is reported.

3.2.1 Photocatalytic Activity of Nano-TiO$_2$ Semiconductor

In the context of environmental science, nano-TiO$_2$ is the most used photoactive semiconductor since it is readily available, inexpensive, chemically and thermally stable and has low toxicity.

The photocatalytic activity of TiO$_2$-based materials is affected by different parameters that can be tuned following specific synthetic routes: crystal structure, crystal morphology, crystal size, specific surface area, average pore size, pore volume and its distribution, and amount and nature of dopants [36]. Titanium dioxide crystallizes in three different phases: anatase, rutile, and brookite. Titanium (Ti^{4+}) atoms are octahedrally coordinated to six oxygen (O^{2-}) atoms, and in particular, anatase and rutile have a tetragonal structure and brookite an orthorhombic structure [37]. Anatase and rutile are more photoactive and stable than brookite; therefore they are the two most used polymorphs in photocatalysis. In both crystalline phases, with the absorption of a photon with energy equal or greater than their bandgap (3.0 and 3.2 eV in rutile and anatase, respectively), displacement of electrons from the valence to the conduction band occurs and an electron-hole pair is generated, separating the pair in a free electron and a free hole. Both the electron and the hole are trapped onto the TiO$_2$ surface, and they can react with acceptor or donor molecules, respectively, or recombine at surface trapping sites in a nanosecond time span. The electron and the hole can be confined in trap sites at the bulk and lead to recombination with the emission of heat [38]. The photogenerated electrons can react with molecular oxygen (O$_2$), producing superoxide radical anions ($\bullet$O$_2^-$), and the holes react with water to produce hydroxyl radicals ($\bullet$OH) [39]. The two reactive radicals promote the decomposition of organic molecules and the oxidation or reduction of inorganic molecules.

Compared to other semiconductors, nano-TiO$_2$ exhibits another determinant photoinduced phenomenon which occurs simultaneously to photocatalytic activation, which is known as superhydrophilic behavior [40]. With the latter phenomenon, oxygen vacancies are formed and water molecules occupy them, with the adsorption of OH groups leading to surface hydrophilicity. In addition, the chemisorbed water molecules can adsorb water by van der Waals forces and hydrogen bond formation, thus preventing the contact between surface and adsorbed contaminants, resulting in the easy removal of soiling from the surface and making the considered surface "self-cleaning" [41].

TiO$_2$-based nanomaterials are characterized by wide bandgaps and they mainly adsorb photons in the ultraviolet region. Nevertheless, continuous research efforts aim at finding strategies to improve the photocatalytic efficiency of nano-TiO$_2$ with sunlight, which include morphological modifications (e.g., increase of surface area and porosity) and chemical modifications, with incorporation of dopants to extend the spectral sensitivity of photocatalysts from UV to visible light [37, 42]. TiO$_2$ nanoparticles should have morphology with a controlled diameter in order to have high surface-to-volume ratio that increases the active sites and to reduce undesired bulk recombination [37, 43]. Encouraging results in shifting the absorption threshold of TiO$_2$ into the visible range were obtained by doping TiO$_2$ with nonmetal atoms, like nitrogen [44]. The use of carbon, phosphorous, and sulfur atoms and the combination of different nonmetal atoms and oxygen-rich TiO$_2$ modifications are promising strategies to achieve visible light absorption photocatalysis [37]. Other synthetic routes provide the modification of TiO$_2$ with different coupled semiconductors such as ZnO/TiO$_2$, CdS/TiO$_2$, and Bi$_2$S$_3$/TiO$_2$ or with transition metals such as Fe, Cu, Co, Ni, Cr, V, Mn, Mo, Nb, W, Ru, Pt, and Au [37]. The combination of noble metals like Ag, Au, Pt, and Pd with TiO$_2$ improves the photocatalytic efficiency under visible light by acting as an electron trap [45]. Moreover, dye photosensitization has been considered as a powerful way to extend the photoactivity of TiO$_2$ into the visible region [37, 42].

3.2.2 Applications in the Field of Architectural Heritage

Two different strategies can be identified for the preparation of TiO$_2$-based self-cleaning products for the protection of marble and compact limestone (very low open porosity sedimentary stones that can be easily polished and are often called "marble") substrates. The first is comprised of hydrophilic nano-TiO$_2$ dispersions, whereas the second includes functionalized nanocomposites, where nanoparticles are introduced inside the polymeric or inorganic matrix. The first category is represented by nano-TiO$_2$ dispersions in different solvents (e.g., water, alcohol, ethylene glycol) that have been applied on the surface of compact limestones, marble, and travertine [46–48]. The setup of nanocomposites follows another approach by including nano-TiO$_2$ in polymeric dispersions often used for the protection of stone surfaces such as alkyl silane and alkyl aryl polysiloxane [49, 50], fluorinated or partially fluorinated polymers [51, 52], or acrylic copolymers [52, 53] to convey hydrophobic and superhydrophobic properties. In addition, hybrid organic and inorganic TiO$_2$-based nanocomposites were developed in order to obtain treatments with both hydrophobic and consolidating properties [54–56]. Both protective approaches include materials with self-cleaning behavior, changing the wettability of the treated surface accordingly with either a hydrophilic (or superhydrophilic) or hydrophobic (or superhydrophobic) mechanism. Nano-TiO_2 dispersions applied to marble surfaces by brush or spray show the advantage of not significantly affecting the natural hydrophilic character of the substrate. However, the adhesion of nanoparticles is poor due to the low chemical affinity and relatively low roughness of the stone, easily washed away by rainfall thus decreasing their effectiveness as "self-cleaning" agents and durability [57, 58].

The formulations of nanocomposites are created in order to overcome the problem of adhesion of the NPs to the substrate; favor a good dispersion of NPs in the polymeric and inorganic matrices, avoiding NPs aggregation and precipitation; grant a certain degree of water repellency due to the organic alkyl groups; and favor the positioning and stability of NPs toward the external surface of the coating, creating nanoroughness and enhancing photoactivity, ensuring long shelf-life and on-site durability [24, 33]. However, an unavoidable drawback of polymeric nanocomposites is that the organic matrix can be subject to photocatalytic degradation. For this reason, a compromise between high photocatalytic activity and stability should be achieved in the formulations.

According to the literature [47], the photoactivity of nano-TiO_2 materials applied on natural stone has been evaluated by methods based on the discoloration of organic dye dispersions (e.g., methylene blue, methyl orange or red, rhodamine B) applied on treated and untreated stone surfaces after exposure to UV light. As reported by Munafò et al. [47], the results obtained by different researchers are hardly comparable since different parameters can be selected for the tests (i.e., concentration and amount of dye dispersions, total irradiance, and wavelength range of light source), and therefore only a relative comparison among a homogeneous set of experimental data can be done. Photoactivity is enhanced by the homogenous NPs distribution, avoiding aggregation phenomena, while the application of successive layers of nano-TiO_2 dispersions does not increase the activity, which is only related to the NPs total specific area exposed to irradiation [46]. A high amount of nano-TiO_2 can cause the formation of cracks and loss of adhesion of nanocomposites to the stone substrate [59, 60].

In addition to photocatalytic activity, some nano-TiO_2 treatments show antifouling and antibacterial properties, especially regarding the removal of biofilm present on stone surface before the application of coating [55] and in the inhibition in fungal cell growth [61, 62].

The preparation of Ag-doped nano-TiO_2 was proposed by different researchers to enhance the photoactivity of the treatment by extending the wavelength absorption toward VIS light [63] and to develop antifouling coatings [64]. In particular, La Russa et al. demonstrated that TiO_2 doped with Ag alone or in combination with Fe or with Sr could be applied in the prevention of stone biodeterioration [64]. During the synthesis of TiO_2-doped materials, a crucial aspect to be taken into account is to perfectly tune the concentration of the dopants, especially metal ions, because too high of a concentration can create a non-acceptable surface color change [63, 64]. Finally, a narrowing of the nano-TiO$_2$ bandgap, with the enhancement of photoactivity and self-cleaning behavior, was obtained by N-doped nanoparticles [65].

Another important aspect to be taken into account is in regard to the durability of the TiO$_2$ treatments applied on natural stone under real working and outdoor exposure conditions. Experience and published data about the long-term effectiveness of photocatalytic treatments are still limited. The durability of the treatments has been evaluated in laboratory by means of peeling tests to assess the adhesion of the coatings to the stone substrate [54, 60] and in an irradiation chamber and accelerated weathering cycles to investigate the photostability and the resistance to the mechanical stress induced by rain [53, 57, 58]. The overall lab results suggest that only treatments based on the use of nano-TiO$_2$ in combination with polymeric or inorganic matrices show good adhesion to the stone substrates and still have photocatalytic properties after ageing. Field exposure tests of treated stone specimens and pilot areas were conducted, aiming to evaluate the durability of nanocomposites in outdoor conditions; however, the collected data are related to a short timeframe of monitoring [50, 66].

3.2.3   Research Open Challenges and Future Perspectives for Nano-TiO$_2$ Treatments

The development of modified TiO$_2$ nanoparticles with enhanced properties, tailored based on the need for protection of very compact natural stone substrates, will be an important challenge in the near future.

A determinant factor of success will be obtaining more on-site evaluation and monitoring data, especially about the real durability of nano-TiO$_2$ on surfaces and the specific rate of deposit accumulation (i.e., blackening) in the outdoors, comparing treated and untreated surfaces.

As already stated, this research should concentrate on the possible extension to visible-light range of nano-TiO$_2$ photoactivity, on the improvement of adhesion to stone surfaces and, therefore, to the development of grafting agents or suitable binders able to fix the nanoparticles to the surface. Another important aspect is the improvement of the stability of the matrix in nano-TiO$_2$ composites. According to the literature, it may be possible to reduce the degradation of the organic binders by adding suitable organic stabilizers in nanocomposite formulations or by covering nano-TiO$_2$ with surface capping with inorganic compounds such as Al$_2$O$_3$, SiO$_2$, or ZnO [67].

Finally, little information about the life cycle assessment (LCA) of this class of treatments, with an evaluation of human and environmental impacts, is available [68]. To deepen this aspect, it will be important to ensure the development of sustainable and compatible materials as well as methods for large surfaces in architecture.

3.3   Application of Nano-TiO$_2$ Dispersions and Nanocomposites: In Lab and In Situ Experiences

In the last 3 years, research has been conducted on the setup and evaluation of the effectiveness of nanomaterials for the conservation of paint [69, 70], stone surfaces, and particularly marbles [48, 66, 71], based on innovative TiO$_2$ nanoparticles. The objective of the experimental work was to assess the activity of TiO$_2$ nanoparticles once applied on marble specimens and compared them with similar commercial nanosystems. Moreover, the possibility of jointing to the photocatalytic properties of the nanoparticles, the coating ability of protecting marble surfaces from water action, has been explored both in laboratory and in situ.

3.3.1   Application of Nano-TiO$_2$ Dispersions on Carrara Marble

TiO$_2$ nanoparticles were synthesized according to Niederberger et al. [72] following a nonaqueous route, by using benzyl alcohol as solvent and titanium (IV) tetrachloride as precursor [73, 74]. The selected nanocrystals contain pure phase anatase covered by residuals of benzyl alcohol deriving from the specific synthesis, anchored on the surface [73]. The surface capping is providing the extension of photoactivity under solar light irradiation. In addition, the obtained TiO$_2$ nanoparticles are particularly transparent [73], and they allow for highly stable dispersions in an aqueous system while respecting important safety and green chemistry requirements as well as preserving the color and texture of the substrates. As confirmed by TEM analysis, they are almost spherical crystals approximately 5–6 nm in size, aggregated in elongated structures whose longest axis measures about 40 nm [48]. UV-visible spectrum obtained from nano-TiO$_2$ powder is characterized by an absorption partially detectable in the visible wavelength region. Benzyl alcohol groups bound to the surface of nano-TiO$_2$ generate defects within the TiO$_2$ energy gap due to the overlapping of the oxide orbitals with those of the anchored alcohol group, allowing the red shift of the nano-TiO$_2$ absorption curve [48].

Water and ethylene glycol dispersions of nano-TiO$_2$ at different concentrations were prepared (1%, 2%, and 3% by weight in water and of 1%, 2%, and 4% by weight in ethylene glycol, named WA1, WA2, WA3, EGA1, EGA2, and EGA4, respectively). For comparison, commercial TiO$_2$ nanoparticles widely used in the photocatalysis field (AEROXIDE® P25 by Evonik) were dispersed in water (1, 2, and 3% by weight, labeled as CA1, CA2, and CA3).

The innovative and commercial dispersions of nano-TiO$_2$ were applied by brush on Carrara marble specimens. The amount, the morphology and aesthetic characteristics of the treatment, the behavior toward liquid water, and the photocatalytic activity of the different treatments were evaluated to assess the difference between the two nanomaterials [48, 71]. The morphology and the penetration depth of the innovative nano-TiO_2 treatments on marble specimens were investigated by ESEM-EDX, proving that the treatment (even at the highest concentration EGA4) is homogenously distributed onto the surface without nanoparticles aggregation, whereas CA1 dispersion tended to create clusters of nanocrystals. By observing the cross section of marble specimens, it was detected that the ethylene glycol dispersion of nano-TiO_2 did not penetrate into the porosity of the stone, and the nanocrystals remained well distributed without any aggregation on the surface, only penetrating tens of microns into the crystalline matrix of the marble [48].

At the same time, the innovative nano-TiO_2 dispersions did not affect the aesthetic appearance and the color of the marble surfaces (the color change ΔE^* is calculated from the average values of L^*, a^*, and b^* of the CIE $L^*a^*b^*$ color space, and in this case, it ranges between 1 and 2), while P25 dispersions containing nano-TiO_2 higher than 1% by weight (CA2 and CA3) caused an intense whitening of the surfaces, leading to significant and not acceptable color alteration ($\Delta E^* > 5$) [75].

The results obtained from static contact angle tests proved the increase in marble wettability and hydrophilicity after the application of the innovative dispersions, even without exposing the specimens to UV light (Fig. 3.4). This behavior is associated with the well-known superhydrophilic effect granted by the presence of small size dispersed TiO_2 nanoparticles, which is promoted by the ability of these nanoparticles to absorb light in the VIS region [76]. On the contrary, CA1 dispersion did not increase the surface wettability of the marble since the nanoparticles tended to agglomerate and needed UV irradiation to reduce the surface free energy [48].

The increase in the wettability of treated marble specimens was not matched with the enhancement of the water absorption by capillarity; this is a property not affected by the application of the nanoparticles dispersions with the exception of the

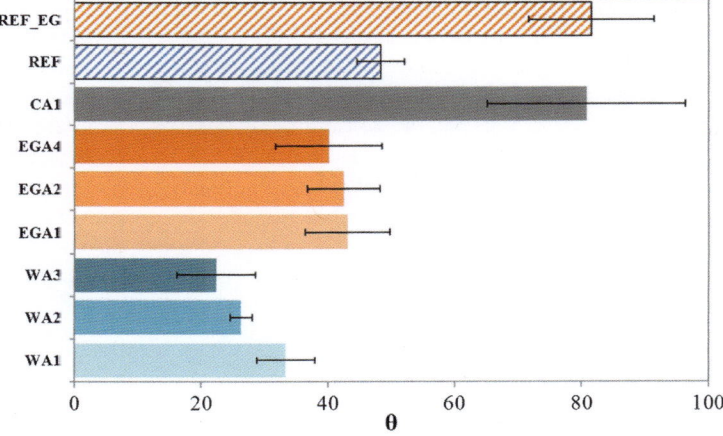

Fig. 3.4 Static contact angle values (θ) for Carrara marble specimens untreated (REF) and treated with ethylene glycol (REF_EG) and with innovative nano-TiO_2 (WA1, WA2, WA3, EGA1, EGA2, EGA4) and commercial nano-TiO_2 (CA1) dispersions

Table 3.1 Values of relative capillary index (CIrel) for Carrara marble specimens treated with nano-TiO$_2$ dispersions

	WA1	WA2	WA3	EGA1	EGA2	EGA4	CA1
CIrel	1.04	1.04	1.00	0.74	0.94	1.01	0.74

Table 3.2 Ratio between the values of stain discoloration D^* of samples treated (D^*_T) and the untreated reference (D^*_{REF}), after 15 and 150 min of irradiation

D^*_T/D^*_{REF}	WA1	WA2	WA3	EGA1	EGA2	EGA4	CA1
15 min	4.00	3.90	5.34	5.16	5.40	6.10	2.91
150 min	1.16	1.16	1.23	1.24	1.25	1.29	1.03

slight decrease in the relative capillary index value obtained by specimens treated with EGA1 and CA1 dispersions (CIrel) (Table 3.1).

Finally, it should be highlighted that this kind of treatments, with low nano-TiO$_2$ concentrations confined at the very surface, did not significantly alter the behavior of the stone toward liquid water, which can be absorbed with an unaltered kinetic.

The evaluation of the photocatalytic activity of treated marble specimens has been carried out by the rhodamine B test [48]. By observing the results of the ratio D^*/D^*_{REF}, which expresses the ability of the treatment of decomposing rhodamine, D^* is the extent of discoloration, and it is evaluated according to the formula, $D^*(\%) = [(a_t^* - a_0^*)/(a_{ns}^* - a_0^*) \, 100]$, where a_t^* and a_0^* stand for the a^* value at time t and 0, respectively, while a_{ns}^* refers to the specimen before the deposition of the rhodamine solution; it is clear that the innovative nano-TiO$_2$-based dispersions promoted a more efficient degradation of the red colorant, especially after 15 min, indicating a higher photoactivity compared to the commercial one (Table 3.2). In addition, the efficiency of the photocatalytic activity was not affected by the type of solvent in the dispersion and only slightly by the concentration of nanoparticles. Further investigations about the threshold value of surface nanoparticles amount related to the maximum photocatalytic effectiveness are in progress. One can imagine that the optimal surface distribution of nanoparticles and, therefore, the "optimal nanostructure" for the photocatalytic activity can be achieved with a specific NPs concentration, depending on the solvent quality. Other authors already observed some similar phenomena, where the superhydrophobic effect is maximized by a specific nanoparticle concentration in different solvents on the surface of marble [26].

Untreated and treated Carrara marble specimens were exposed to unsheltered outdoor conditions by rain washing on the laboratory roof, in Milan urban and polluted environment, for 9 months. The photocatalytic and self-cleaning activity of the dispersions applied on marble surfaces was monitored by color measurements, following up the blackening of the specimens. The results clearly indicated that the innovative dispersions experienced only a slight decrease in particulate matter accumulation on the surface compared to the untreated reference specimens, and, therefore, this kind of treatment could not lead to a satisfactory self-cleaning effect, even on low-porosity and very compact stones [71].

3.3.2 Application of TiO₂-Based Nanocomposites in Lab on Carrara Marble

Starting from the innovative nano-TiO_2, photocatalytic nanocomposites were prepared by mixing nano-TiO_2 water dispersion (WA3 with 3% w/w) with commercial products used as stone protective treatments [71]. The selected commercial formulations for the conservation of natural stones were from two different classes of materials: oligomeric siloxanes and functionalized silica. These materials were carefully selected to achieve durable, water-repellent, and self-cleaning treatments. The first one, labeled as S REF, is an aqueous dispersion of organosiloxanes (10% w/w) (*Silo 112*, *CTS srl*), a well-known oligomeric alkyl-siloxane widely used in the restoration field. The second, labeled FS REF, is a suspension of SiO_2 functionalized by silicon alkoxides in isopropyl alcohol (20% by weight) (*SIOX-5S*) obtained by sol-gel process from a small Italian spin-off company (Siltea srl, Italy).

Nanocomposites were set up by adding, upon stirring, different amounts of water dispersion of TiO_2 nanoparticles to the commercial products, without any further dilution. Starting from the polysiloxane-based sample (S REF), three different stable and transparent emulsions were obtained, labeled S16, S28, and S44 with 16%, 28%, and 44% concentrations of nanoparticles by weight, respectively. In the case of the functionalized SiO_2-based product (FS REF), only one stable composite, labeled FS16, was obtained with 16% by weight of nano-TiO_2 [71].

TEM images obtained from the polysiloxane nanocomposites indicated the presence of elongated nanostructures of about 40 nm well dispersed in the matrix, independent from the nanoparticles concentration (Fig. 3.5a–c), confirming the results obtained from water dispersions of the innovative nano-TiO_2 [48]. In the case of the functionalized SiO_2-based treatment, the nanoparticles were also homogeneously dispersed, but some slightly larger aggregates were also observed (Fig. 3.5d).

Once applied on Carrara marble, the nanocomposites proved to be perfectly compatible with the substrate, as the obtained color change (ΔE^*) values were lower than 3. In addition, the nanocomposites showed lower ΔE^* values and therefore better color compatibility compared to the reference protective products S REF and FS REF, likely due to the slight whitening effect of nano-TiO_2, favorable in the marble applications.

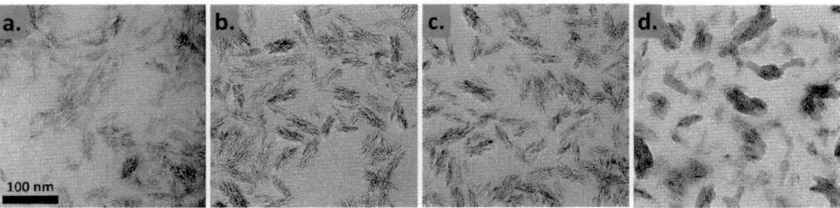

Fig. 3.5 TEM images of the nanocomposites based on polysiloxane: (**a**) S16, (**b**) S28, and (**c**) S44; on functionalized SiO_2: (**d**) FS16

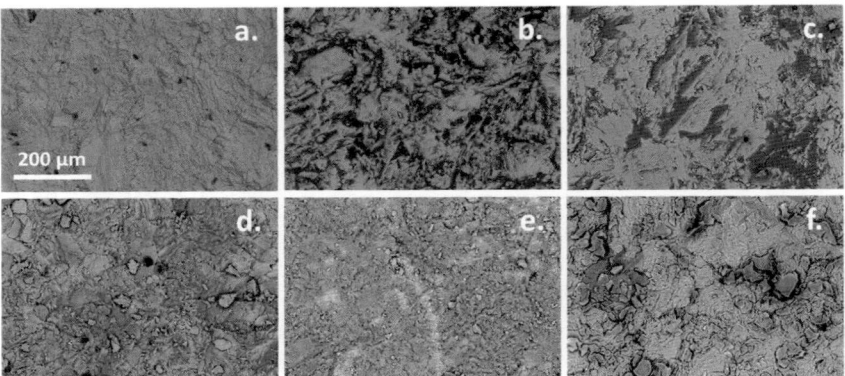

Fig. 3.6 ESEM images of Carrara marble. (**a**) Untreated; treated with (**b**) S REF; (**c**) FS REF; (**d**) S16 nanocomposite; (**e**) S44 nanocomposite; (**f**) FS16 nanocomposite

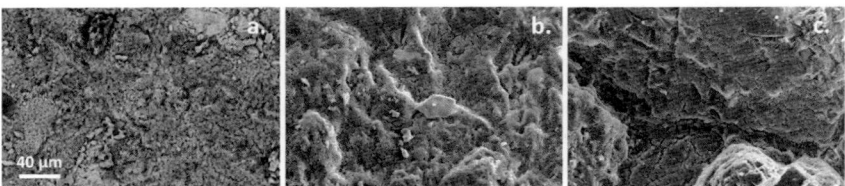

Fig. 3.7 ESEM images of different marbles treated with S44: (**a**) unaged Carrara marble treated in lab conditions; (**b**) Candoglia marble treated on-site; (**c**) Crevoladossola marble treated on-site. Typical nanostructured surface morphology can be observed in the three different surfaces

The reference products (S REF and FS REF) can be perceived on the marble surface since they accumulate in the surface micropores and cavities, giving rise to a slight roundish effect of the sharp mineral edges (Fig. 3.6). The nanocomposites covered the crystal grains of Carrara marble with a very thin layer of material, preserving the surface roughness, as evidenced by ESEM observations (Fig. 3.6) and by EDX mapping of the main elements (Si and Ti) which proved that the treatments were homogenously distributed onto the surface. In particular, a typical nanorough porous pattern can be observed on the S44 treatment at higher magnification (Fig. 3.7a) [71].

A better reduction of the water absorption by capillarity was obtained with both nanocomposites as compared to the reference treatments (S REF and FS REF) (Table 3.3), indicating that the addition of nano-titania in the formulations plays a role in the reduction of the water absorption inside the pores.

Interesting results were obtained by static contact angle test, and they indicated that the application of polysiloxane-based nanocomposites (S16, S28, and S44)

Table 3.3 Values of relative capillary index (CIrel) and static contact angle values (σ) for Carrara marble specimens either untreated or treated with the commercial reference products (S REF and FS REF) and nanocomposites (S16, S28, S44, and FS 16)

	S REF	S16	S28	S44	FS REF	FS16	Untreated
CIrel	0.78	0.55	0.44	0.48	0.55	0.51	–
σ	95	129	138	142	60	34	48

Table 3.4 Ratio between the values of stain discoloration D^* of samples treated (D^*_T) and the untreated reference (D^*_{REF}), after 15 and 150 min of irradiation

D^*_T/D^*_{REF}	S REF	S16	S28	S44	FS REF	FS16
15 min	0.85	10.73	17.42	16.06	0.69	6.17
150 min	1.34	2.31	2.45	2.53	0.94	2.04

confers a significant water-repellent characteristic to the marble's surface compared to the reference polymeric product (S REF) (Table 3.3). In particular, by increasing the nanoparticles' concentration in the composites, higher contact angle values were measured. This result should be associated to the surface nanoroughness created by the presence of nano-TiO$_2$ homogeneously distributed in the treatment with the reduction of the surface free energy. Atomic force microscopy (AFM) measurements of the average nanoroughness of the treated specimens support this conclusion [71].

The specimens treated with functionalized SiO$_2$ (FS REF) and functionalized SiO$_2$-based nanocomposite (FS16) showed a different behavior, revealing values of static contact angle similar or lower to those obtained by the untreated marble specimen. The treatment did not grant any water-repellent characteristic to the marble surface. This result can be explained again by AFM observations that showed low comparable surface roughness values measured on specimens either untreated or treated with FS REF and FS16 [71]. The specimens treated with FS16 showed lower contact angle values compared to those treated with FS REF, concluding that the addition of nano-TiO$_2$ to a hydrophilic coating increased the wettability of the surface [77].

The polymeric and inorganic matrix in the nanocomposites did not compromise the photoactivity of nano-TiO$_2$, as the specimens treated with the nanostructured blends showed higher values of stain discoloration in the rhodamine test compared to the untreated ones and to those treated with reference commercial products (S REF and FS REF). The best results were obtained by the polysiloxane-based nanocomposite with the highest nano-TiO$_2$ content (S44) (Table 3.4).

The overall results demonstrated that the addition of nano-TiO$_2$ improved the performances of the reference commercial products as protective treatments. Thus, an in situ testing campaign was carried out on the pilot yard of the marble façade of the Renaissance Cathedral of Monza (Italy).

3.3.3   On-Site Evaluation of the Effectiveness of Nanostructured Treatments

The application of treatments on-site, in real conditions, should be regarded as necessary because laboratory testing protocol [17] only considers fresh-quarried, sound specimens for experimental work. Therefore, the most important difference in the case of marble surface is the condition of the deteriorated surface and the higher surface porosity induced by weathering. By transferring this technology and the assessment of the treatments on-site, the application method changes, and therefore the related parameters (e.g., solvents, concentration, and amount of products) should be tuned according to the surface characteristics. In general, protective coatings for natural stones are applied by brush or by spray. In this study the selected nanocomposites based on polysiloxane (S44) and functionalized SiO_2 (FS16) were applied by brush, till refuse on coarse grain marbles, showing surface decohesion and corrosion. Two different marbles from northern Italian regions, Candoglia and Crevoladossola, were treated. The context, the stone materials, and the conservation state are thoroughly described in a previous publication [66]. A 12-month monitoring protocol of nondestructive tests, based on microscopic on-site observation, color, and water absorption by capillarity measurements, was set up and carried out every 3 months. Moreover, from both untreated and treated areas, small stone fragments were collected, and the surface morphology was studied by microscopic observations (ESEM-EDX).

One month after the application, the reference and nanocomposite treatments showed good color compatibility, with color change values (ΔE^*) lower than 5; for both marbles, the highest color change was related to the commercial polysiloxane-based product (S REF) (Table 3.5) [66].

After 12 months of monitoring, all the treatments demonstrated good color stability on both marble substrates (Table 3.5). In particular, on Candoglia the total color variation of the nanocomposites (S44 and FS16) can be attributed to the increase of the lightness due to the slight whitening induced by nano-TiO_2 applied on a pinkish substrate. On Crevoladossola, due to the higher surface porosity caused by weathering and better penetration of the treatment, this did not occur, and the color change was almost negligible (Table 3.5). It must be pointed out that the

Table 3.5 ΔE^* values measured on untreated reference area (NT) and on areas with the following treatments: commercial polysiloxane (S REF), polysiloxane-based nanocomposite (S44), commercial functionalized SiO_2 (FS), functionalized SiO_2-based nanocomposite (FS16) measured after 1 and 12 months on Candoglia and Crevoladossola marble

Marble surface	Monitoring time	ΔE^* S REF	ΔE^* S44	ΔE^* FS REF	ΔE^* FS16	ΔE^* NT
Candoglia	1 month	2.58	1.00	0.96	2.13	0.91
	12 months	2.16	3.30	0.59	2.13	0.59
Crevoladossola	1 month	3.46	1.99	2.71	0.75	0.58
	12 months	4.46	1.82	2.30	0.70	0.62

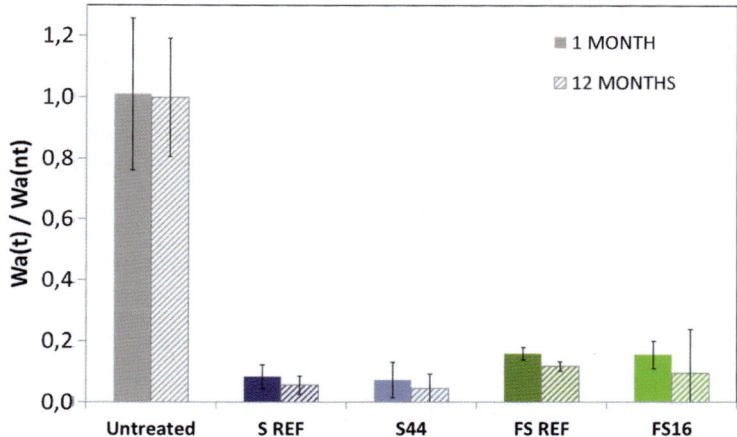

Fig. 3.8 Water absorption ratios by contact sponge method Wa(t)/Wa(nt), measured on Crevoladossola marble surfaces: untreated reference area (untreated); treated with commercial polysiloxane (S REF); treated with polysiloxane-based nanocomposite (S16); treated with commercial functionalized SiO_2 (FS REF); treated with functionalized SiO_2-based nanocomposite (FS16)

condition of the treated area was only partially sheltered by rainfall and wind, not allowing for the evaluation of the efficacy of these treatments in preventing the soiling of marble surfaces during the short monitoring period. As such, no conclusions can be drawn about the ability of the nanocomposites to act as self-cleaning treatments in real conditions.

The analyses of the micro-samples collected from the treated areas proved that S REF polysiloxane penetrated into the surface micropores and was able to cover the crystalline matrix with a very thin layer, as also observed in lab experiments. The surfaces treated with the S44 nanocomposite created a rough and porous layer with a spongelike morphology, as evidenced also in fresh, laboratory-treated Carrara marble specimens (Fig. 3.7) and characterized by a good distribution of both the polymeric matrix and nano-TiO_2 on the stone surface, as assessed by EDX spectrometry. A glass-like, rather smooth morphological feature for both substrates can be observed on the samples collected from the areas treated with both the functionalized SiO_2 product (FS REF) and the nanocomposite (FS16), similarly to what was observed for the same treatments in the lab.

These observations confirmed that the products can penetrate into the permeable micropores creating a sort of barrier and, therefore, a reduction of the water uptake by capillarity. Both the reference commercial products and the nanocomposites lead to a significant reduction of the water absorption compared to the untreated reference areas, obtaining comparable values after 12 months (Fig. 3.8). It was not possible with the noninvasive selected methodology (contact sponge method) applied to low-porosity marbles (concerning very small amount of liquid water) to clearly remark the small absorption difference in the behavior of the nanocomposites compared to the reference treatments, which were pointed out in the lab tests (water absorption by capillarity).

The overall preliminary results from the on-site evaluation are important, because the on-site experience is quite different from the results gathered in laboratory controlled conditions. These results clearly suggest the importance of a long-term monitoring activity (about 3–5 years) for the assessment of innovative nanostructured conservation treatments.

3.4   Conclusion

The protection of marble surfaces from aggressive polluted atmosphere in changing climatic conditions is a hard challenge and is far from being won. The chemical-physical phenomena in a low-porosity crystalline solid are complex. The interaction of mineral surfaces with materials and nanomaterials, the interaction of mineral surfaces and nanomaterials with electromagnetic radiations, and the interactions of nanomaterials with water and pollutants are scarcely investigated and known. This work analyzed the major directions of the research for the development of suitable, efficient, and durable nanostructured materials for the protection of marble, offering some specific examples of innovative nanomaterial evaluation, in laboratory and on-site. Nanocomposites, after a careful selection of both the matrix and the nanoparticles, can display very interesting and suitable performances. Unfortunately, limited attention and economic support by public and private entities are addressed to the specific application field (the protection of historical architecture). It would also be necessary to perform larger studies and experimental activity directly on real surfaces, with adequate knowledge and instruments, in close collaboration with architects, conservators, and authorities.

Acknowledgments   The authors would like to express their deep gratitude to Dr. A. Colombo and Prof. R. Simonutti of the University of Milano-Bicocca for the development of innovative nanoparticles, their collaboration, and fruitful discussions.

References

1. Siegesmund S, Snethlage R, editors. Stone in architecture: properties, durability. Berlin Heidelberg: Springer; 2014.
2. Cantisani E, Pecchioni E, Fratini F, Garzonio CA, Malesani P, Molli G. Thermal stress in the Apuan marbles: relationship between microstructure and petrophysical characteristics. Int J Rock Mech Min. 2009;46:128–37.
3. Malaga-Starzec K, Åkesson U, Lindqvist JE, Schouenborg B. Microscopic and macroscopic characterization of the porosity of marble as a function of temperature and impregnation. Constr Build Mater. 2006;20(10):939–47.
4. Steiger M, Charola AE, Sterflinger K. Weathering and deterioration. In: Siegesmund S, Snethlage R, editors. Stone in architecture: properties, durability. Berlin Heidelberg: Springer; 2014. p. 227–316.

5. Vergès-Belmin V, editor. Illustrated glossary on stone deterioration patterns. Paris: ICCMOS; 2008.
6. Dohene E, Price C. Stone conservation—an overview of current research. Los Angeles: The Getty Conservation Institute; 2010.
7. Snethlage R. Stone conservation. hydrophobic treatment. In: Siegesmund S, Snethlage R, editors. Stone in architecture: properties, durability. Berlin Heidelberg: Springer; 2014. p. 515–28.
8. Poli T, Toniolo L. The challenge of protecting outdoor exposed monuments from atmospheric attack: experience and strategy. In: Kourkoulis SK, editor. Fracture and failure of natural building stones: applications in the restoration of ancient monuments. Dordrecht: Springer; 2006. p. 553–64.
9. Casadio F, Toniolo L. Polymeric treatments for stone materials conservation: the problem of penetration depth. J Am Inst Conserv. 2004;43:1–19.
10. Brugnara M, Degasperi E, Della Volpe C, Maniglio D, Penati A, Siboni S, Toniolo L, Poli T, Invernizzi S, Castelvetro V. The application of the contact angle in monument protection: new materials and methods. Colloid Surf A Physicochem Eng Asp. 2004;241:299–312.
11. Della Volpe C, Brugnara M, Maniglio D, Siboni S, Wangdu T. About the possibility of experimentally measuring an equilibrium contact angle and its theoretical and practical consequences. In: Mittal KL, editor. Contact angle wettability and adhesion, vol. 4; 2006. p. 79–99.
12. Peruzzi R, Poli T, Toniolo L. The experimental test for the evaluation of protective treatments: a critical survey of the Capillary Absorption Index. J Cult Herit. 2003;4(3):93–6.
13. Favaro M, Mendichi R, Ossola F, Russo U, Simon S, Tomasin P, Vigato PA. Evaluation of polymers for conservation treatments of outdoor exposed stone monuments. Part I: photo-oxidative weathering. Polym Degrad Stabil. 2006;91(12):3083–96.
14. Favaro M, Mendichi R, Ossola F, Simon S, Tomasin P, Vigato PA. Evaluation of polymers for conservation treatments of outdoor exposed stone monuments. Part II: photo-oxidative and salt-induced weathering of acrylic–silicone mixtures. Polym Degrad Stabil. 2007;92(3):335–51.
15. Charola AE. Water repellents and other 'protective' treatments: a critical review. Int J Rest Build Mon. 2003;9(1):3–21.
16. Alvarez de Buergo M, Fort R. A basic methodology for evaluating and selecting water-proofing treatments applied to carbonatic materials. Progr Org Coat. 2001;43:258–66.
17. EN 16581:2014. Conservation of Cultural Heritage—Surface protection for porous inorganic materials—Laboratory test methods for the evaluation of the performance of water repellent products, European Committee for Standardization; 2014.
18. Chiantore O, Lazzari M. Photo-oxidative stability of paraloid acrylic protective polymers. Polymer. 2001;42(1):17–27.
19. Doherty B, Pamplona M, Selvaggi R, Miliani C, Matteini M, Sgamellotti A, Brunetti B. Efficiency and resistance of the artificial oxalate protection treatment on marble against chemical weathering. Appl Surf Sci. 2007;253:4477–84.
20. Meloni P, Manca F, Carcangiu G. Marble protection: an inorganic electrokinetic approach. Appl Surf Sci. 2013;273:377–85.
21. Alessandrini G, Aglietto M, Castelvetro V, Ciardelli F, Peruzzi R, Toniolo L. Comparative evaluation of fluorinated and unfluorinated acrylic copolymers as water-repellent coating materials for stone. J Appl Polym Sci. 2000;76:962–77.
22. Poli T, Toniolo L, Chiantore O. The protection of different Italian marbles with two partially flourinated acrylic copolymers. Appl Phys A Mater Sci Process. 2004;79:347–51.
23. Tsakalof A, Manoudis P, Karapanagiotis I, Chryssoulakis I, Panayiotou C. Assessment of synthetic polymeric coatings for the protection and preservation of stone monuments. J Cult Herit. 2007;8(1):69–72.
24. Sierra-Fernandez A, Gomez-Villalba LS, Rabanal ME, Fort R. New nanomaterials for applications in conservation and restoration of stony materials: a review. Mater Constr. 2017;67(325):e107.
25. Manoudis P, Karapanagiotis I, Tsakalof A, Zuburtikudis I, Kolinkeová B, Panayiotou C. Superhydrophobic films for the protection of outdoor cultural heritage assets. Appl Phys A. 2009;97(2):351–60.

26. Aslanidou D, Karapanagiotis I, Panayiotou C. Tuning the wetting properties of siloxane-nanoparticle coatings to induce superhydrophobicity and superoleophobicity for stone protection. Mater Design. 2016;108:736–44.
27. Karapanagiotis I, Pavlou A, Manoudis PN, Aifantis KE. Water repellent ORMOSIL films for the protection of stone and other materials. Mater Lett. 2014;131:276–9.
28. Fatorić S, Seekamp E. Are cultural heritage and resources threatened by climate change? A systematic literature review. Clim Chang. 2017;142(1/2):227–54.
29. Brimblecombe P. Past, present and future damage to materials and building surfaces in the polluted urban environment. In: Brimblecombe P, editor. Urban pollution and changes to materials and building surfaces, vol. 5. London: Imperial College Press; 2015. p. 1–18.
30. Bonazza A, Messina P, Sabbioni C, Grossi CM, Brimblecombe P. Mapping the impact of climate change on surface recession of carbonate buildings in Europe. Sci Total Environ. 2009;407(6):2039–50.
31. Bonazza A, Sabbioni C, Messina P, Guaraldi C, De Nuntiis P. Climate change impact: mapping thermal stress on Carrara marble in Europe. Sci Total Environ. 2009;407(15):4506–12.
32. Grossi CM. Soiling and discolouration in Urban pollution and changes to materials and building surfaces. In: Brimblecombe P, editor. Urban pollution and changes to materials and building surfaces, vol. 5. London: Imperial College Press; 2015. p. 127–41.
33. Munafò P, Goffredo GB, Quagliarini E. TiO_2-based nanocoatings for preserving architectural stone surfaces: an overview. Constr Build Mater. 2015;84:201–18.
34. Ditaranto N, Loperfido S, van der Werf I, Mangone A, Cioffi N, Sabbatini L. Synthesis and analytical characterisation of copper-based nanocoatings for bioactive stone artworks treatment. Anal Bioanal Chem. 2011;399(1):473–81.
35. Bellissima F, Bonini M, Giorgi R, Baglioni P, Barresi G, Mastromei G, Perito B. Antibacterial activity of silver nanoparticles grafted on stone surface. Environ Sci Pollut Res. 2014;21(23):13278–86.
36. Chen X, Mao SS. Titanium dioxide nanomaterials: synthesis, properties, modifications, and applications. Chem Rev. 2007;107(7):2891–959.
37. Pelaez M, Nolan NT, Pillai SC, Seery MK, Falaras P, Kontos AG, Dunlop PSM, Hamilton JWJ, Byrne JA, O'Shea K, Entezari MH, Dionysiou DD. A review on the visible light active titanium dioxide photocatalysts for environmental applications. Appl Catal B Environ. 2012;125:331–49.
38. Fujishima A, Zhang X. Titanium dioxide photocatalysis: present situation and future approaches. C R Chim. 2006;9(5–6):750–60.
39. Hoffmann MR, Martin ST, Choi W, Bahnemann DW. Environmental applications of semiconductor photocatalysis. Chem Rev. 1995;95(1):69–96.
40. Fujishima A, Rao TN, Tryk DA. Titanium dioxide photocatalysis. J Photoch Photobio C. 2000;1(1):1–21.
41. Guan K. Relationship between photocatalytic activity, hydrophilicity and self-cleaning effect of TiO_2/SiO_2 films. Surf Coat Tech. 2005;191(2–3):155–60.
42. Fujishima A, Zhang X, Tryk DA. TiO_2 photocatalysis and related surface phenomena. Surf Sci Rep. 2008;63(12):515–82.
43. Shon H, Phuntsho S, Okour Y, Cho DL, Kim KS, Li HJ, Na S, Kim JB, Kim JH. Visible light responsive titanium dioxide (TiO_2). J Korean Ind Eng Chem. 2008;19(1):1–16.
44. Asahi R, Morikawa T, Ohwaki T, Aoki K, Taga Y. Visible-light photocatalysis in nitrogen-doped titanium oxides. Science. 2001;293(5528):269–71.
45. Seery MK, George R, Floris P, Pillai SC. Silver doped titanium dioxide nanomaterials for enhanced visible light photocatalysis. J Photoch Photobio A. 2007;189(2–3):258–63.
46. Quagliarini E, Bondioli F, Goffredo GB, Licciulli A, Munafò P. Smart surfaces for architectural heritage: preliminary results about the application of TiO_2-based coatings on travertine. J Cult Herit. 2012;13(2):204–9.
47. Quagliarini E, Bondioli F, Goffredo GB, Licciulli A, Munafò P. Self-cleaning materials on architectural heritage: compatibility of photo-induced hydrophilicity of TiO_2 coatings on stone surfaces. J Cult Herit. 2013;14(1):1–7.

48. Gherardi F, Colombo A, D'Arienzo M, Credico B, Goidanich S, Morazzoni F, Simonutti R, Toniolo L. Efficient self-cleaning treatments for built heritage based on highly photo-active and well-dispersible TiO$_2$ nanocrystals. Microchem J. 2016;126:54–62.
49. Manoudis PN, Tsakalof A, Karapanagiotis I, Zuburtikudis I, Panayiotou C. Fabrication of super-hydrophobic surfaces for enhanced stone protection. Surf Coat Technol. 2009;203:1322–8.
50. Cappelletti G, Fermo P, Camiloni M. Smart hybrid coatings for natural stones conservation. Prog Org Coat. 2015;78:511–6.
51. Colangiul D, Calia A, Bianco N. Novel multifunctional coatings with photocatalytic and hydrophobic properties for the preservation of the stone building heritage. Constr Build Mater. 2015;93:189–96.
52. La Russa MF, Rovella N, Alvarez de Buergo M, Belfiore CM, Pezzino A, Crisci GM, Ruffolo SA. Nano-TiO$_2$ coatings for cultural heritage protection: the role of the binder on hydrophobic and self-cleaning efficacy. Prog Org Coat. 2016;91:1–8.
53. Scalarone D, Lazzari M, Chiantore O. Acrylic protective coatings modified with titanium dioxide nanoparticles: comparative study of stability under irradiation. Polym Degrad Stabil. 2012;97(11):2136–42.
54. Pinho L, Mosquera MJ. Titania-silica nanocomposite photocatalysts with application in stone self-cleaning. J Phys Chem C. 2011;115(46):22851–62.
55. Kapridaki C, Maravelaki-Kalaitzaki P. TiO$_2$–SiO$_2$–PDMS nano-composite hydrophobic coating with self-cleaning properties for marble protection. Prog Org Coat. 2013;76(2–3):400–10.
56. Kapridaki C, Pinho L, Mosquera MJ, Maravelaki-Kalaitzaki P. Producing photoactive, transparent and hydrophobic SiO$_2$-crystalline TiO$_2$ nanocomposites at ambient conditions with application as self-cleaning coatings. Appl Catal B Environ. 2014;156:416–27.
57. Munafò P, Quagliarini E, Goffredo GB, Bondioli F, Licciulli A. Durability of nano-engineered TiO$_2$ self-cleaning treatments on limestone. Constr Build Mater. 2014;65(0):218–31.
58. Calia A, Lettieri M, Masieri M. Durability assessment of nanostructured TiO$_2$ coatings applied on limestones to enhance building surface with self-cleaning ability. Build Environ. 2016;110:1–10.
59. Licciulli A, Calia A, Lettieri M, Diso D, Masieri M, Franza S, Amadelli R, Casarano G. Photocatalytic TiO2 coatings on limestone. J Sol-Gel Sci Technol. 2011;60(3):437–44.
60. Pinho L, Mosquera MJ. Photocatalytic activity of TiO$_2$–SiO$_2$ nanocomposites applied to buildings: influence of particle size and loading. Appl Catal B Environ. 2013;134–135:205–21.
61. La Russa MF, Ruffolo SA, Rovella N, Belfiore CM, Palermo AM, Guzzi MT, Crisci GM. Multifunctional TiO$_2$ coatings for cultural heritage. Prog Org Coat. 2012;74(1):186–91.
62. Goffredo GB, Accoroni S, Totti C, Romagnoli T, Valentini L, Munafò P. Titanium dioxide based nanotreatments to inhibit microalgal fouling on building stone surfaces. Build Environ. 2017;112:209–22.
63 Pinho L, Rojas M, Mosquera MJ. Ag–SiO$_2$–TiO$_2$ nanocomposite coatings with enhanced photoactivity for self-cleaning application on building materials. Appl Catal B Environ. 2015;178:144–54.
64. La Russa MF, Macchia A, Ruffolo SA, De Leo F, Barberio M, Barone P, Crisci GM, Urzì C. Testing the antibacterial activity of doped TiO$_2$ for preventing biodeterioration of cultural heritage building materials. Int Biodeter Biodegr. 2014;96:87–96.
65. Bergamonti L, Predieri G, Paz Y, Fornasini L, Lottici PP, Bondioli F. Enhanced self-cleaning properties of N-doped TiO$_2$ coating for cultural heritage. Microchem J. 2017;133:1–12.
66. Gherardi F, Gulotta D, Goidanich S, Colombo A, Toniolo L. On-site monitoring of the performance of innovative treatments for marble conservation in architectural heritage. Herit Sci. 2017;5(1):4.
67. Allen NS, Edge M, Ortega A, Sandoval G, Liauw CM, Verran J, Stratton J, McIntyre RB. Degradation and stabilisation of polymers and coatings: nano versus pigmentary titania particles. Polym Degrad Stabil. 2004;85(3):927–46.
68. Ferrari A, Pini M, Neri P, Bondioli F. Nano-TiO$_2$ coatings for limestone: which sustainability for cultural heritage? Coatings. 2015;5(3):232.

69. Colombo A, Gherardi F, Goidanich S, Delaney JK, de la Rie ER, Ubaldi MC, Toniolo L, Simonutti R. Highly transparent poly(2-ethyl-2-oxazoline)-TiO_2 nanocomposite coatings for the conservation of matte painted artworks. RSC Adv. 2015;5(103):84879–88.
70. Beccaria C, Colombo A, Gherardi F, Mombrini V, Toniolo L. Use of nanocoatings for the restoration of matte paintings. Stud Conserv. 2016;61(suppl 2):265–6.
71. Gherardi F. Nano-structured coatings for stone and paint surfaces of cultural heritage. PhD thesis in materials engineering. Milan: Politecnico di Milano; 2016.
72. Niederberger M, Bartl MH, Stucky GD. Benzyl alcohol and titanium tetrachloride—a versatile reaction system for the nonaqueous and low-temperature preparation of crystalline and luminescent titania nanoparticles. Chem Mater. 2002;14(10):4364–70.
73. Colombo A, Tassone F, Mauri M, Salerno D, Delaney JK, Palmer MR, de la Rie ER, Simonutti R. Highly transparent nanocomposite films from water-based poly(2-ethyl-2-oxazoline)/TiO_2 dispersions. RSC Adv. 2012;2(16):6628–36.
74. Simonutti R, Colombo A, Beccaria C, Mombrini V. Nanoparticle dispersions in polymer matrices. Patent N. WO2012/160525. Universita' degli Studi Milano—Bicocca and Fondazione Cassa di Risparmio delle Province Lombarde; 2012.
75. García O, Malaga K. Definition of the procedure to determine the suitability and durability of an anti-graffiti product for application on cultural heritage porous materials. J Cult Herit. 2012;13(1):77–82.
76. Wang R, Hashimoto K, Fujishima A, Chikuni M, Kojima E, Kitamura A, et al. Light-induced amphiphilic surfaces. Nature. 1997;388(6641):431–2.
77. Kazuhito H, Hiroshi I, Akira F. TiO_2 photocatalysis: a historical overview and future prospects. Jpn J Appl Phys. 2005;44(12R):8269–85.

Chapter 4
A Hybrid Consolidant of Nano-Hydroxyapatite and Silica Inspired from Patinas for Stone Conservation

Pagona Maravelaki and Anastasia Verganelaki

4.1 Introduction

The increasing amount of pollution has resulted in the intensification of weathering phenomena in building materials, especially in the case of pure carbonate materials which generally exhibit low durability and mechanical resistance. Therefore, the development of consolidants that enhance the robustness and durability of carbonate stones is of primary importance.

Recently, research has focused on the development of materials based on hydroxyapatite (HAp), with encouraging results [1–3]. The treatment with HAp has been found to provide remarkable consolidation without altering the porosity of the substrate while preserving stone from hydrophilic behavior [1]. Sassoni et al. [2] studied the effectiveness of HAp formed by the reaction of a diammonium hydrogen phosphate (DAP) solution with calcite in limestone, proving that the mechanical properties of the treated substrate significantly increased after just 2 days of reaction. Moreover, the same research group that reported the durability of HAp-treated limestone when subjected to salt crystallization demonstrated that HAp-treated samples exhibited less micro-cracking and higher resistance to mechanical damage than TEOS-treated limestones [3].

Scherer et al. [4] applied DAP solutions to limestone samples by capillarity. Although the sorptivity of the DAP-treated substrate was negatively affected, the absence of toxic byproducts and the improvement in mechanical properties and retained water uptake had positively affected the consolidation. Liu et al. [5] prepared a protective film of apatite on a surface-modified marble through a biomimetic method using a solution of $CaCl_2$ and $(NH_4)_2HPO_4$.

P. Maravelaki (✉) • A. Verganelaki
Lab of Materials for Cultural Heritage and Modern Building, School of Architecture,
Technical University of Crete, Akrotiri campus, Chania, Crete 73100, Greece
e-mail: pmaravelaki@isc.tuc.gr

© Springer International Publishing AG 2018
M. Hosseini, I. Karapanagiotis (eds.), *Advanced Materials for the Conservation of Stone*, https://doi.org/10.1007/978-3-319-72260-3_4

Finally, layer patinas detected on several well-preserved Mediterranean monuments have shown HAp as one of the main constituents. These patinas, called "scialbatura" by several authors, were often observed in Roman and other monuments of the same period, are usually hard, well adhered to the marble surface, maintained the relief details, and protected the surface from further weathering [6, 7].

The choice of HAp as a potential consolidant was based on its low solubility and slow dissolution rate. In addition, HAp has a crystal structure and lattice parameters similar to those of calcite (differing by only ~5%), indicating that it can be strongly bonded with calcareous substrates, thus overcoming one of the basic drawbacks of commercial consolidants, that is, an insufficient bonding with calcareous substrates [8, 9].

As it is well known, tetraethyl orthosilicate (TEOS)-based alkoxysilanes are popular commercial products for the protection of building materials. The low viscosity of the initial sol enables a deeper penetration of TEOS into the substrate, polymerizing in the pores of the building materials by using atmospheric moisture, thus creating a silica matrix similar to the siliceous minerals of the stone that stabilizes the structure [10]. However, TEOS products are characterized by important drawbacks, such as inefficient chemical bonding to calcite and a tendency to crack during their drying process [11, 12]. Cracking is generated by the high capillary pressure exerted by the gel network during drying. It has been proven that the surfactant n-octylamine ($CH_3(CH_2)_7NH_2$) plays an important role during polymerization and condensation in providing crack-free gels by acting as the structure-directing agent and decreasing solvent surface tension [13, 14]. Therefore, the contribution of n-octylamine in the crack-free gel enhanced the consolidant effectiveness of siloxane interactions, while it can be easily removed via its hydrogen bonding with silica [15, 16].

In this chapter, the synthesis of a novel nanocomposite based on TEOS and HAp is reported, with amylamine ($CH_3(CH_2)_4NH_2$) used as a surfactant for the first time, providing an efficient means of protecting gels from cracking by reducing the capillary pressure. The use of amylamine in this work aims to propose a surfactant with similar properties to n-octylamine but without inducing any deposition between the initial reagents. Amylamine is characterized by a shorter chain amine molecule than n-octylamine. Therefore, the amphiphilicity increases by the hydrophilic functional groups and hydrophobic tail groups without causing the formation of any amphiphilic molecule clusters [17]. For better understanding and comparing, the role of HAp as well as amylamine into the silica structure and the adherence to the calcareous substrate, two other nanocomposites, were synthesized, containing (1) TEOS and HAp and (2) TEOS and amylamine as basic reagents.

The synthesized products have been characterized and evaluated for their effectiveness as strengthening agents on a limestone that is widely used in historic and modern architectural structures in the Mediterranean basin. Finally, for comparison purposes, commercial TEOS was also characterized and evaluated for its performance on the same building materials.

4.2   Experimental Section

4.2.1   Materials

TEOS (Sigma-Aldrich), calcium hydroxide powder (CH, Fluka), and phosphoric acid (H_3PO_4, 85%, Mallinckrodt) were used as raw materials, while amylamine was used as surfactant for the synthesis of the nanocomposite. Isopropanol (ISP, Sigma-Aldrich) and deionized water were used as solvents.

4.2.2   Synthesis of Nanocomposites

The sol was prepared by following a two-step synthesis route. The first step includes the synthesis of a colloidal solution of HAp produced by the reaction of CH with H_3PO_4. In the second step, the synthesized HAp was added slowly into a mixture containing TEOS and water. A small quantity of amylamine was also added to the final mixture, after examining that neither reaction nor deposition occurred between amylamine and the reagents of the synthesis. This synthesis led to the formation of the final colloidal nanocomposite, defined as SiHAp.

More specifically, an aqueous solution of HAp was prepared by using aqueous solutions of CH and H_3PO_4 [18]. The diluted H_3PO_4 solution was added to the CH suspension drop by drop (1 mL/min) at room temperature, under ultrasonic conditions. When the titration is finished, the final solution was maintained under ultrasonic conditions for 30 min and then was subjected to vigorous stirring for 24 h. According to the literature, the presence of ultrasound in the synthesis process promotes the chemical reaction and contributes to the formation of nanoparticles [19]. Careful control of pH is very important because allowing the pH of solution to decrease below 9 could cause the formation of Ca-deficient HAp [20].

Afterward, the synthesized HAp was dried at 100 °C and re-dispersed in isopropanol (ISP) under vigorous stirring and ultrasonic agitation alternatively for 6 h, producing a 1% w/v colloidal solution, defined as HAp-s. According to the literature, the re-dispersion of HAp particles in alcoholic solvents (e.g., ISP) significantly effects the chemical composition, crystallinity degree, and particle size of HAp-s [18]. The synthesized HAp-s was added slowly to a mixture of TEOS/H_2O. Phosphoric acid (0.01 M) was used to adjust the pH to 2–3. Finally, a small quantity of amylamine was added to the mixture since a higher concentration promotes the instantaneous gelation of TEOS [15]. It is assumed that amylamine functions as both a base catalyst and surfactant. The final sol, defined as SiHAp, was subjected to ultrasonic agitation for 15 min and then was maintained under vigorous stirring for 48 h. The SiHAp sol has a molar ratio of TEOS/ISP/H_2O/HAp/amylamine equal to 1/8.8/2.5/0.01/0.02.

Three different sets of formulations were also prepared in order to gain insight into the role of the HAp and amylamine in the xerogel structure and the mechanism

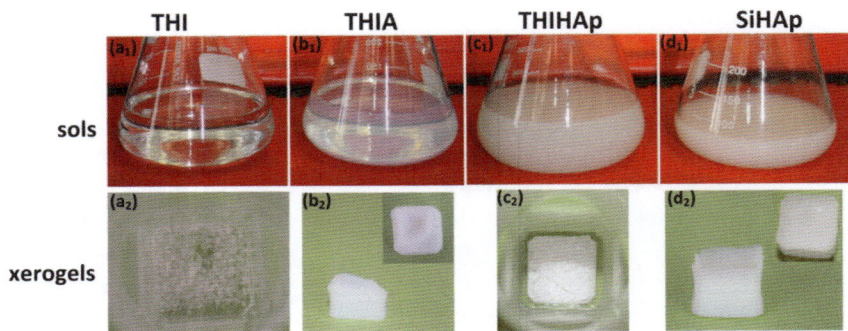

Fig. 4.1 THI (a_1), THIA (b_1), THIHAp (c_1), and SiHAp (d_1) sols obtained after 48 h under magnetic stirring; THI (a_2), THIA (b_2), THIHAp (c_2), and SiHAp (d_2) xerogels after 40 days of drying in laboratory conditions

Table 4.1 Molar ratio and gelation time of the sols under study

Sample code	TEOS	ISP	H_2O	HAp	Amylamine	pH	Gelation time (days)
THI	1	8.8	2.5	–	–	7.3	42
THIA	1	8.8	2.5	–	0.02	9.9	25
THIHAp	1	8.8	2.5	0.01	–	5.5	40
SiHAp	1	8.8	2.5	0.01	0.02	8.8	26

of reaction: (1) the first formulation was synthesized containing TEOS, H_2O, and ISP, defined as *THI*; (2) the second one consisted of TEOS, ISP, H_2O, and amylamine, defined as *THIA*; and the third one was formed by using TEOS, H_2O, ISP, and HAp, termed as *THIHAp*. The proportions of the reagents in the three sols were maintained according to the molar ratio which was previously described for SiHAp nanocomposite.

The final sols under study were each placed on a cubic bottom flask covered with perforated parafilm to allow gentle evaporation of the solvent. Gelation and drying occurred by simple exposure of the cast sols to laboratory conditions (RH = 60 ± 5%, T = 20 ± 2 °C) until a constant weight was reached. Gel times were determined by a visual inspection of the gel transition inside the transparent mold. After the polymerization and drying processes, xerogels were produced in a period ranging of approximately 25–40 days (Fig. 4.1).The experimental formulation for the synthesized sols expressed in moles is listed in Table 4.1.

4.2.3   Evaluation of Effectiveness on Calcareous Stones

The effectiveness of the products under study was evaluated on a calcareous stone, called Alfas, which is frequently found in historic and modern architectural structures in the Mediterranean basin. It is a homogeneous texture stone, with yellow-cream color, consisted of approximately 95% calcium carbonate with small amounts of quartz and plagioclase [21]. Its porosity ranges from 25 to 35%, while the pore size distribution is located in the range from 1 to 10 μm [21].

The stone samples were cut in the form of cylinder with 50 mm diameter and 28 mm height. After shaping, the stone specimens were rinsed with water and dried in an oven at 80 °C until a constant weight was reached. The sols under study were applied by brushing to all the faces of the specimens. The treatment was completed when the surfaces remained wet for 1 min. The quantity of each sol that was applied on the stone samples was the minimum possible, in order to avoid the creation of a layer on the stone surface. After treatment, all the test specimens were allowed to dry in laboratory conditions (RH = 60 ± 5%, T = 20 ± 2 °C) until their weight was stabilized. This was achieved within 40–45 days of drying; then, the amount of the xerogels deposited in the stone specimens in terms of g/cm^2, denoted as dry matter, was calculated.

All of the results reported correspond to average values obtained from three stone samples.

4.2.4   Characterization

The synthesized nanocomposites were assessed with the analytical techniques described below. Initially, the crystallinity of the nanocomposites was studied by X-ray diffraction analysis (XRD) using a Bruker D8 Advance Diffractometer with a Ni-filtered Cu Kα radiation (35 kV 35 mA) and a Bruker Lynx Eye strip silicon detector. Fourier transform infrared spectroscopy (FTIR) spectra were recorded on a Perkin-Elmer 1000 spectrometer in the spectral range of 400–4000 cm^{-1}. For the analysis of the sols and xerogels, two different experimental procedures were followed. The sol samples were placed into AgBr cells, while pellets with KBr and powder from xerogels were prepared in a ratio of 100/1. Twenty scans were recorded for each spectrum. The spectra of sols were recorded at various time intervals during the synthesis of the final product, starting from the mixing of all the reagents up to 12 days, in order to gain insights into the sol-gel process, hydrolysis of TEOS, and role of amylamine into the sol-gel process. The spectra are obtained at intervals of 1 h from the mixing of all the reagents, as well as at intervals of 1, 5, 7, and 12 days from the completion of the synthesis. Finally, powders derived from the xerogels cured for 1.5 and 2.5 months were also studied by FTIR.

The thermal behavior of the nanocomposites along with the synthesized HAp powder was determined by thermogravimetric analysis (DTA-DSC-TG) using a Setaram LabSys Evo 1600 °C at a heating rate of 10 °C/min under nitrogen atmosphere from 27 to 1000 °C. From the comparison of TG and DTA curves, significant observations were recorded concerning weight loss, endotherm, and exotherm effects.

The surface morphology of the xerogels under study was examined under scanning electron microscopy (SEM) using an FEI Quanta Inspect D8334 instrument operating at 25 kV, after 1.5 months of curing. Energy-dispersive spectroscopy (EDX) was used to characterize the elemental composition of the xerogels.

4.2.5 Limestone Treated with the Nanocomposites

The characterization of the treated specimens was assessed 1.5 months after the treatment process with the following techniques.

The coloring effect of the applied treatment on the treated specimens was evaluated with reflectance spectra measured by a spectrophotometer (CM-2600d, Konica Minolta). L^*, α^*, b^*, and ΔE^* are derived from random areas from stone specimens.

The ability of the treated specimens to allow water vapor evaporation was evaluated by carrying out water vapor permeability tests according to UNI EN 15803:2010 [22].

The mechanical properties of the treated samples were evaluated by comparing the indirect tensile strength (Brazilian test) according to ASTM D3967-86 and the drilling resistance properties of the treated and untreated specimens. The mechanical robustness was studied by using the drilling resistance measuring system (DRMS) (Sint Technology). For DRMS measurements, a drill bit of 4.8 mm diameter was used with a rotation speed of 600 rpm and penetration rate of 20 mm/min. Moreover, through drilling resistance profiles, the penetration depth of the nanocomposites into the calcareous substrate was determined.

4.3 Results and Discussion

4.3.1 Nanocomposites

The structural evaluation of the four xerogels after curing for 2.5 months was initially conducted by XRD analyses. The X-ray diffractograms of the xerogels are shown in Fig. 4.2. A broad peak, which is typical of amorphous silica, ranging from 15° to 25°, is shown in all the XRD patterns. In the SiHAp_x and THIHAp_x patterns, two main peaks at around 26° and 32°, accompanied with smaller peaks at 34°, 40°, 42°, and 49.5° are observed, all corresponding to HAp.

Figure 4.3 illustrates the infrared spectra of sols under study, derived from different time intervals, starting from 1 h after mixing of all the ingredients/reagents up to 12 days. The THI and THIHAp sols showed no new peaks or changes in the intensity, indicating that the hydrolysis process had not yet started. Conversely, in THIA and SiHAp, the hydrolysis has been started 1 day after the completion of the sol synthesis as new peaks that correspond to silanols appeared. The new peak at 1408 cm^{-1} at B (b) and D (d) spectra assigned to stretching vibration of Si–O–H

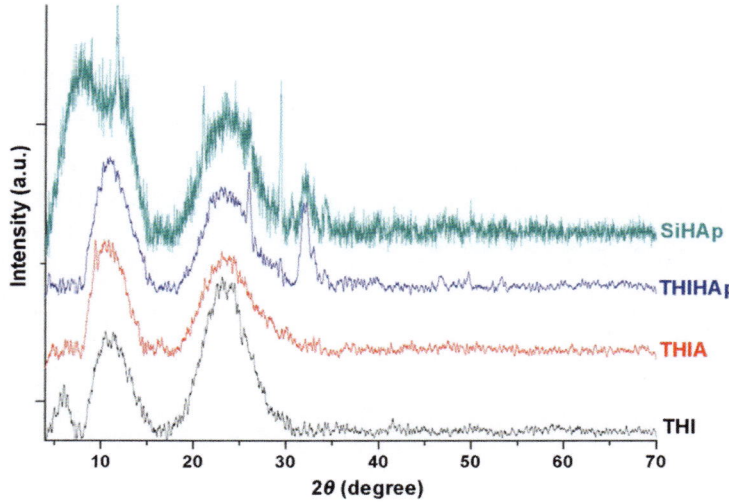

Fig. 4.2 XRD patterns of the (a) THI_x, (b) THIA_x, (c) THIHAp_x, and (d) SiHAp_x xerogels after curing for 1.5 months

hydrogen bonded to water evidenced that the hydrolysis process started [23]. Indeed, the intensity of this new peak slowly increased up to the 12th day, suggesting the ongoing hydrolysis of TEOS. Further supporting evidence of the hydrolysis of TEOS is the gradual decrease of the SiO_4 ring structure of TEOS at 1106, 1090, and 795 cm^{-1}, and the new absorptions at 1051 and 881 cm^{-1} attributed to the production of EtOH during the hydrolysis of TEOS [24, 25]. These results were expected since in THI and THIHAp, without base or acid catalysts and at neutral pH, no hydrolysis took place (Table 4.1) [10]. On the other hand, in THIA and SiHAp sols, amylamine provides the base-catalyzed environment for TEOS and functions as a template to form mesostructured materials with ordered pores [26].

In all of the spectra of Fig. 4.3A–D, the absorptions in the spectral regions at 1470–1300 cm^{-1} are associated with the C–H asymmetric and symmetric bending, respectively, from TEOS and ISP.

The identification of the HAp by its characteristic peaks at 1090 and 1037 cm^{-1} in the spectra of Fig. 4.3C, D was not possible, due to overlapping with the TEOS absorption in the same spectral area [27]. Similarly, the presence of amylamine in the spectra of THIA and SiHAp was not feasible because its characteristic peaks at 3316, 3263, 2928, 2925, and 2858 cm^{-1} were overlapped by TEOS and ISP absorptions in the same region.

The spectra of THI, THIA, and SiHAp final xerogels illustrated in Fig. 4.4 show characteristic peaks of Si–O–Si bonds in the spectral ranges 1000–1100 cm^{-} and 450–480 cm^{-1}, indicating that the polymerization process of TEOS had taken place [24, 27]. The peak at 798 cm^{-1}, which corresponds to Si–O–Si symmetric stretching, indicates the ongoing condensation process. In the THI xerogels, the peaks at

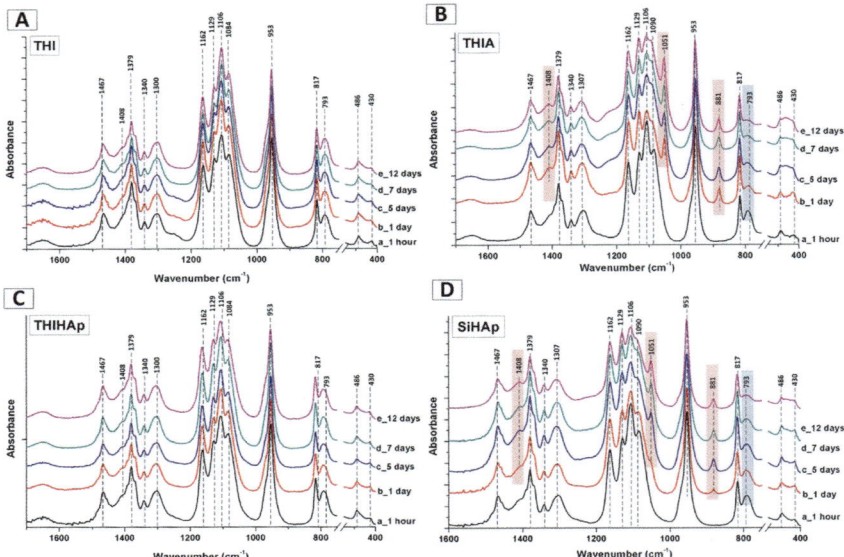

Fig. 4.3 FTIR spectra of sols under study at various time intervals: (**A**) THI sol (a) 1 h, (b) 1 day, (c) 5 days, (d) 7 days, and (e) 12 days after mixing the reagents; (**B**) THI sol (a) 1 h, (b) 1 day, (c) 5 days, (d) 7 days, and (e) 12 days after mixing the reagents; (**C**) THIA sol (a) 1 h, (b) 1 day, (c) 5 days, (d) 7 days, and (e) 12 days after mixing the reagents; and (**D**) spectra (a), (b), (c), (d), and (e) derived after 1 h, 1 day, 5 days, 7 days, and 12 days from the addition of amylamine into SiHAp, respectively

2980 and 2930 cm⁻¹ correspond to residual ISP, to EtOH, and to the ethyl group derived from the hydrolyzed TEOS, suggesting that the hydrolysis continued. This evidence is enhanced by the presence of the absorption band at 1169 cm⁻¹, which is related to the rocking of the C–H bond in -CH₃ of TEOS [28]. Finally, the characteristic double peak of HAp (604 and 565 cm⁻¹ assigned to bending vibrations of P–O mode) is presented in the spectra of SiHAp [18, 29].

Observing the spectra e_1 and f_1 in Fig. 4.4 that correspond to the THIHAp cured for 1.5 and 2.5 months, respectively, a totally different situation is revealed. In these spectra, the absorptions of HAp predominate. Apart from the peaks at 604 and 565 cm⁻¹, the absorptions at 1090 and 1040 cm⁻¹, which correspond to P–O asymmetrical stretching mode, and 3570 cm⁻¹, which is assigned to OH– group, are also attributed to HAp [18]. The weak absorptions at 878 and 960 cm⁻¹, which are attributed to Si–OH hydrogen bond, indicated that the hydrolysis process of TEOS had taken place, while the absorption at 1230 cm⁻¹, which is indicative of Si–O–Si asymmetric stretching vibration, implies the condensation process. The low intensity of the peaks which related to TEOS is due to its evaporation during the drying process.

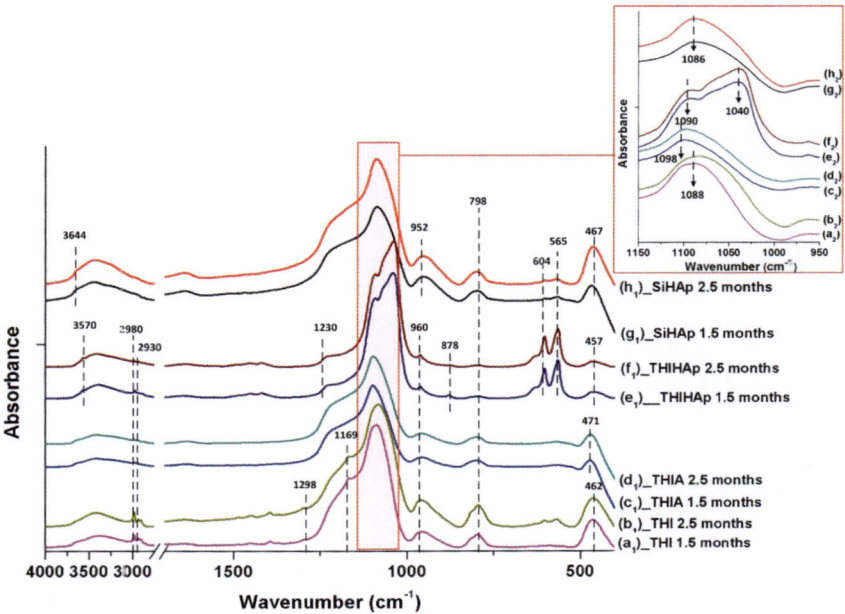

Fig. 4.4 FTIR spectra of xerogels under study at two different time intervals: (a_1, a_2) THI, (c_1, c_2) THIA, (e_1, e_2) THIHAp, and (g_1, g_2) SiHAp cured for 1.5 months; (b_1, b_2) THI, (d_1, d_2) THIA, (f_1, f_2) THIHAp, and (h_1, h_2) SiHAp cured for 2.5 months

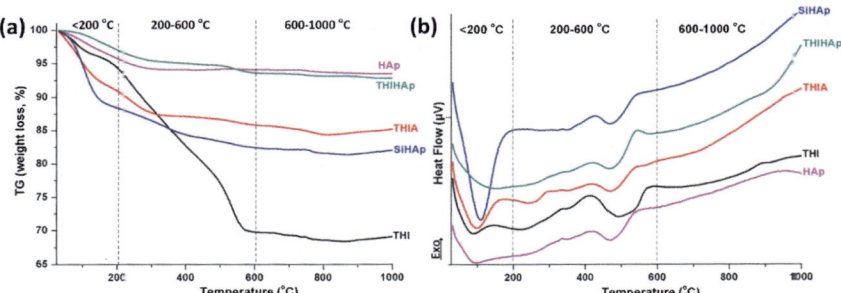

Fig. 4.5 TG (**a**) and DTA (**b**) curves of THI, THIA, THIHAp, and SiHAp xerogels, after 2.5 months of curing

All the samples under study (xerogels) were examined by thermal analysis in order to get information about their thermal stability and to cross the results of the previous analyses (XRD, FTIR). Figure 4.5 shows the TG (Fig. 4.5a) and DTA (Fig. 4.5b) diagrams of xerogels and powder of hydroxyapatite, while Table 4.2 lists

Table 4.2 Weight loss (%) of THI, THIA, THIHAp, SiHAp xerogels, and HAp powder in various temperature ranges

Sample code	Weight loss (%)			
	30–200 (°C)	200–600 (°C)	600–1000 (°C)	30–1000 (°C)
HAp	4.23	1.66	0.58	6.47
THI	5.36	24.75	0.64	30.75
THIA	9.00	5.14	0.76	14.9
THIHAp	2.90	3.42	0.78	7.10
SiHAp	11.43	5.95	0.59	17.97

the weight losses in various temperature ranges as well as the total weight loss of each sample.

The thermal decomposition of all samples was studied in 3 temperature ranges. The first range is from 30 to 200 °C, denoted by an endothermic peak, which corresponds to the absorbed water, as well as the volatilization and thermal decomposition of the remaining organic solvents [21]. In this area all xerogels show weight loss, with the greatest loss being observed in SiHAp (approximately 11.5%) due to the incomplete hydrolysis of TEOS (as proven by FTIR) and the production of ethanol, which thermally decomposes up to 200 °C. THIHAp presents the lower weight loss (2.9%) due to the small quantity of silica in the final xerogel, as also proved by the FTIR. The weight loss indicated in the case of HAp is due to the moisture residuals which had not evacuated during the drying process.

In the second stage of thermal degradation (200–600 °C), a significant loss of mass (24.75%) occurred in the THI xerogel, linked to both thermal decomposition of organic components and loss of water molecule generated from the condensation of silanol (Si–OH) groups [30]. The incomplete hydrolysis of TEOS in THI resulted in a large weight loss because of the continuous production of Si–OH groups and EtOH. In THIA where hydrolysis process was completed, the mass loss is due to the dehydroxylation of silanol groups. The weight loss of THIHAp and SiHAp is due to the vaporization of the water of crystallization of HAp [31]. In the third stage of thermal decomposition (600–1000 °C), the minor weight losses can be correlated to further condensation of silanol groups, while for the THIHAp and SiHAp xerogels, this can be attributed to the breakage of HPO_4^- in HAp.

Figure 4.6 provides important information about the microstructure and porosity of the four materials under study. The THI xerogel (Fig. 4.6a) presents a uniform dense structure, without pores, where the characteristic spherical shape of SiO_2 particles is not distinctive, due to the aggregation phenomena. Unlike the THIs structure, the THIHAp (Fig. 4.6c) exhibits pores as a result of the addition of HAp nanoparticles in silica substrate and aggregates with sizes ranging from 0.5 up to 2 μm. The xerogels with amylamine THIA and SiHAp, illustrated in Figs. 4.6b and d, show completely different structures in relation to the previous materials. In both cases, the aggregation phenomena have declined significantly, particularly in the case of SiHAp xerogel where both the characteristic spherical shape of SiO_2 particles and the rod-like shape of HAp particles are no longer distinctive (Fig. 4.6d) as a result of the

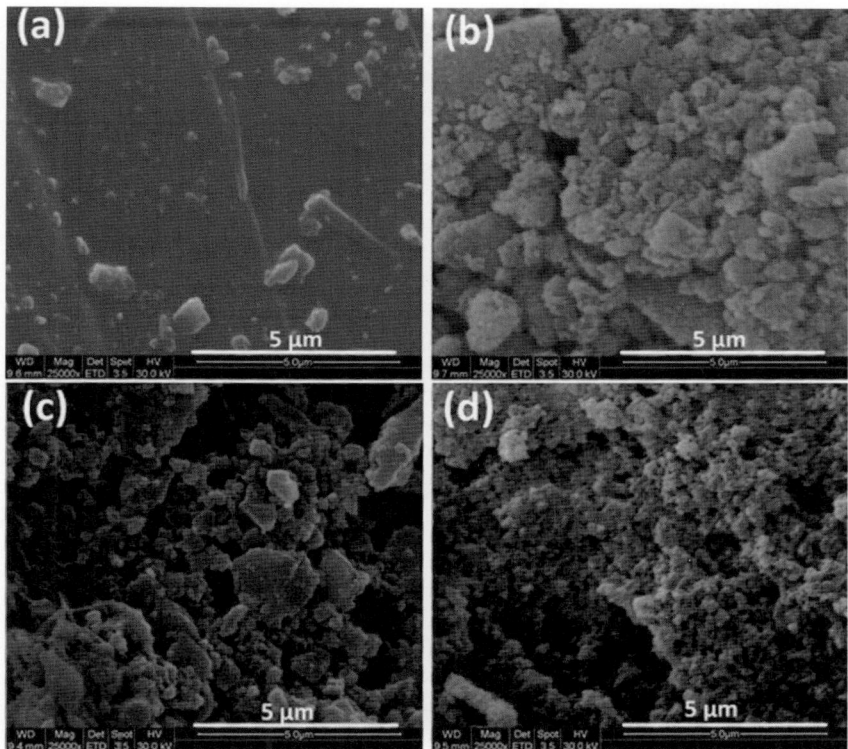

Fig. 4.6 SEM images for (**a**) THI, (**b**) THIA, (**c**) THIHAp, and (**d**) SiHAp after curing for 1.5 months

incorporation of HAp particles within the silica matrix [18]. It is worth highlighting that the amylamine-based xerogels THIA and SiHAp are characterized by a more uniform microstructure derived by the amylamine function as a structure template similarly to what is observed and well-established for n-octylamine [15].

4.3.2   Performance Evaluation of the Nanocomposites

The effectiveness of the nanocomposites under study as strengthening and/or protective agents was evaluated in a calcareous substrate by comparing hygric, microstructural, and mechanical properties before and after the treatment process. Initially, the dry matter values of the materials under study were recorded and are presented in Table 4.3. As expected, significant variations are noted in the values of dry matter. Calcareous samples treated with THIA and SiHAp contain almost double the quantity of dry matter compared to those which are treated with THI and

Table 4.3 Dry matter of the absorbed nanocomposites and coefficient of WCA, TWCA, and WVP of untreated and treated with THI, THIA, THIHAp, and SiHAp specimens

Sample code	Dry matter (g cm^{-2})	WCA (g cm^{-2} s$^{-1/2}$)	TWCA (%)	WVP (g cm^{-2} h^{-1})
Alfas_un	–	0.0126 (±0.0020)	13.40 (±0.06)	0.0010 (±0.0001)
Alfas_tr_THI	0.011 (±0.001)	0.0095 (±0.0009)	12.58 (±0.10)	0.0008 (±0.0001)
Alfas_tr_THIA	0.028 (±0.002)	0.0105 (±0.0001)	13.30 (±0.15)	0.0007 (±0.0001)
Alfas_tr_THIHAp	0.015 (±0.001)	0.0103 (±0.0000)	12.84 (±0.09)	0.0005 (±0.0001)
Alfas_tr_SiHAp	0.018 (±0.002)	0.0100 (±0.0002)	12.73 (±0.13)	0.0009 (±0.0002)

Table 4.4 Chromatic parameters (L^*, α^*, and b^*) and total color variation (ΔE^*) of untreated and treated with THI, THIA, THIHAp, and SiHAp specimens

Sample code	Chromatic parameters			ΔE^*
	L^*	α^*	b^*	
Alfas_un	82.63 (±0.09)	1.47 (±0.01)	13.29 (±0.05)	–
Alfas_tr_THI	81.72 (±0.42)	1.7 (±0.17)	14.62 (±0.69)	1.66 (±0.76)
Alfas_tr_THIA	81.87 (±0.25)	1.80 (±0.26)	13.41 (±1.10)	1.36 (±0.17)
Alfas_tr_THIHAp	84.03 (±0.39)	1.31 (±0.08)	8.24 (±0.70)	5.26 (±0.75)
Alfas_tr_SiHAp	83.22 (±0.57)	1.67 (±0.06)	10.60 (±0.83)	2.80 (±0.88)

THIHAp. This is due to the large amount of TEOS, in the case of THI and THIHAp sols, that was not hydrolyzed and evaporated during drying process, as proven by FTIR analyses.

The hygric properties of the treated samples presented in Table 4.3 as mean values of three samples showed that no significant changes are noticed on the stone's microstructure. Specifically, the coefficient of water capillary absorption (WCA) decreased by 16–24%, with samples treated with THI showing the highest decrease. Reductions to a smaller extent are also observed in the total water absorbed by capillarity (TWCA) measurements. Similarly, THI-treated specimens presented higher changes than the other cases. This result in conjunction with the reduction of TWCA indicates that THI had probably blocked some small pores. Additionally, the coefficient of water vapor permeability (WVP) was measured after the treatment process in order to check the circulation and the evaporation of water vapor. THIHAp-treated samples showed a significant reduction in WVP coefficient, further supporting that HAp particles were deposited on the surface of the stone samples, thus preventing the substrate from "breathing." The changes in WVP coefficient which are observed in the cases of THI-, THIA-, and SiHAp-treated samples were insignificant, indicating that the microstructure of the stone remained unchanged after treatment.

One of the basic criteria in consolidation refers to the color alterations which are noticed in the treated surface after the consolidants application. The color changes were evaluated as L^* (brightness), α^* (redness color), and b^* (yellowness color) coordinates, and the total color variation (ΔE^*), equal to $\Delta E* = \sqrt{\left(\Delta L*\right)^2 + \left(\Delta \alpha*\right)^2 + \left(\Delta b*\right)^2}$, was measured after the consolidation process and is presented in Table 4.4. According to literature, a total color variation expressed with $\Delta E^* < 2.5$ cannot be detected by the naked eye [32]. According to the ΔE^*

values, the application of THI, THIA, and SiHAp nanocomposites caused insignificant color alterations, while THIHAp induced a $\Delta E^* = 5.26$ (± 0.75), which corresponds to visible intense decrease of the b^* parameter (approximately 38%), which is related to yellow along with a slight increase of L^* parameter which expresses the luminosity, indicating that their surface had become slightly brighter after treatment. The color changes in the treated surface of THIHAp may be due to the heterogeneous nature of the sol. As it was mentioned before, THIHAp is separated into two phases during drying process (Fig. 4.1c$_2$). Similarly, TEOS penetrated enough into the stone during the treatment procedure, while HAp particles were deposited on the sample surface causing color alterations (Fig. 4.7).

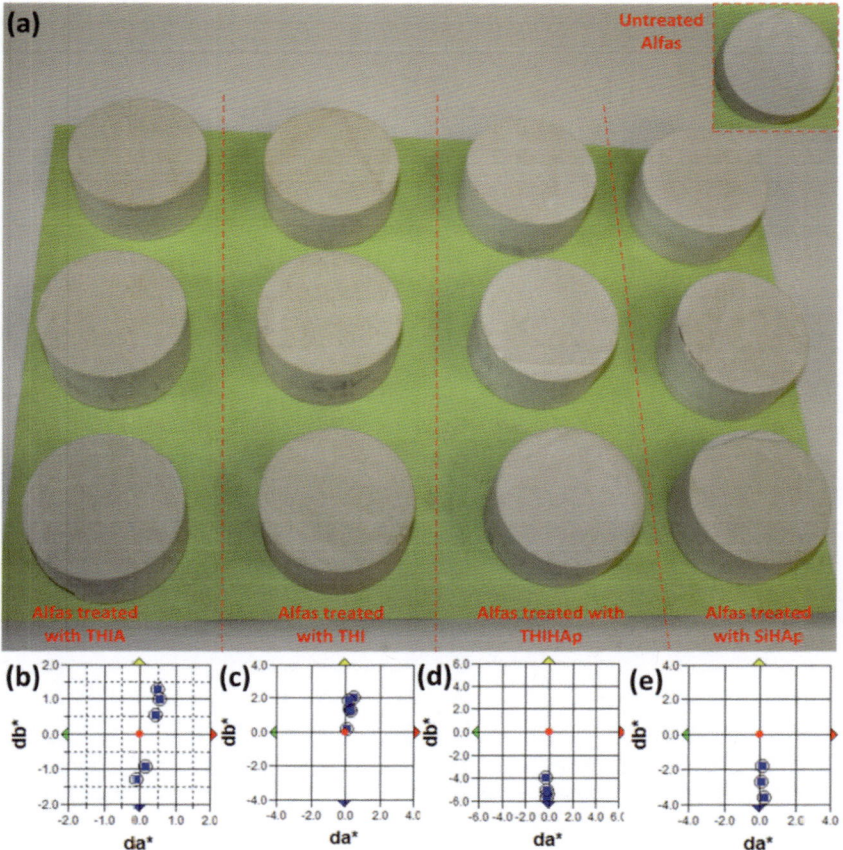

Fig. 4.7 Specimens treated with THIA, THI, THIHAp, and SiHAp nanocomposites cured for 1.5 months (**a**). The stone sample which is shown as inset figure in **a** indicates the untreated sample. Scatter plot of the Δb^* chromatic parameter against the Δa^* chromatic parameter for treated (blue spots) and untreated (red spots) surfaces of the Alfas stone treated with THIA (**b**), THI (**c**), THIHAp (**d**), and SiHAp (**e**)

Table 4.5 Tensile strength and drilling resistance force measurements for untreated and treated with THI, THIA, THIHAp, and SiHAp limestones

Sample code	Tensile strength (MPa)	Drilling resistance (N)
Alfas_un	3.68 (±0.66)	11.13 (±0.45)
Alfas_tr_THI	4.48 (±0.23)	12.10 (±0.41)
Alfas_tr_THIA	3.78 (±1.22)	11.69 (±0.25)
Alfas_tr_THIHAp	3.40 (±0.91)	12.07 (±0.12)
Alfas_tr_SiHAp	4.10 (±0.43)	12.32 (±0.30)

It is widely accepted that the penetration depth of consolidants is one of the most important parameters that must be considered in the evaluation of consolidation effectiveness. The microdrilling measurements used for the indirect evaluation of the product penetration depth as well as for the direct mechanical examination of the robustness of the stone. Table 4.5 presents the tensile strength (MPa) and drilling resistance force (N) measurements for untreated and treated with THI, THIA, THIHAp, and SiHAp limestones. The plots illustrated in Fig. 4.8 revealed that all the nanocomposites increased the stone's resistance up to a depth of 20 mm, thus indicating an adequate penetration depth. It can be stated that all the materials under study penetrated at least 20 mm into the substrate, thus enhancing the robustness of the stone. However, the SiHAp product increased the resistance by at least 11% compared to the increase by approximately 7% achieved with the rest of products.

As mentioned above, by using drilling resistance measuring system, the improvement in mechanical properties can be measured. All the materials increased the stone's resistance up to their respective penetration depths. The tensile strength of the treated stone samples with THI and SiHAp increased by 21% and 11%, respectively, indicating that THI was more homogeneously distributed in the treated samples. However, the SiHAp-treated samples achieved a considerable resistance to the tensile strength in accordance with the increased resistance measured during the microdrilling test.

4.4 Conclusion

The synthesis and characterization of a novel nanocomposite, SiHAp, based on TEOS and HAp were achieved; amylamine was used for the first time as a surfactant and template, providing an efficient means of preventing gels from cracking by forming a homogeneous microstructure. Amylamine increased the amphiphilicity by the hydrophilic functional groups and hydrophobic tail groups without causing the formation of any amphiphilic molecule clusters. The hydroxyapatite was selected on the grounds of its remarkable weathering resistance often encountered in patinas and attributed to the combination of hydroxyapatite, with calcium oxalate and silica. When examining and comparing the synthesized consolidant with other nanocomposites containing TEOS, TEOS and HAp, and TEOS and amylamine as

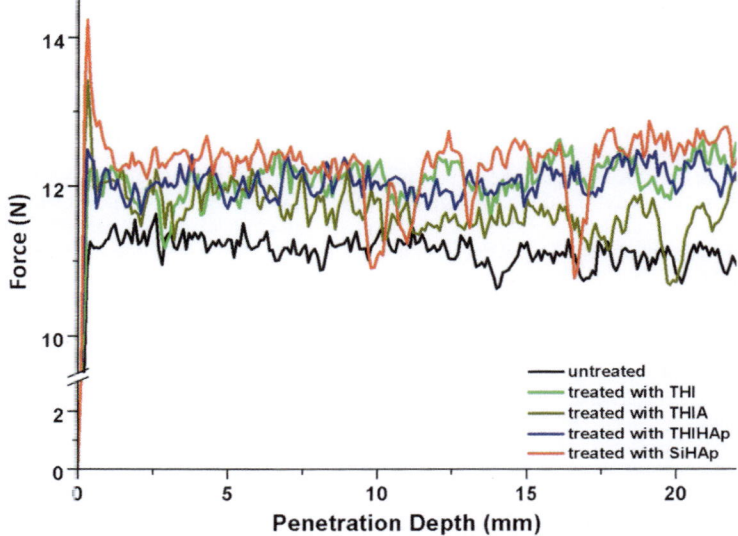

Fig. 4.8 Drilling resistance force measurements for untreated and treated with THI, THIA, THIHAp, and SiHAp limestones

basic reagents, the beneficial effect of HAp and amylamine into the silica structure was proven. More specifically, FTIR, thermal analysis, and SEM observations of the synthesized nanocomposites indicated that amylamine contributed to the completion of the hydrolysis of TEOS, the incorporation of HAp within the silica matrix, and the formation of a homogeneous gel microstructure acting as a surfactant and template. The treated calcareous stone with SiHAp preserved their microstructural integrity and exhibited adequate water repellency. A penetration depth of the consolidant at more than 15 mm is related to the increase of the robustness of the treated stone and the resistance to the tensile strength. The improvement of the hygric properties, drilling resistance, and tensile strength were attributed to the integrity and crack-free structure of the synthesized nanocomposite.

References

1. Sassoni E, Franzoni E, Pigino B, Scherer GW, Naidu S. Consolidation of calcareous and siliceous sandstones by hydroxyapatite: comparison with a TEOS-based consolidant. J Cult Herit. 2013;14:103–8.
2. Sassoni E, Naidu S, Scherer GW. The use of hydroxyapatite as a new inorganic consolidant for damaged carbonate stones. J Cult Herit. 2011;12:346–55.

3. Sassoni E, Franzoni E, Graziani G, Sagripanti F. Limestone resistance to sodium sulfate degradation after consolidation by hydroxyapatite and TEOS. In: 3rd International Conference on Salt Weathering of Buildings and Stone Sculptures; 2014.
4. Naidu S, Liu C, Scherer GW. New techniques in limestone consolidation: hydroxyapatite-based consolidant and the acceleration of hydrolysis of silicate-based consolidants. J Cult Hert. 2015;16(1):94–101.
5. Liu Q, Zhang B. Synthesis and characterization of a novel biomaterial for the conservation of historic stone building and sculptures. Mater Sci Forum. 2011;675–677:317–20.
6. Martín-Gil J, Martín-Gil FJ, del Carmen Ramos-Sánchez M, Martín-Ramos P. The orange-brown patina of Salisbury Cathedral (West Porch) surfaces: evidence of its man-made origin. Environ Sci Pollut Res Int. 2005;12(5):285–9.
7. Lazzarini L, Salvadori O. A reassessment of the formation of the patina called scialbatura. Stud Conserv. 1989;34:20–6.
8. Sassoni E, Graziani G, Franzoni E. An innovative phosphate-based consolidant for limestone. Part 1: effectiveness and compatibility in comparison with ethyl silicate. Constr Build Mater. 2016;102:918–30.
9. Sassoni E, Franzoni E. Consolidation of Carrara marble by hydroxyapatite and behaviour after thermal ageing. In: Built heritage: monitoring conservation management. Cham: Springer International Publishing; 2015. p. 379–389.
10. Brinker CJ, Scherer GW. Sol-gel science: the physics and chemistry of sol-gel processing. Boston, MA: Academic Press; 1990.
11. Kim EK, Won J, Do J, Kim SD, Kang YS. Effects of silica nanoparticle and GPTMS addition on TEOS-based stone consolidants. J Cult Herit. 2009;10:214–21.
12. Scherer GW, Wheeler GS. Silicate consolidants for stone. Key Eng Mater. 2009;391:1–25.
13. Mosquera MJ, de los Santos DM, Montes A, Valdez-Castro L. New nanomaterials for consolidating stone. Langmuir. 2008;24:2772–8.
14. Mosquera MJ, Montes A, de los Santos DM. Method of strengthening stone and other construction materials. US2008/0209847 A1; 2008.
15. Mosquera MJ, de los Santos DM, Valdez-Castro L, Esquivias L. New route for producing crack-free xerogels: obtaining uniform pore size. J Non-Cryst Solids. 2008;354:645–50.
16. Pinho P, Mosquera MJ. Titania-silica nanocomposite photocatalysts with application in stone self-cleaning. J Phys Chem C. 2011;115(46):22851–62.
17. Yener DO, Sindel J, Randall CA, Adai JH. Synthesis of nanosized silver platelets in octylamine-water bilayer systems. Langmuir. 2002;18:8692–9.
18. Guo X, Xiao P. Effects of solvents on properties of nanocrystalline hydroxyapatite produced from hydrothermal process. J Eur Ceram Soc. 2006;26:3383–91.
19. Poinern GJE, Brundavanam R, Thi Le X, Djordjevic S, Prokic M, Fawcett D. Thermal and ultrasonic influence in the formation of nanometer scale hydroxyapatite bio-ceramic. Int J Nanomedicine. 2011;6:2083–95.
20. Liu DM, Troczynski T, Tseng WJ. Water-based sol-gel synthesis of hydroxyapatite: process development. Biomaterials. 2001;22:1721–30.
21. Verganelaki A, Kapridaki C, Maravelaki-Kalaitzaki P. Modified tetraethoxysilane with nano-calcium oxalate in one-pot synthesis for protection of building materials. Ind Eng Chem Res. 2015;54(29):7195–206.
22. UNI EN 15803:2010. Conservation of cultural property—test methods—determination of water vapour permeability (dp). Official Italian Version of EN 15803:2009 European Standard Emitted by CEN, Technical Body CEN/TC 346—Conservation of Cultural Property. Date of Availability (DAV) 2009-12-09.
23. Hench LL, West JK. The sol-gel process. Chem Rev. 1990;90:33–72.
24. Rubio F, Rubio J, Oteo JL. A FT-IR study of the hydrolysis of tetraethylorthosilicate (TEOS). Spectrosc Lett. 1998;31:199–219.
25. Lucovsky G, Wong CK, Pollard WB. Vibrational properties of glasses: intermediate range order. J Non-Cryst Solids. 1983;59-60:839–46.

26. Prado AC, Airoldi C. Different neutral surfactant template extraction routes for synthetic hexagonal mesoporous silicas. J Mater Chem. 2002;12:3823–6.
27. Orcel G, Phalippou GJ, Hench LL. Structural changes of silica xerogels during low temperature dehydration. J Non-Cryst Solids. 1986;88:114–30.
28. Téllez L, Rubio J, Rubio F, Morales E, Oteo JL. FT-IR study of the hydrolysis and polymerization of tetraethyl orthosilicate and polydimethyl siloxane in the presence of tetrabutyl orthotitanate. Spectrosc Lett. 2004;37:11–31.
29. Alobeedallaha H, Ellis JL, Rohanizadehc R, Costera H, Dehghania F. Preparation of nanostructured hydroxyapatite in organic solvents for clinical applications. Trends Biomater Artif Organs. 2011;25(1):12–9.
30. Pramanik A, Bhattacharjee K, Mitra MK, Das GC, Duari B. A mechanistic study of the initial stage of the sintering of sol-gel derived silica nanoparticles. Int J Mod Eng Res. 2013;3(2):1066–70.
31. Zhang X, Li Y, Lv G, Zuo Y, Mu Y. Thermal and crystallization studies of nano-hydroxyapatite reinforced polyamide 66 biocomposites. Polym Degrad Stab. 2006;91:1202–7.
32. Sasse HS, Snethlage R. Methods for the evaluation of stone conservation treatments. In: Baer NS, Snethlage R, editors. Report of Dahlem workshop on saving our architectural heritage, Berlin; 1996.

Chapter 5
Compatible Mortars for the Sustainable Conservation of Stone in Masonries

M. Apostolopoulou, E. Aggelakopoulou, A. Bakolas, and A. Moropoulou

5.1 Introduction

The deterioration of stone monuments and historical buildings, as well as their restoration, has been the aim of much research, especially in the past several decades. Many existing masonry buildings require repair or strengthening to address long-term deterioration effects, structural deficiencies, or concerns regarding seismic performance. This is attributed not only to an increased interest in the world's cultural heritage and its preservation but also to the higher rate of deterioration in recent years due to atmospheric pollution and its increasing effects, to climate change and the induced environmental alterations, as well as to the extensive use of incompatible restoration materials and their detrimental effects, especially in environmentally loaded conditions [1–4].

The deterioration of stone is dependent on intrinsic and extrinsic factors. Material properties such as microstructural characteristics, density, hygric properties, thermal expansion coefficient, modulus of elasticity, and compressive strength play an important role in the susceptibility of stone to decay factors such as environmental pollutants, extreme environmental conditions, moisture ingress, salts, dynamic stresses, biological attack, and the use of incompatible materials. The susceptibility level of a stone to deterioration from certain decay factors, such as salt decay and frost, is directly related to its microstructural and mechanical characteristics [5–7].

The application of restoration mortars during masonry conservation works is common practice in the case of masonry units bound together with mortar. The masonry's deteriorated mortars demand substitution in order to ensure continuity of the masonry and enhance masonry performance. Furthermore, plastic repair presents

M. Apostolopoulou • E. Aggelakopoulou • A. Bakolas • A. Moropoulou (✉)
Department of Materials Science and Engineering, School of Chemical Engineering,
National Technical University of Athens (NTUA), Athens, Greece
e-mail: mair_apostol@hotmail.com; eaggela@gmail.com;
abakolas@mail.ntua.gr; amoropul@central.ntua.gr

© Springer International Publishing AG 2018
M. Hosseini, I. Karapanagiotis (eds.), *Advanced Materials for the Conservation of Stone*, https://doi.org/10.1007/978-3-319-72260-3_5

97

the advantage of lower cost, as the application of new compatible mortars is less costly than the substitution of carved stones or bricks, and moreover offers the advantages of reapplicability and retreatability [1, 3].

Restoration mortars of cement and polymer, highly incompatible with historical building materials, were used extensively in the last century. Apart from the damaging effects that their application inferred on historical building materials, they also presented a limited lifetime and obscured retreatability. The damaging effects of incompatible restoration mortars are attributed to their different physicochemical characteristics and potentials in comparison to the original building materials, intensified by environmental factors and dynamic stresses [1]. The lack of ductility of cement mortars and low chemical and physical affinity with traditional building materials negates the advantages that led to its extensive use, mainly its quick setting time and high bearing capacity [1, 4]. In recent years, the incompatibility of Portland cement with many natural stones was manifested through the irreversible damage it inflicted.

The detrimental effects of the use of cement and the resulting strategy of the conservation of the restoration ("restauro di restauro") found a dramatic paradigm in recent efforts to conserve Evans' 1920s restoration of the Knossos Palace in Crete, which had, until then, survived for almost four millennia [1]. Cement mortars, apart from altering the aesthetic characteristics of the monument, lead to structural and surface failures due to the different chemical and mechanical properties of cement and historical building materials. Recent efforts to remove the cement mortars and degradation by-products from the gypsum building elements of the monument through mechanical means highlighted the irreversibility accompanying the application of incompatible mortars, as it led to partial loss of authentic material [8]. Research is now involved with the development of appropriate techniques of cement removal [8]. The use of incompatible restoration mortars in past restorations in the Medieval City of Rhodes, and the inferred damages to the historical masonries, also demonstrates the damaging effects of incompatibility. The decay of the bio-calcareous, highly porous, and low-strength local stone, mainly comprising the monument, evolved through a combination of intrinsic and extrinsic factors in mechanisms mainly concerned with transfer phenomena of aquatic solutions. Thus, the presence of incompatible cement mortars led to advanced cavernous alveolar damage of the building stones [9–11].

Studies on the degradation phenomena of architectural monuments constructed using tuffeau stones from Val de Loire region of France revealed that they are often related to the use of mortars incompatible with the tuffeau stones [12]. Further examples of the damage inflicted on various lithologies due to the use of incompatible materials can be found on sandstones, porous limestones, and tuff in Hungary and Germany, where the improper selection and use of repair mortars have led to salt crystallization damage, cracking, scaling, and partial loss of mortar and stone [13]. The use of incompatible mortars can affect even stones which are considered durable, such as granite, as demonstrated during conservation work carried out at Trinity College, Dublin, where granite weathering was associated with damaging salts originating from incompatible joint mortars and exhibited through fracturing and chemical alteration of minerals [14].

Thus, through the examination of such cases, where the damaging effects of incompatible mortars are evident in real time and real conditions, the achievement of compatibility becomes the focal point of conservation works related to restoration mortar application.

5.2 Compatibility and Performance Through a Reverse Engineering Methodological Approach

Compatibility is a complex term, as the parameters governing its achievement are multiple and interconnected. In a general approach, it can be perceived as the opposite of incompatibility; therefore, a conservation material is one that will not harm the in-place historical materials in any manner, directly or indirectly. Thus, a compatible restoration mortar does not induce chemical alterations to the historical materials, does not initiate or intensify deterioration phenomena, and ensures a homogenous hygric and mechanical behavior of the masonry. International research is involved with the development of methodologies aiming to assist in the assessment of compatibility, through general directions, case studies, or the establishment of incompatibility indexes regarding the restoration mortar characteristics in relation to the substrate [1, 15–18]; similarity with the substrate and/or the original mortars is usually the aim. Many studies involved with new mortar design also address the issue of compatibility for use in cultural heritage conservation works [17–23].

When compatibility is ensured, maximum durability can be guaranteed, as well as retreatability of the intervention. Compatibility, in regard to the substrate it is applied to, is assessed in chemical, dimensional (e.g., modulus of elasticity, thermo-mechanical, volume shrinkage), hygric adhesion to the substrate, as well as historical and aesthetical terms; furthermore, the use of bioecological materials enhances compatibility and sustainability (Fig. 5.1).

Chemical compatibility demands that the restoration mortar must not introduce hazardous compounds or compounds which could interact in a negative manner with a compound or with the decay products of the materials of the masonry on which it is applied. Dangerous compounds include types of salts which in the presence of moisture could induce decay typologies. For example, soluble salts such as calcium sulfates and sodium salts sometimes present in Portland cement can leach out over time, inflicting damage to the surrounding materials of the masonry [24]. Soluble salts and their respective cations, through a continuous sequence of evaporation-crystallization, can result in crystallization and hydration pressures in the material's pores [11].

A homogenous hygric behavior within the masonry must be ensured. Moister transfer between the building materials of the masonry should take place in a

Fig. 5.1 Parameters which
determine restoration
mortar compatibility in
relation to the substrate

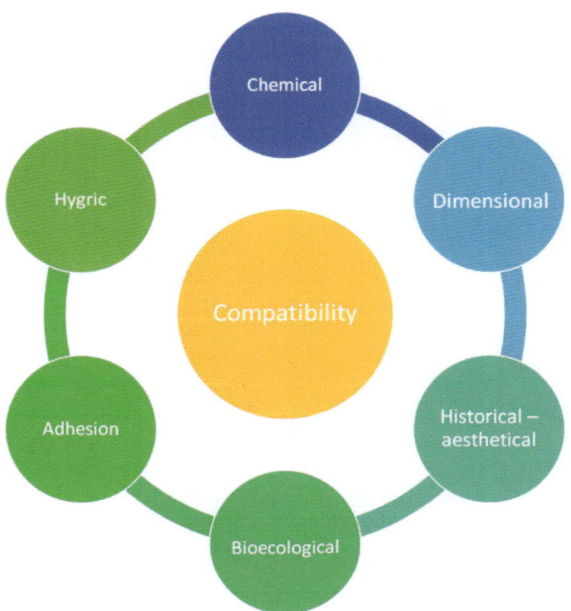

homogenous manner, ensuring that moisture is not concentrated in a specific type of material, especially historical ones. Also, uptake of moisture through capillary rise must not be intensified with the application of the new restoration mortar [25]. Capillary rise plays an important role, as soluble salts can enter the masonry building materials through rising damp and moisture plays a basic role in the course of various decay reactions. Permeability must guarantee protection from corrosive environmental factor, such as salts, gaseous pollutants, etc., but at the same time allow normal transpiration of the masonry [26]. Capillary rise and permeability characteristics are correlated with the amount and distribution of pores of the restoration mortar, as well as the dimensions of the pores; a proper microstructure guarantees an effective moisture transfer within the structure, correct transpiration and high endurance in freeze-thaw cycles, and salt crystallization.

Dimensional compatibility is governed by several parameters. Modulus of elasticity or otherwise the rigidity of the restoration mortar in terms of possible deformation is important, especially in seismic hazard areas [27, 28]. A difference in the modulus of elasticity between the stone substrate and the restoration mortar can result in the concentration of stresses in the less deformable material and can obscure deformation of the residing materials. Creep coefficient expresses the amount of deformation of a material in relation to loading. Therefore, how much it can remain constant under a specific load must also be taken into account [29, 30]. Thermal expansion coefficient, which expresses the variation of dimensions in relation to the variation of temperature, must be similar between materials as different thermal expansion coefficients between the restoration mortar and the stone substrate can lead to the development of stresses on the interface of the two materials [17]. Volumetric shrinkage is related to two phenomena: plastic contraction, which is active in the beginning of the hardening process after setting, and hygrometric shrinkage, which takes place after hardening is complete. Lime mortars present only the first type of contraction, while hydraulic mortars present both types due to the loss of excess humidity present in the mortar over time [23, 31, 32]. Plastic shrinkage leads to the development of stresses on the interface of the two materials, while hygrometric shrinkage can create stresses both in the matrix of the restoration mortar and on the interface of the two materials. Hygrometric contraction, modulus of elasticity, creep coefficient, and tensile strength are interconnected parameters which affect the durability of a restoration mortar in time.

Compatibility in terms of adhesion is related with both the restoration mortar and the substrate characteristics. The higher adhesion is, the better the cooperation of the restoration mortar and the substrate, as it will result in an enhanced performance of the composite system [33, 34]. If adhesion is too low, detachment of the restoration mortar from the substrate could occur due to contraction phenomena or stresses originating from the presence of salts or moisture. Conditions which can contribute to a good cooperation between the restoration mortar and the substrate are related to the correct preparation of the substrate, adequate workability of the restoration mortar, and adequate water retention capability of the restoration mortar.

Historical and aesthetical compatibility are also significant factors when designing mortar for restoration purposes. This is important especially in the case where the restoration mortar is visible after the restoration works, as within an intervention, the authenticity of the built heritage must be understood and preserved as much as possible [35]. The restoration mortar must have the same appearance as the authentic mortar. The color, texture, symbolisms, and aesthetic must create an overall appearance similar to the original. The shape and design, materials used, and workmanship are often addressed as key heritage values [18].

The assessment of a repair mortar as bioecological is related with compatibility but also contributes to sustainability [36]. The assessment of a restoration mortar in terms of bioecology is related to more than one aspect. The first aspect is related to the origin of the raw materials used for its production; bioecological mortars are composed of recycled materials, materials which originate from renewable sources, etc. The second aspect is in relation to the production procedure of the raw materials and the mortar and is assessed in terms of required energy consumption and emitted pollution due to the production. The third aspect is related to the mortar's life-cycle assessment, its impact on the environment and on humans. The final aspect is related to the mortar's durability as a restoration mortar that is durable and at the same time compatible does not demand frequent substitution. Raw materials used in traditional mortars (e.g., lime, gypsum, pozzolans) demand low energy consumption for their production due to low firing temperatures while emitting a low amount of emissions to the environment with no toxic or dangerous residues that require disposal. The high durability that traditional mortars have exhibited over time and exposed to the environment, as evident in many historical buildings and

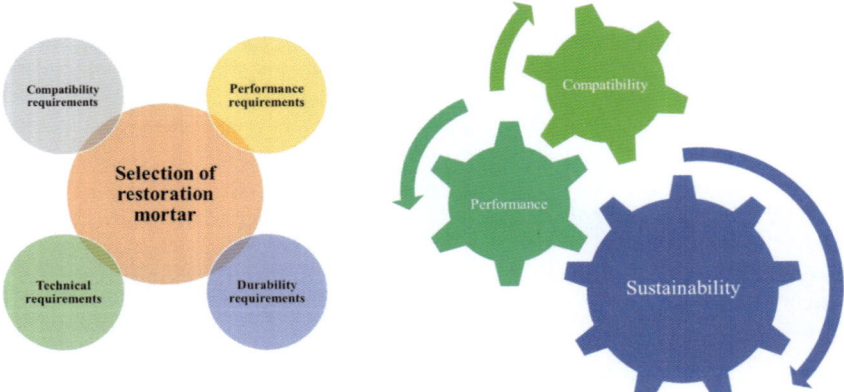

Fig. 5.2 Requirements that must be taken into account for the selection of the appropriate restoration mortar (left); the interaction of compatibility, performance, and sustainability (right)

monuments, guarantees a very satisfactory life-cycle assessment and contributes to sustainability.

Performance of the restoration mortars is related to the mechanical behavior of the masonry after their application while under static and dynamic stresses. In order to apply the optimum restoration mortar, compatibility and performance must be ensured in order to reinstate structural integrity (many times along with other measures). Limited compatibility will eventually negatively affect performance due to the degradation of the building material properties. Computational methods through numerical analysis, such as finite element model analysis and fragility curves, are a valuable tool for the assessment of performance, especially in seismically prone areas [37–39]. Thus, either the analysis can set the demand for the minimum restoration mortar strength required for the minimization of damages, or different restoration mortars can be compared to evaluate their impact on the behavior of the masonry. Sustainability of a restoration work can only be ensured when both compatibility and performance are secured (Fig. 5.2). Furthermore, technical requirements related to a specific restoration project must also be taken into account, for example, in terms of project implementation limitations.

The exemplary longevity and durability exhibited by most traditional mortars have led in recent years to an increased interest in traditional mortar production technologies and materials, aiming to design compatible mortars through a reverse engineering methodological approach. The reverse engineering methodology aims to design and apply new compatible and performing restoration mortars on monuments and historical buildings based on the characteristics of the historical mortars [1, 40]. In cases where the historical mortar (in place) served adequately in the masonry, the design of the restoration mortar should be implemented in the direction of achieving the characteristics of the historical mortar; if however the historical mortar did not serve adequately, the restoration mortar design should be optimized.

Table 5.1 Range of acceptability limits for different types of restoration mortars connected to thermal analysis results (TG/DTA)—physicochemical compatibility; (%) refers to mass loss in relation to the total sample mass (mg/mg); inverse hydraulicity ratio: hydraulic water content (%)/ CO_2 content (%)

Restoration mortar type	Mass loss % (mg/mg) attributed to loss of hygroscopic water $H_2O_{ph.w.}$ content (%)	Mass loss % (mg/mg) attributed to loss of hydraulic water $H_2O_{ch.w.}$ content (%)	Mass loss % (mg/mg) due to decarbonation CO_2 content (%)	Inverse hydraulicity ratio $CO_2/H_2O_{ch.w.}$
Lime	<1	2–4	>30	>8.5
Crushed brick-lime	1.5–4.5	2.3–5.3	<20	3.2–6 5
Hot lime	0.7–1.5	2–4.6	>25	6–15
Hydraulic	1–2.5	4–7.2	<25	1.8–6 1
Lime-pozzolan	2–4	3.3–5.4	<22	1.3–5 1
Rubble masonry	–	5.6–5.9	<30	3.36–5.13

Table 5.2 Range of acceptability limits for different types of restoration mortars through mercury intrusion porosimetry (MIP) results —microstructural compatibility

Restoration mortar type	Cumulative volume (mm³/g)	Apparent density (g/ cm³)	Average pore radius (μm)	Specific surface area (m²/g)	Total porosity (%)
Lime	170–320	1–1.8	0.8–3.3	1.3–3.3	30–45
Crushed brick-lime	170–290	1.5–1.9	0.1–0.8	3.5–15	32–43
Hot lime	110–180	1.7–1.9	0.3–0.8	2.5–4.7	20–30
Hydraulic	90–230	1.7–2.1	0.1–3.5	2.5–13.5	18–40
Lime-pozzolan	160–265	1.6–1.9	0.1–1.5	3–14	30–42
Rubble masonry	117–220	1.8–2.1	0.2–20.6	1.2–4.7	25–39

Furthermore, the study of various historical mortars which have exhibited a high degree of compatibility with traditional natural substrates and adequate performance under static and dynamic loads contributes to establish acceptability limits a restoration mortar should fulfill in terms of chemical and physicomechanical compatibility. Through the analysis, characterization, and classification of historical mortars presenting high durability, Moropoulou et al. defined the range of acceptability limits of characteristic mortar parameters in order to assist in the assessment of new mortars simulating historical ones through the reverse engineering methodological approach as stated in Tables 5.1 and 5.2 [41, 42]. The assessment of a restoration mortar as compatible for use in a restoration project, in addition to the characteristics of the historical mortar, must also take into account the characteristics of the other building elements (stones and/or bricks), as well as the environmental

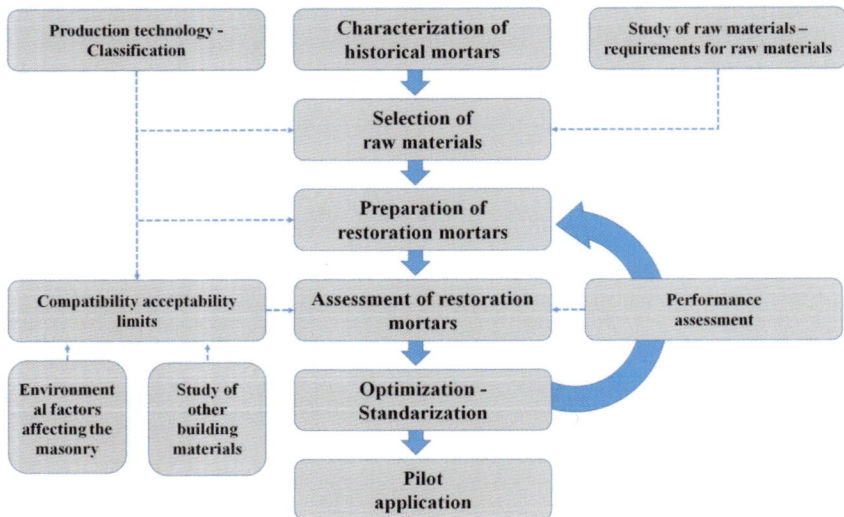

Fig. 5.3 Integrated reverse engineering methodological approach

and deterioration factors, especially in the case of new loads imposed on the structure throughout its lifetime. The supplementary but necessary performance assessment through laboratory masonry mechanical tests and/or computational methods leads to an integrated assessment of the repair mortar for use in a restoration project (Fig. 5.3).

5.3 Application of Methodological Approach: Evaluation of Restoration Mortar Compatibility and Performance

Herein, three case studies are presented in order to illustrate the proposed methodological approach for the selection of the optimum compatible and performing restoration mortar among restoration mortars designed through reverse engineering for various monuments. The characteristics of restoration mortars, as studied through thermal analysis (DTA/TG), mercury intrusion porosimetry (MIP), capillary rise tests, and mechanical tests [23, 43, 44], are evaluated in terms of compatibility and performance for three different monuments. The evaluation takes into account the historical mortars of each monument, technical requirements, as well as the characteristics of the stones and bricks comprising the masonry and the environmental loads the repair mortars must successfully serve in. The syntheses of the different restoration mortars are presented in Table 5.3 and cover a wide range of restoration mortar types [23, 43, 44].

Table 5.3 Mix design of restoration mortars

	Lime powder	Lime putty	Hydraulic lime NHL2	Ceramic powder	Natural pozzolan	Metakaolin	Siliceous sand 0–2	Siliceous sand 0–4
Hydraulic lime mortars								
NHLFS			25					75
Lime mortars								
L	30						70	
Lpu		40					60	
Lime-metakaolin mortars								
MK1	15					15	70	
MK05	20					10	70	
MK5	25					5	70	
MK2.5	27.5					2.5	70	
Lime natural pozzolan mortars								
EM1	15				15		70	
EM2	10				20		70	
EM3	7.5				22.5		70	
EM4	6				24		70	
Lime ceramic powder mortars								
CP1	15			15			70	
CP2	10			20			70	
CP3	7.5			22.5			70	
CP4	6			24			70	

5.3.1 The Byzantine Monastery of Kaisariani in Athens, Greece

The Catholicon of the Kaisariani Monastery is a typical mid-byzantine Athenian church structure, built in the late eleventh or early twelfth century. Originally it was a complex cross-in-square four-column-domed church. Throughout the centuries a narthex was added to the west, and a chapel was added to the south (Fig. 5.4a). The building materials consist of carved stones, bricks, and mortars. The east façade is a typical cloisonné-type masonry, considered to be original (Fig. 5.4b) [45–47].

Samples of structural and joint mortars were taken from the masonries corresponding to the main church of the Catholicon, considered as authentic, and two brick unit samples were taken from the east façade cloisonné masonry [37]. The sampling of stones was not permitted; however, during an in situ investigation using nondestructive techniques, compressive strength of the fossiliferous stone (most typical stone in the masonry) was estimated through Schmidt hammer rebound test at 8.2 MPa average value with a standard deviation of 1.72 [48]. This value sets the upper limit regarding the compressive strength of the restoration mortar, in order to ensure mechanical compatibility. The good state of preservation displayed by the

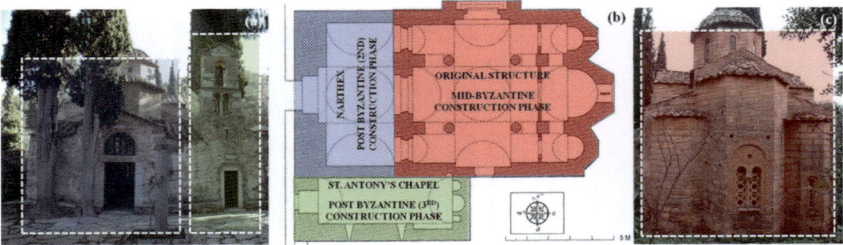

Fig. 5.4 Image of west façade, which is a later addition (**a**); ground plan of the church [47], with probable construction phases (**b**); image of east cloisonné-type façade, considered original (**c**)

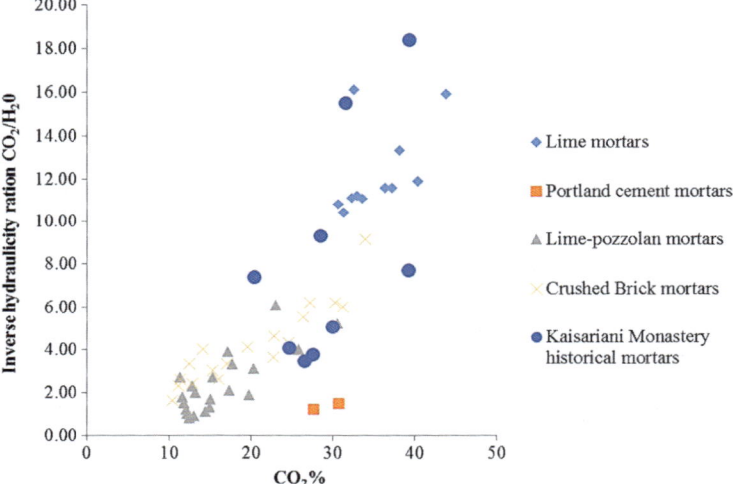

Fig. 5.5 Classification of Kaisariani Monastery mortars

building elements of the east façade cloisonné masonry allows for the capillary rise coefficient of the brick to set the determining value for the hygric compatibility of the restoration mortar.

Thermal analysis (TG/DTA) of the historical mortars allowed for their classification as lime-based mortar with varying hydraulicity ratios (Fig. 5.5). Analysis of the mortars' and ceramics' microstructure was conducted through MIP on certain samples (Table 5.4). The brick samples presented characteristics similar to handmade byzantine bricks [37, 48]. Therefore, the masonry system has most probably served adequately for almost ten centuries. The microstructural characteristics of the mortars are typical of lime-based mortars [49]. The capillary rise coefficient of the brick units was estimated through water absorption by capillarity test at $C.R.C._{\cdot brick} = 16.5 \ mg/(cm^2 \ s^{1/2})$ [37]. Therefore, any restoration mortar must present a similar value in order to be assessed as compatible in terms of achieving a

Table 5.4 Mercury intrusion porosimetry results of Kaisariani historical mortars and bricks

Historical mortar/brick sample	Total cumulative volume (mm³/g)	Bulk density (g/cm³)	Average pore radius (μm)	Specific surface area (m²/g)	Total porosity (%)
Brick 1	131.24	1.89	0.81	5.82	24.85
Brick 2	167.68	1.82	0.85	1.03	30.54
Mortar 6	233.75	1.61	0.44	3.27	37.70
Mortar 8	255.81	1.62	0.47	3.40	36.55

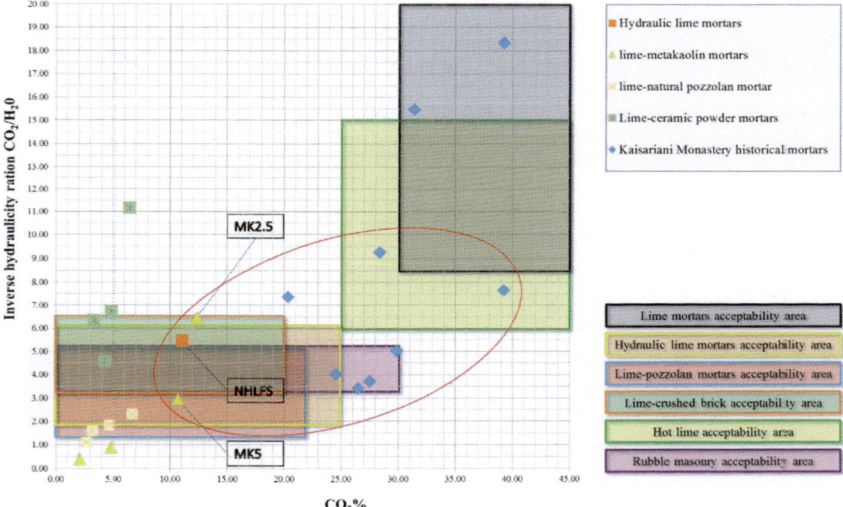

Fig. 5.6 Graphical representation of acceptability limits according to thermal analysis results and position of historical and restoration mortars in graph

homogenous hygric behavior of the masonry. A high amount of soluble salts were detected in all samples, ranging from 3.2 to 8.28%, attributed to rising damp phenomena. Hydraulic mortars show greater durability in the presence of soluble salts than non-hydraulic ones [50].

After the characterization of the historical mortars and taking into account measurements conducted on the other building elements, as well as the estimation of the environmental factors affecting the monument, the restoration mortars were assessed in relation to their compatibility with the structural materials, as well as in regard to durability and performance. In Fig. 5.6, the compliance of the restoration mortars' characteristics with the set acceptability limits is graphically represented in relation to the historical mortars in regard to thermal analysis results and physicochemical compatibility. Environmental factors affecting the monument demand the use of a hydraulic mortar. The correlation shows that the two lime-metakaolin mortars (MK5, MK2.5), as well as the hydraulic lime mortar (NHLFS), are within

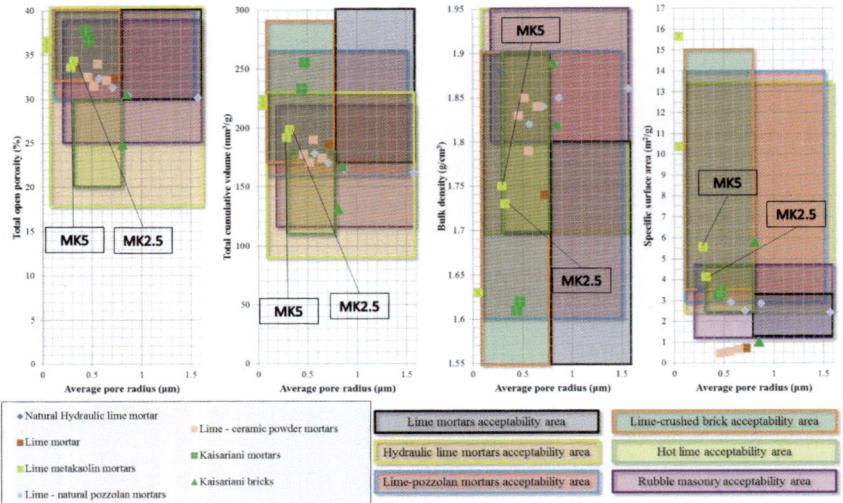

Fig. 5.7 Graphical representation of acceptability limits regarding microstructural characteristics and position of historical and restoration mortars in graphs

acceptability limits and at the same time are in close proximity with the historical mortars in place.

In Fig. 5.7, the compliance of the restoration mortars with the set acceptability limits is graphically represented, in relation to the historical mortars as well, regarding microstructural compatibility. The lime-metakaolin mortars MK5 and MK2.5 again are within acceptability limits in relation to all microstructural characteristics and fall within the range of characteristics of the historical mortars and bricks of the Kaisariani Monastery.

The hygric compatibility of the brick and the restoration mortars was assessed through comparison of the measured capillary rise coefficients, and similarity is the aim, allowing a 25% range of acceptable values (Fig. 5.8). Lime-metakaolin mortars presented the best values.

An important parameter in selecting a restoration mortar is the achievement of mechanical performance while complying with compatibility limits. All mortars except for MK1 presented compressive strength values below the fossiliferous stones and are thus considered acceptable in term of mechanical compatibility (Fig. 5.9). In a previous study, the performance of Kaisariani Monastery was evaluated through fragility analysis and demonstrated that even the use of a 5 MPa compressive strength restoration mortar can decrease the possibility of damage occurring during an earthquake stress [37].

Thus, in the case of selecting the lime-metakaolin mortar MK5, all demands regarding compatibility and performance are fulfilled, as it is within acceptability limits in regard to physicochemical and microstructural characteristics. It presents a

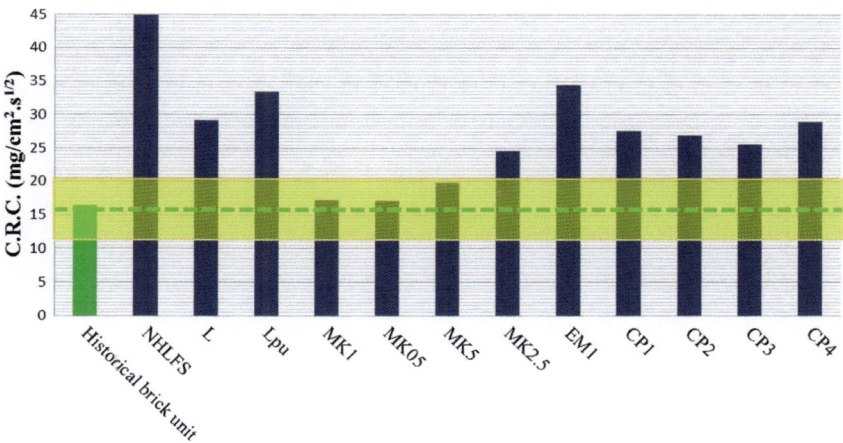

Fig. 5.8 Capillary rise coefficients of the Kaisariani Monastery masonry brick and the restoration mortars; the green-dashed line states the desired value

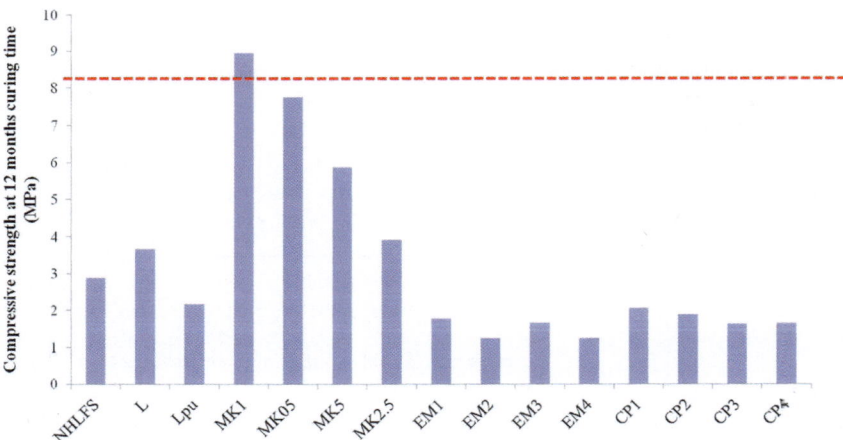

Fig. 5.9 Compressive strength (MPa) of restoration mortars at 12 months—the dashed line is indicative of the upper limit of acceptable compressive strength

Fig. 5.10 Image of the Plaka Bridge before (**a**) and after (**b**) collapse

similar value of capillary rise coefficient (CRC) to the historical brick and at the same time presents mechanical strength values compatible with the stone while enhancing mechanical performance of the structure.

5.3.2 The Traditional Bridge of Plaka in Epirus, Greece

An example of the area's traditional architecture and an emblematic monument of the region, the traditional bridge of Plaka in Epirus, Greece, built in 1866, collapsed partially in February 2015 (Fig. 5.10). An interdisciplinary study was initiated by the National Technical University of Athens (NTUA) to examine the building materials and the structure, taking into account historical references and images, in order to ascertain the factors leading to its collapse and propose new restoration materials and techniques [51]. The materials science and engineering laboratory of NTUA undertook the study of the historical mortars in order to discover their production technology as well as to propose a compatible and performing restoration mortar so that it may be used in the restoration project of the bridge [52].

This interdisciplinary study revealed the multiple causes leading to its collapse through a synergistic action. Structural deficiencies, high river velocity, incompatible repair actions in order to address damages, and high moisture content of the materials resulting in loss of performance with the additional negative mechanical action of plant roots growing on the bridge led to the collapse after the heavy rainfalls of January 2015. Mortar samples were taken during an in situ visit from the parts of the bridge, which had detached and collapsed into the river as large fragments (K1–K3), as well as from the remaining parts (K4–K5, K7) [51].

Due to the complexity of the structure's geometry, important technical requirements from the restoration mortar are low shrinkage, high acquirement of mechanical strength, and good creep behavior. The volume shrinkage of the different restoration mortars is represented in Fig. 5.11. Lime-metakaolin mortars exhibit the best behavior in terms of volume shrinkage, as they present the lowest values, especially MK1 and MK05 (Fig. 5.11). As demonstrated earlier, in Fig. 5.9, lime-metakaolin mortars present the highest values of compressive strength, while the addition of metakaolin to mortars is reported to enhance creep behavior [29, 30].

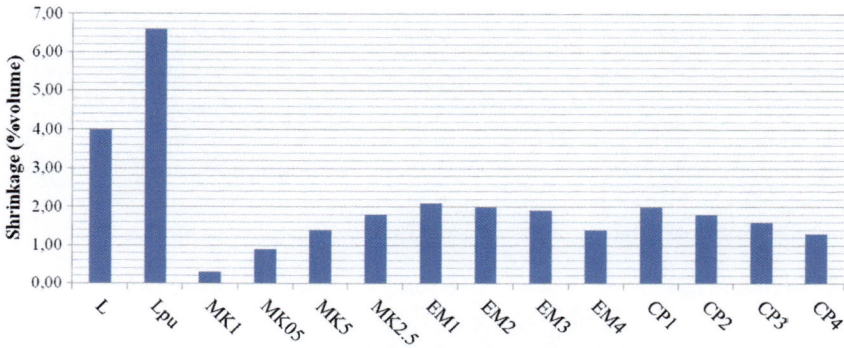

Fig. 5.11 Shrinkage (%) of the volume of different restoration mortars

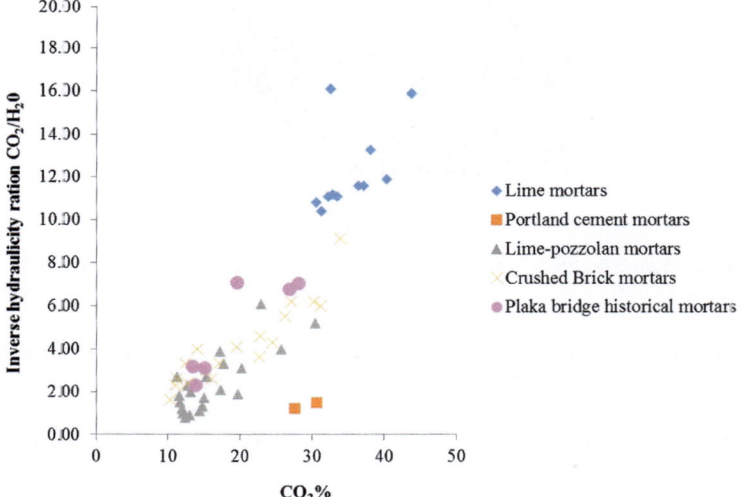

Fig. 5.12 Classification of Plaka Bridge historical mortars

Thermal analysis (TG/DTA) of the historical mortar samples was conducted, and the results allowed for the classification of the historical mortars in place (Fig. 5.12). The mortars of the Plaka Bridge exhibit high hydraulicity and are classified as lime-pozzolan mortars. The microstructural characteristics of the historical mortars were analyzed through MIP (Table 5.5). The results indicate a certain degree of deterioration of the mortars due to the continuous wetting-drying cycles that they have endured.

A study conducted by the school of civil engineering of NTUA measured the historical mortars' tensile strength in a range from 0.09 to 1.04 MPa, with an average value of 0.36 MPa and standard deviation between samples' tensile strength

Table 5.5 Microstructural characteristics of the historical mortars of Plaka Bridge (MIP)

Historical mortar sample	Total cumulative volume (mm³/g)	Bulk density (g/cm³)	Average pore radius (μm)	Specific surface area (m²/g)	Total porosity (%)
K1	280.9	1.48	0.047	19.47	41.8
K2	193.8	1.63	0.013	36.01	31.6
K3	387.5	1.24	0.046	29.69	48.1
K4	389.3	1.19	0.017	49.49	46.6
K5	418.1	1.11	0.025	41.66	46.6
K7	333.8	1.32	0.017	45.05	44.0

0.31 MPa [53]. This is an indication of anisotropy due to different levels of deterioration. The results of the building stone characterization through petrographic analysis, X-ray diffraction, and scanning electron microscopy coupled with microanalysis show that it is a micaceous calcareous sandstone, with quartz, feldspars, and calcite as principal mineralogical phases [52]. They presented a considerably lower value of compressive strength values ranging from 49.60 to 68.47 MPa [53]. However, a considerably lower value should be set as the upper limit for the restoration mortar's compressive strength to ensure mechanical compatibility due to statistical inadequacy of sampling, as well as the presence of lower-strength porous stones measured in the core of the structure as additional filling material.

The results from the analysis of the historical materials present the guidelines for the design and selection of the appropriate restoration mortar. The high moisture environment of the monument demands the use of a strongly hydraulic lime-pozzolan restoration mortar in order to harden and serve adequately under the environmental loads. Special attention must be given to achieve a low inverse hydraulicity ratio and the compliance of microstructural characteristics to acceptability limits, as well as to the historical mortars in place.

In general lime-metakaolin mortars and lime-natural pozzolan mortars present the optimum hydraulicity values and correlate with the historical mortars of Plaka Bridge (Fig. 5.13). Regarding microstructural characteristics, lime-metakaolin mortars present the best correlation with the historical mortars and the acceptability limits of lime-pozzolan mortars (Fig. 5.14). Although marginally not within preset acceptability limits, MK1 and MK05 present the optimum compliance with the historical mortars in place; this is an illustration of why the preset acceptability limits can only act as a general guide and correlation with the historical mortars of each specific monument is necessary in order to assist in decision-making.

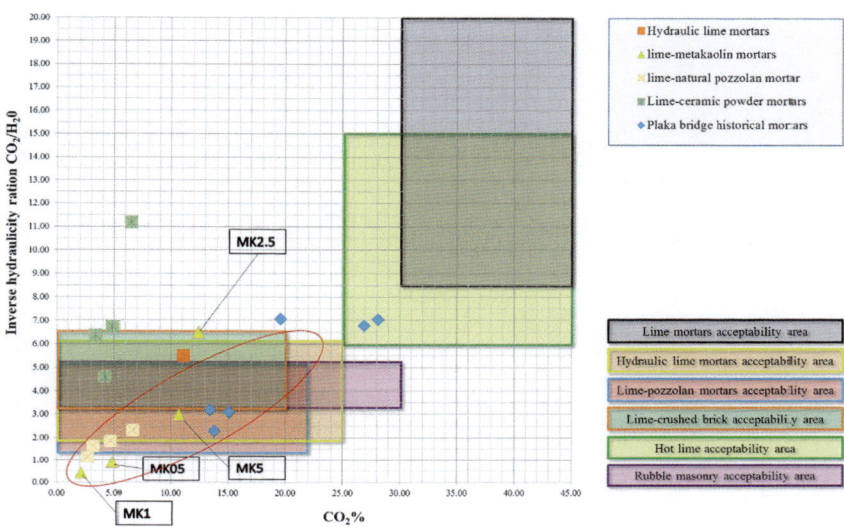

Fig. 5.13 Graphical representation of acceptability limits according to thermal analysis results and position of historical and restoration mortars in graph

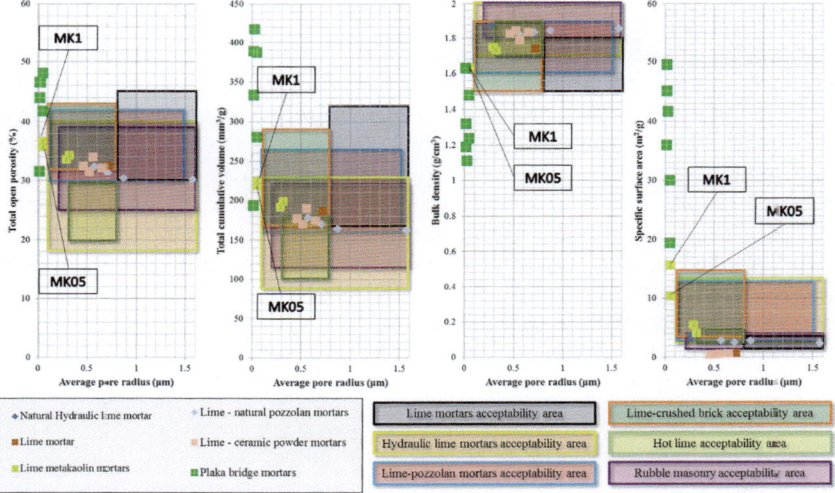

Fig. 5.14 Graphical representation of acceptability limits according to thermal analysis results and position of historical and restoration mortars in graph

Fig. 5.15 Image of the Holy Aedicule before interventions: front view (**a**), top view (**b**); intense buckling of the structure (**c**)

Table 5.6 Mercury intrusion porosimetry results of Holy Aedicule historical mortars [59]

Historical mortar sample	Total cumulative volume (mm³/g)	Bulk density (g/cm³)	Average pore radius (µm)	Specific surface area (m²/g)	Total porosity (%)
JHS_1fa_m1	240.54	1.54	0.82	7.97	37.10
JHS_1fa_m2	166.69	1.70	0.10	7.72	28.42
JHS_1fa_m3	570.51	0.98	0.24	15.25	56.03
JHS_fe_m	450.66	1.14	0.57	2.45	51.53
JHS_intsf_m1	387.45	1.24	0.50	6.24	48.17
JHS_intss_m	432.43	1.18	0.78	6.24	50.90
JHS_t_pl	215.82	1.55	0.47	1.95	33.45

5.3.3   The Holy Aedicule of the Holy Sepulchre in Jerusalem

The Holy Aedicule of the Holy Sepulchre in Jerusalem is the most important site of Christianity as, according to tradition, it is the place of Christ's burial and resurrection (Fig. 5.15a, b). The Holy Aedicule, embedding the Holy Tomb, is a complex structure with a 1700-year history, throughout which it was destroyed, reconstructed, and restored many times [54–56]. The rehabilitation of the Holy Aedicule [57], completed by NTUA in March 2017, aimed to address serious deformation problems (Fig. 5.15c) and was based on the results of an NTUA diagnostic study [58].

Prior to the rehabilitation project, masonry core samples, as well as mortar samples from different areas of the structure, were selected for laboratory analysis [59, 60]. Thermal analysis (TG/DTA) results of the mortar samples allowed for their classification as lime-based mortars, exhibiting hydraulicity (Fig. 5.16). Microstructural characteristics of the historical mortars (MIP results shown in Table 5.6) are characteristic of disintegration from intense moisture, entrapped in the masonry due to the lead sealing of the stone facades enveloping it.

The building stones of the Holy Aedicule presented low values of specific surface area (between 0.1 and 0.9 m²/g), while the values of total porosity showed a wide variation from 0.3 to 28%; this is indicative of a highly anisotropic structure.

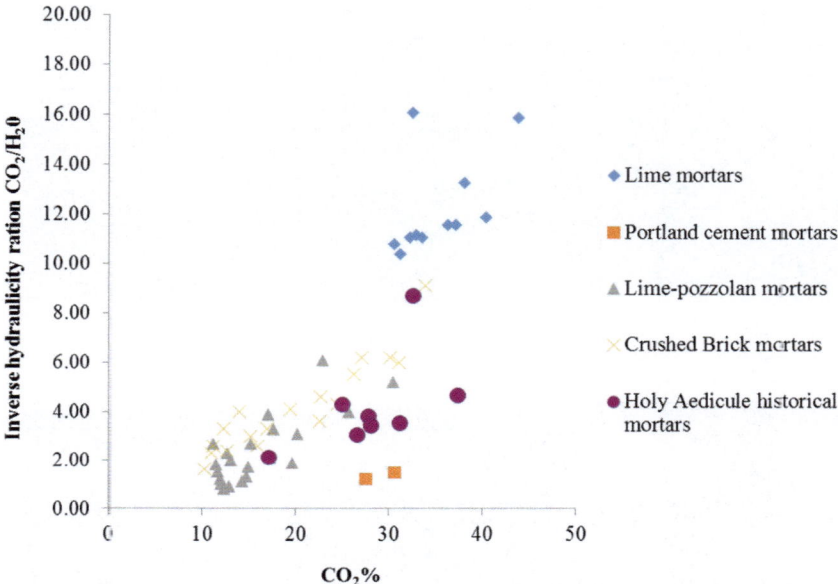

Fig. 5.16 Classification of the historical mortars of the Holy Aedicule

This is attributed to the use of different building stones throughout the evolution of the structure with one historical phase embedded and merging with the other. The range of mechanical properties of the stones was also wide; compressive strength ranged from 10.3 to 77.8 MPa, while static modulus of elasticity ranged from 0.8 to 38.2 GPa [51]. Due to the serious damage exhibited and in order to ensure structural integrity, it was decided that the limit for compressive strength should be determined by static and dynamic numerical analysis of the structure. A finite element modeling and analysis of the structure was conducted by Spyrakos and Maniatakis [38]. The analysis took into account different restoration materials and intervention scenarios proposed by the interdisciplinary team. The results suggested the application of a restoration mortar of at least 15 MPa compressive strength in order to ensure structural integrity of the structure. Taking into consideration the anisotropy of the building materials and the wide range of values exhibited, it was decided that the restoration project would also involve the injection of grouts to homogenize the structure.

As the deterioration and swelling of the mortars was considered one of the main causes for the observed buckling and deformations of the Aedicule, the design and proposal of a compatible and performing restoration mortar for the rehabilitation interventions were necessary. In order to select the optimum mortar synthesis, different restoration mortars were examined. Lime-metakaolin mortars (except for weaker MK2.5) presented the optimum hydraulicity values and correlation with the historical mortars of the Holy Aedicule (Fig. 5.17) while presenting the best correlation of microstructural characteristics as well (Fig. 5.18).

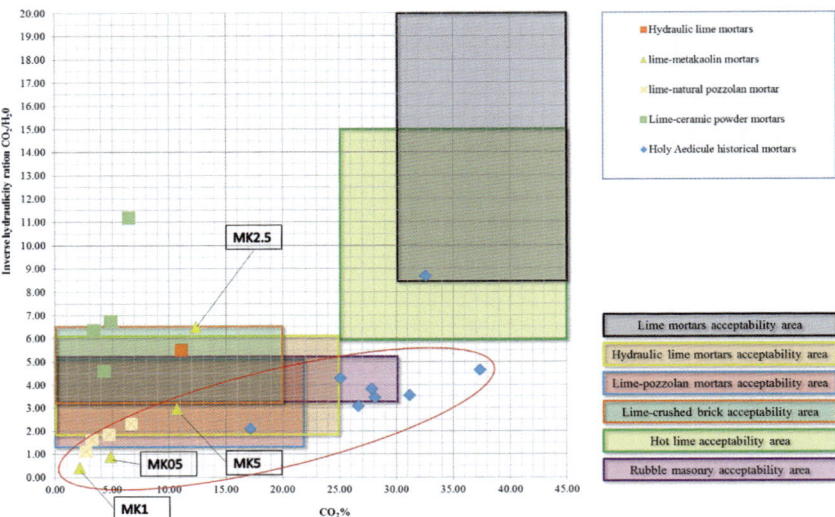

Fig. 5.17 Graphical representation of acceptability limits according to thermal analysis results and position of historical and restoration mortars in graph

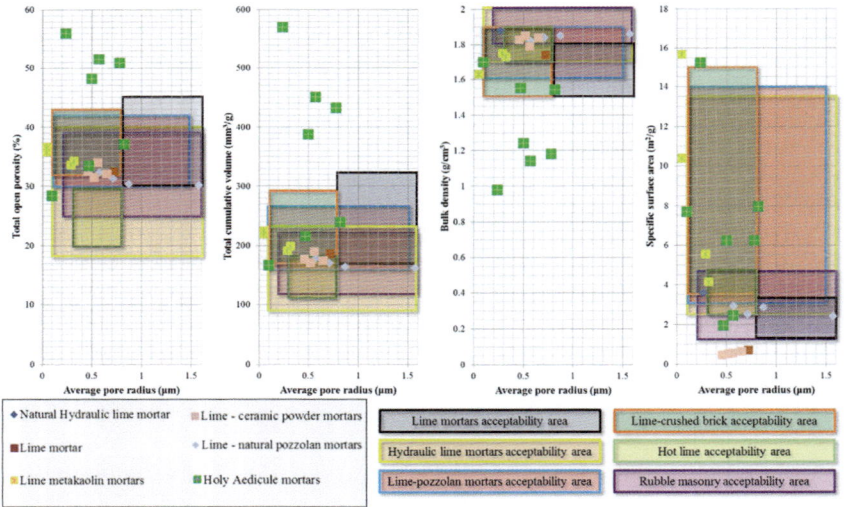

Fig. 5.18 Graphical representation of acceptability limits regarding microstructural characteristics and position of historical and restoration mortars in graph

Further technical requirements were derived from the geometry and layering of the structure; the masonry was enclosed within marble exterior and interior facades, with a layer of filling mortar between the facings and the masonry. The steps of the restoration project regarding the application of the restoration mortar to the masonry included the removal of certain stone panels that made up the exterior façade, the removal of the disintegrated filling mortar layer, and the application of the restoration mortar on the masonry. Subsequently, after a relatively short period of time, the panels were reassembled with the parallel use of a restoration "concrete-type" mortar to serve as filling mortar in order to fill the gap between the masonry and the stone façade [62]. The small timeframe in which the project had to be concluded (10 months) and consequently the small amount of time during which the masonry would remain visible before being enveloped again demanded the selection of a restoration mortar with a high rate of carbonation, early acquirement of mechanical properties, and adequate hydraulicity. The prerequisite that the Holy Aedicule would remain open to pilgrims throughout the rehabilitation project, and therefore safety of visitors would be an issue, further highlighted the importance of the above-mentioned requirements. Furthermore, the high percentage of soluble salts measured in the mortar samples, ranging from 4.46 to 11.28%, also demanded the use of a hydraulic lime-pozzolan restoration mortar. In Figs. 5.19 and 5.20, the superior behavior of lime-metakaolin mortars in relation to early acquirement of mechanical strength can be observed, as well as the good mechanical strength final values. The carbonation rate is related to $Ca(OH)_2$ consumption in relation to time, and lime-metakaolin mortars presented extremely satisfactory rates (Fig. 5.21). Lime-metakaolin mortars fulfilled all criteria of compatibility, as well as criteria deriving from technical requirements. Thus, a commercial lime-metakaolin mortar with 15 MPa of compressive strength was sought out and selected in order to ensure compatibility and performance regarding all requirements.

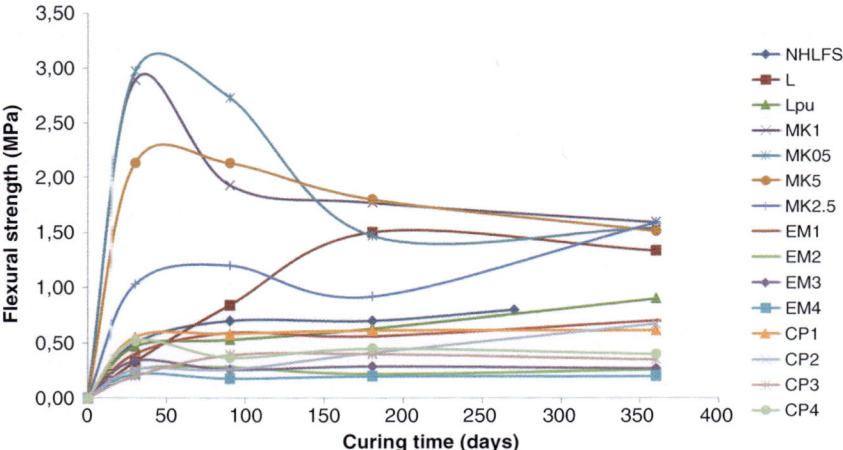

Fig. 5.19 Flexural strength (MPa) in relation to curing time (days)

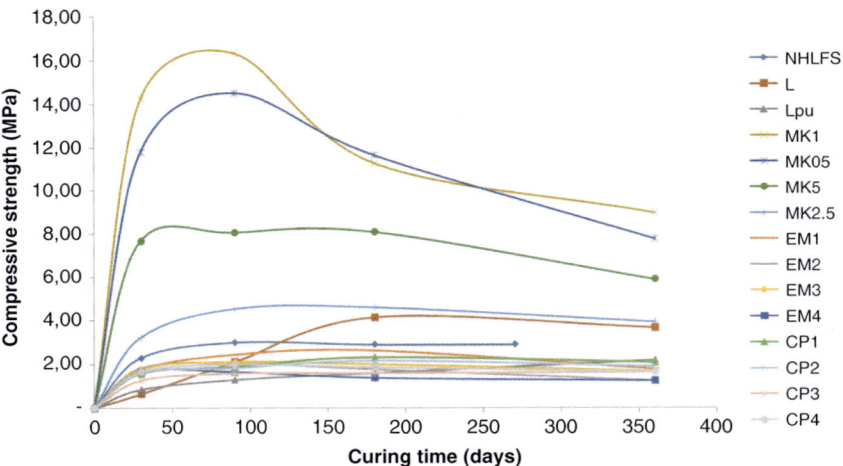

Fig. 5.20 Acquirement of compressive strength (MPa) in relation to curing time (days)

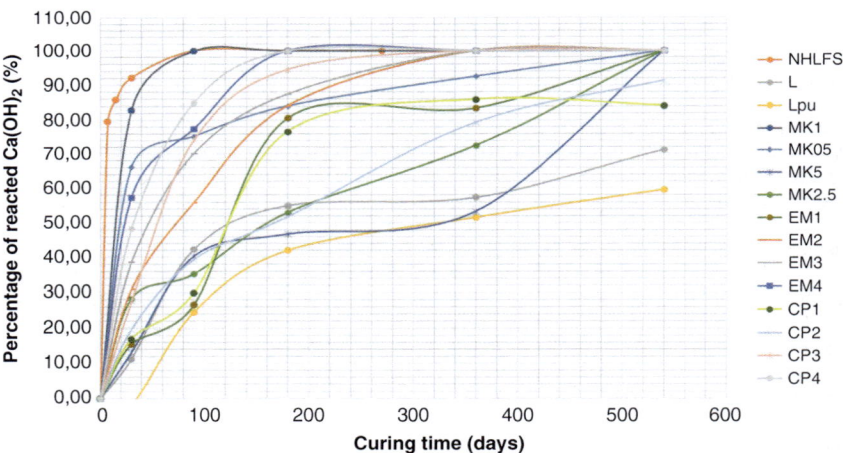

Fig. 5.21 Percentage of Ca(OH)₂ consumption (%) in relation to curing time (days)

5.4 Conclusion

The assessment of compatibility between a new restoration mortar and the stone substrate it is to be applied to during the conservation or restoration of a monument or historical building is an important process in order to protect the stone units and at the same time ensure sustainability of the structure. The proposed methodology aims to use the acceptability limits deriving from the study of a great number of compatible historical mortars as a tool to assess different traditional restoration mortars' characteristics. The use of acceptability limits, in correlation with the historical mortars of the specific monument to be restored, can assist in decision-making in regard to physicochemical compatibility and compatibility regarding microstructural characteristics. Further selection criteria arise from performance evaluation, technical requirements which have to do with each unique monument or historical building conservation/restoration details, as well as from the characteristics of the other building materials of the masonry, such as the stones and/or bricks. Environmental conditions also demand specific qualities from the restoration mortar. The optimum selection must take all parameters into account. It is important to stress that in the field of monument preservation, sampling is many times limited or even restricted for many materials, especially when dealing with important monuments. It is therefore necessary to always keep in mind that the statistical sample may not be adequate; however, any obtained data regarding any of the materials of the masonry is important and can assist in decision-making regarding the optimum restoration mortar selection. The reverse engineering-based methodological approach employed can assist in the selection of the appropriate restoration mortar or provide the guidelines for the optimization of restoration mortar mixes. Restoration mortars produced with traditional raw materials and techniques can fulfill compatibility and performance requirements for use in stone masonries and contribute to sustainability.

Acknowledgments The study and the rehabilitation project of the Holy Aedicule were possible and executed under the governance of His Beatitude Patriarch of Jerusalem, Theophilos III. The Common Agreement of the Status Quo Christian Communities provided the statutory framework for the execution of the project; His Paternity the Custos of the Holy Land, Archbishop Pierbattista Pizzaballa (until May 2016—now the Apostolic Administrator of the Latin Patriarchate of Jerusalem), Fr. Francesco Patton (from June 2016), and His Beatitude the Armenian Patriarch of Jerusalem, Nourhan Manougian, authorized His Beatitude the Patriarch of Jerusalem, Theophilos III, and NTUA to perform this research and the project. Contributions from all over the world secured the project's funding. Worth noting Mica Ertegun's and Jack Shear's donations through WMF, Aegean Airlines et al., the interdisciplinary NTUA team for the Protection of Monuments, Em. Korres, A. Georgopoulos, A. Moropoulou, C. Spyrakos, and Ch. Mouzakis were responsible for the rehabilitation project, and A. Moropoulou, as Chief Scientific Supervisor, was responsible for its scientific supervision.

Part of this investigation regarding the Kaisariani Monastery was performed within the research project "Seismic Protection of Monuments and Historic Structures— SEISMO," which was cofinanced by the Greek Ministry of Education and Religions and the European Union under the action "Thales" within the context of the Operational Program—Education and Lifelong Learning, NSRF 2007–2013. The authors would like to thank the employees of the first Ephorate of Byzantine Antiquities, responsible for the Kaisariani Monastery, for their cooperation.

The authors would like to thank the Ministry of Culture for the research conducted in the framework of the NTUA Scientific Committee of the Plaka Bridge Restoration, as well as all members of the interdisciplinary scientific team responsible for the study and proposals.

Finally, the authors would like to commemorate the late Dr. Petros Moundoulas, an esteemed member of the research group, and dedicate this research to him.

References

1. Moropoulou A, Bakolas A, Moundoulas P, Aggelakopoulou E. Reverse engineering: a proper methodology for compatible restoration mortars. In: Proceedings of the workshop repair mortars for historic masonry, TC RMH. Delft: RILEM; 2005. p. 25–8.
2. Franzoni E, Sassoni E. Correlation between microstructural characteristics and weight loss of natural stones exposed to simulated acid rain. Sci Tot Environ. 2011;412–413:278–85.
3. Torney C, Forster AM, Kennedy CJ, Hyslop EK. "Plastic" repair of natural stone in Scotland: perceptions and practice. Struct Surv. 2012;30(4):297–311.
4. Juhasova E, Sofronie R, Bairrao R. Stone masonry in historical buildings—ways to increase their resistance and durability. Eng Struct. 2008;30:2194–205.
5. Amoroso G, Fassina V. Stone decay and conservation. New York: Elsevier; 1983.
6. Moropoulou A, Kouloumbi N, Haralampopoulos G, Konstanti A, Michailidis P. Criteria and methodology for the evaluation of conservation interventions on treated porous stone susceptible to salt decay. Prog Org Coat. 2003;48:259–70.
7. Cardell C, Benavente D, Rodríguez-Gordillo J. Weathering of limestone building material by mixed sulfate solutions. Characterization of stone microstructure, reaction products and decay forms. Mater Charact. 2008;59:1371–85.
8. Grammatikakis G, Demadis KD, Melessanaki K, Pouli P. Laser-assisted removal of dark cement crusts from mineral gypsum (selenite) architectural elements of peripheral monuments at Knossos. Stud Conserv. 2015;60(Supplement 1):S3–11.
9. Avdelidis NP, Moropoulou A. Applications of infrared thermography for the investigation of historic structures. J Cult Herit. 2004;5:119–27.
10. Moropoulou A, Labropoulos KC, Delegou ET, Karoglou M, Bakolas A. Non-destructive techniques as a tool for the protection of built cultural heritage. Constr Build Mater. 2013;48:1222–39.
11. Moropoulou A, Koui M, Kourteli C, Zezza F, et al. Techniques and methodology for the preservation and environmental management of historic complexes—the case of the Medieval City of Rhodes. In: Moropoulou A, Zezza F, Kollias E, Papachristodoulou I, editors. Proceedings of the 4th international symposium on the conservation of monuments in the Mediterranean Basin, vol. 4. Rhodes: Technical Chamber of Greece; 1997. p. 603–34.
12. Beck K, Al-Mukhtar M. Formulation and characterization of an appropriate lime-based mortar for use with a porous limestone. Environ Geol. 2008;56:715–27.
13. Szemerey-Kiss B, Torok A. The effects of the different curing conditions and the role of added aggregate in the strength of repair mortars. Environ Earth Sci. 2017;76:284.
14. O'Brien PF, Bell E, Pavia Santamaria S, Boyland P, Cooper TP. Role of mortars in the decay of granite. Sci Total Environ. 1995;167:103–10.
15. Delgado Rodrigues J, Grossi A. Indicators and ratings for the compatibility assessment of conservation actions. J Cult Herit. 2007;8:32–43.
16. Groot C, Bartos P, Hughes J. Characterisation of old mortars with respect to their repair. In: 12th International Brick/Block Masonry Conference, Madrid; 2000. p. 815–27.
17. Van Balen K, Papayianni I, Van Hees R, Binda L, Waldum A. Introduction to requirements for and functions and properties of repair mortars. Mater Struct. 2005;38:781–5.
18. Schueremans L, Cizer Ö, Janssens E, Serré G, Balen KV. Characterization of repair mortars for the assessment of their compatibility in restoration projects: research and practice. Constr Build Mater. 2011;25(12):4338–50.
19. Faria P, Henriques F, Rato V. Comparative evaluation of lime mortars for architectural conservation. J Cult Herit. 2008;9:338–46.
20. Klisinska-Kopacz A, Tislova R, Adamski G, Kozlowski R. Pore structure of historic and repair Roman cement mortars to establish their compatibility. J Cult Herit. 2010;11(4):404–10.
21. Papayianni I, Pachta V, Stefanidou M. Analysis of ancient mortars and design of compatible repair mortars: the case study of Odeion of the archaeological site of Dion. Constr Build Mater. 2013;40:84–92.

22. Silva BA. Pinto APF, Gomes A. Natural hydraulic lime versus cement for blended lime mortars for restoration works. Constr Build Mater. 2015;94:346–60.
23. Aggelakopoulou E, Bakolas A, Moropoulou A. Properties of lime–metakolin mortars for the restoration of historic masonries. Appl Clay Sci. 2011;53:15–9.
24. Ventola L, Vendrell M, Giraldez P, Merino L. Traditional organic additives improve lime mortars: new old materials for restoration and building natural stone fabric. Constr Build Mater. 2011;25:3313–8.
25. Moropoulou A, Bakolas A, Moundoulas P, Michailidis P. Evaluation of the compatibility between repair mortars and building materials in historic structures by the control of the microstructure of cement-based systems. Conc Sci Eng. 2000;2:191–5.
26. Mosquera MJ, Benitez D, Perry SH. Pore structure in mortars applied on restoration—effect on properties relevant to decay of granite buildings. Cem Concr Res. 2002;32:1883–8.
27. Moropoulou A, Bakolas A, Moundoulas P, Cakmak AS. Compatible restoration mortars, preparation and evaluation for Hagia Sophia earthquake protection. PACT. 1998;56:79–118.
28. Moropoulou A, Aggelakopoulou E, Bakolas A. Earthquakes and monuments—the role of materials in the earthquake protection of monuments. In: Lourenço PB, Roca P, Modena C, Agrawal S, editors. 5th International Conference on Structural Analysis of Historical Constructions, vol. 3, New Delhi, India; 2006. p. 1625–1631.
29. Siddique R, Klaus J. Influence of metakaolin on the properties of mortar and concrete: a review. Appl Clay Sci. 2009;43:392–400.
30. Brook JJ. Johari MMA. Effect of metakaolin on creep and shrinkage of concrete. Cem Concr Compos. 2001;23:495–02.
31. Lanas J, Álvarez-Galindo JI. Masonry repair lime-based mortars: factors affecting the mechanical behavior. Cem Concr Res. 2003;33(11):1867–76.
32. Gameiro A, Silva AS, Faria P, Grilo J, Branco T, Veiga R, et al. Physical and chemical assessment of lime–metakaolin mortars: influence of binder: aggregate ratio. Cem Concr Compos. 2014;45:264–71.
33. Moropoulou A. Reverse engineering to discover traditional technologies: a proper approach for compatible restoration mortars. PACT. 2000;58:81–107.
34. Rosario Veiga M, Velosa A, Magalhaes A. Experimental applications of mortars with pozzolanic additions: characterization and performance evaluation. Constr Build Mater. 2009;23(1):318–27.
35. The Venice Charter. International charter for the conservation and restoration of monuments and sites. In: Second International Congress of Architects and Technicians of Historic Monuments. Venice, Italy; 1964.
36. Zsolnai L, Tencati A, editors. The future international manager: a vision of the roles and duties of management. London: Palgrave; 2009.
37. Moropoulou A, Apostolopoulou M, Moundoulas P, Aggelakopoulou E, Siouta L, Bakolas A, Douvika M, Asteris PG. The role of restoration mortars in the earthquake protection of the Kaisariani Monastery. In: ECCOMAS 2016, VII European Congress on Computational Methods in Applied Sciences and Engineering. Crete Island, Greece; 5–10 June 2016, p. 5340–5358.
38. Spyrakos C, Maniatakis Ch. Assessment of current condition under static and seismic loading for the Holy Aedicule of the Holy Sepulchre. In: Moropoulou, A., Korres Emm, Georgopoulos A, Spyrakos C, editors. Materials and conservation, reinforcement and rehabilitation interventions in the Holy Edicule of the Holy Sepulchre, NTUA; 2016.
39. Spyrakos C, Maniatakis Ch. Retrofitting of a historic masonry building. In: Proceedings of the 10th National and 4th International Scientific Meeting on Planning, Design, Construction and Renewal in the Construction Industry iNDiS 2006; 2006. p. 535–544.
40. Moropoulou A, Bakolas A, Moundoulas P, Aggelakopoulou E, Anagnostopoulou S. Optimization of compatible restoration mortars for the earthquake protection of Hagia Sophia. J Cult Herit. 2013;14:e147–52.

41. Moropoulou A, Polikreti K, Bakolas A, Michailidis P. Correlation of physico-chemical and mechanical properties of historical mortars and classification by multivariate statistics. Cem Concr Res. 2003;33(6):891–8.
42. Moropoulou A, Bakolas A. Range of acceptability limits of physical, chemical and mechanical characteristics deriving from the evaluation of historical mortars. PACT. 1998;56:165–78.
43. Aggelakopoulou E. Criteria and methodology for the evaluation of physicochemical and mechanical characteristics of restoration mortars addressed to historic masonries restoration interventions. PhD thesis, National Technical University of Athens (NTUA); 2006.
44. Moundoulas P. Design and evaluation methodology of compatible restoration mortars for historic monuments according to their mineralogic—physicochemical and physicomechanical characteristics. PhD thesis, National Technical University of Athens; 2004.
45. Charkiolakis N. The monasteries of mount Hymettus, 7 days Kathimerini newspaper; 28 September 1997 (in Greek).
46. Bouras Ch, Boura L. Greek religious architecture in the 12th century. Emporiki Bank of Greece; 2002.
47. Pallis G. Topography of the Athenian field during the post-Byzantine period. Thessalonica: Post-Byzantine Monuments I; 2009.
48. Moropoulou A, Apostolopoulou M, Moundoulas P, Karoglou M, Delegou E, Lampropoulos K, Gritsopoulou M, Bakolas A. The combination of NDTS for the diagnostic study of historical buildings: the case study of Kaisariani Monastery. In: COMPDYN 2015. 5th ECCOMAS Thematic Conference on Computational Methods in Structural. Dynamics and Earthquake Engineering. Crete Island, Greece; 25–27 May 2015. p. 2321–36.
49. Moropoulou A, Bakolas A, Anagnostopoulou S. Composite materials in ancient structures. Cem Concr Compos. 2005;27:295–300.
50. Maravelaki-Kalaitzaki P, Bakolas A, Karatasios I, Kilikoglou V. Hydraulic lime mortars for the restoration of historic masonry in Crete. Cem Concr Res. 2005;35:1577–86.
51. Gkolias I, et al. The bridge of Plaka: the NTUA contribution towards its restoration. Athens: NTUA; 2016.
52. Moropoulou A, Bakolas A, Delegou ET, Moundoulas P, et al. Characterization of the historical mortars of the bridge in Plaka and proposal for a new restoration mortar. In: The bridge of Plaka: the NTUA contribution towards its restoration. Athens: NTUA; 2016.
53. Vintzileou E, Gianellos C, Adami CE, Tsakanika E. Documentation of building materials mechanical characteristics. In: The bridge of Plaka: the NTUA contribution towards its restoration. Athens: NTUA; 2016.
54. Biddle M. The tomb of Christ. Gloucestershire England: Sutton; 1999.
55. Lavas G. The Holy Church of the resurrection in Jerusalem. The Academy of Athens; 2009 (in Greek).
56. Mitropoulos T. The Church of Holy Sepulchre—the work of Kalfas Komnenos. European Centre of for Byzantine and Post-Byzantine Monuments; 2009 (in Greek).
57. Moropoulou A, Korres Emm, Georgopoulos A, Spyrakos C, Mouzakis Ch. Presentation upon completion of the Holy Sepulcher's Holy Aedicule Rehabilitation. Athens: NTUA; 2017.
58. Moropoulou A, Korres Emm, Georgopoulos A, Spyrakos C. Materials and conservation, reinforcement and rehabilitation interventions in the Holy Edicule of the Holy Sepulchre. Athens: NTUA; 2016.
59. Moropoulou A, Bakolas A, Delegou E, Moundoulas P, Alexakis Emm, Apostolopoulou M, Sioula L. Building materials' characterization; conservation reinforcement and rehabilitation materials and interventions. In: Moropoulou A, Korres Emm, Georgopoulos A, Spyrakos C, editors. Materials and conservation, reinforcement and rehabilitation interventions in the Holy Edicule of the Holy Sepulchre. Athens: NTUA; 2016.
60. Moropoulou A, Lampropoulos K, Alexakis E, Delegou E, Moundoulas P, Apostolopoulou M, Bakolas A. Evaluation of the preservation state of the Holy Aedicule in the Holy Sepulchre complex in Jerusalem. In: 13th International Congress on the Deterioration and Conservation of Stone, vol. 2, Glasgow, UK; 2016. p. 1219–26.

61. Alexakis Emm. Diagnostic study of decay and characterization of building materials of the Holy Aedicule of the Holy Sepulchre in Jerusalem. MSc thesis, National Technical University of Athens; September 2016.
62. Moropoulou A, Korres E, Georgopoulos A, Spyrakos C, Mouzakis Ch, Lampropoulos KC, Apostolopoulou M, Delegou ET, Alexakis Em. The rehabilitation of the Holy Aedicule. In: XXXIII Convegno Internazionale Scienza e Beni Culturali, Le Nuove Frontiere del Restauro: Trasferimenti, Contaminazioni, Ibridazioni, Bressanone, 27–30 giugno 2017, Arcadia Ricerche Editore; 2017. p. 1–16.

Chapter 6
Inorganic Nanomaterials for the Consolidation and Antifungal Protection of Stone Heritage

A. Sierra-Fernandez, L.S. Gomez-Villalba, S.C. De la Rosa-García,
S. Gomez-Cornelio, P. Quintana, M.E. Rabanal, and R. Fort

6.1 Introduction

Stone cultural heritage and artworks suffer from weathering due to their interactions with the environment [1, 2]. These significant weathering processes are generally related to an important reduction in mechanical properties and an increase in porosity, pore size, and water absorption [3, 4]. Moreover, the deterioration caused by microorganisms is another critical threat to monuments worldwide [5, 6], especially in tropical areas where high humidity and temperatures encourage the growth of

A. Sierra-Fernandez (✉)
Instituto de Geociencias (CSIC, UCM), Madrid, Spain

Departamento de Ciencia e Ingeniería de Materiales e Ingeniería Química, Avda,
Universidad Carlos III de Madrid e Instituto Tecnológico de Química y Materiales
"Álvaro Alonso Barba" (IAAB), Madrid, Spain
e-mail: arsierra@ucm.es

L.S. Gomez-Villalba • R. Fort (✉)
Instituto de Geociencias (CSIC, UCM), Madrid, Spain
e-mail: rafael.fort@csic.es

S.C. De la Rosa-García
Laboratorio de Microbiología Aplicada, División de Ciencias Biológicas,
Universidad Juárez Autónoma de Tabasco (UJAT), Tabasco, Mexico

S. Gomez-Cornelio
Universidad Politécnica del Centro (UPC), Tabasco, Mexico

P. Quintana
Departamento de Física Aplicada, CINVESTAV-IPN, Unidad Mérida,
Mérida, Yucatán, Mexico

M.E. Rabanal
Departamento de Ciencia e Ingeniería de Materiales e Ingeniería Química, Avda,
Universidad Carlos III de Madrid e Instituto Tecnológico de Química y Materiales
"Álvaro Alonso Barba" (IAAB), Madrid, Spain

© Springer International Publishing AG 2018
M. Hosseini, I. Karapanagiotis (eds.), *Advanced Materials for the Conservation of Stone*, https://doi.org/10.1007/978-3-319-72260-3_6

microorganisms [7]. So, among different stone treatments, consolidation (i.e., restoring materials' integrity by improving cohesion and binding loose grains) [8, 9] and antifungal treatment protection (i.e., reducing and preventing existing fungal growth) are key phases of conservation practice used to safeguard stone heritage [10, 11].

In this context, the application of nanoscience for the conservation of stone may address a significant number of conservation issues. The possibility to manipulate and control materials at the atomic level and the subsequent understanding of fundamental processes at the nanoscale have led to new research challenges. Within nanoscience, the development of nanoparticles specifically designed for cultural heritage preservation purposes has become of interest to researchers. The reason is based on the unique and sometimes unexpected physical and chemical properties that are present in materials at the nanoscale, such as their increased surface-to-mass ratio, diffusivity, and electrical, optical, and thermal properties [12–14]. However, the application of nanoparticles (NPs) in the stone heritage conservation field requires a multidisciplinary approach which combines material science, petrophysics, microbiology, and cultural heritage conservation along with many other scientific disciplines. In this way, many challenges must be overcome such as studies of the physicochemical properties of materials and how they respond to changing environmental conditions, ultimately providing an exciting opportunity to improve the understanding of their action as conservation treatments.

Therefore, both organic and inorganic nanomaterials have been developed in order to recover the mechanical properties of damaged stone heritage and to provide successful antimicrobial coatings [15–18]. It is essential to note that consolidating and antifungal protective products must take into account fundamental conservation and restoration principles, mainly compatibility, efficacy, and durability [19]. Both treatments are the riskiest conservation and restoration practices due to their irreversible nature and threaten to produce undesirable effects. This is why a wide number of studies have been centered on the design of new stone consolidants and protective coatings with antimicrobial action. A detailed description of the different types of nanomaterials currently used to produce conservation treatments can be found in Sierra-Fernandez et al. [20].

In this chapter, a brief overview about the main synthesis methods and the most common analytical techniques employed for the physicochemical characterization of NPs will be presented. Then, the application of nanotechnology for the stone heritage conservation will be discussed. In particular, a significant portion of the discussion will be devoted to inorganic nanomaterials that specifically focus on stone heritage preservation and the main factors that condition their effectiveness.

6.2   Nanomaterials

6.2.1   Synthesis Methods

The term "nanomaterial" can be referred to as materials composed of single elements, such as metals or carbon, or materials made up of several elements, such as metal oxides or composites [21]. In this chapter, when the term "nanomaterial" is used, it is mainly referring to inorganic nanoparticles composed of metals and metal hydroxides and oxides, unless otherwise stated.

The selection of the synthesis method used for obtaining the NPs or the nanostructured materials is a decisive factor used to enhance the suitability of these materials in function of their use for the conservation of artworks. In this way, many synthesis strategies have been developed to obtain inorganic nanocrystals with certain morphologies, particle sizes, agglomeration level, and crystallographic structures. The two commonly used synthetic methods for nanofabrication of NPs or nanostructured materials include "top-down" and "bottom-up" [22]. In the "top-down" approach, the process begins from a bulk piece of material, which is then gradually removed or handled to form materials in the nanometer size range either by milling [23], photo lithography [24], electron beam lithography [25], or anodization [26], among others. The "bottom-up" approach begins with atoms and molecules that get rearranged and assembled to form larger structures [22]. This synthesis approach was the first method used in the field of material science applied to cultural heritage for obtaining calcium hydroxide nanoparticles specifically designed for artwork conservation [27]. Since then, several NPs with different morphologies and structures have been synthesized for application in cultural heritage by using chemical methods mainly due to their potential for scale-up and their lower cost. Examples of widely used chemical methods for synthesis of inorganic NPs include the solvothermal [28], the hydrothermal [29], and the solgel [30, 31] methods. Moreover, alkaline hydroxide NPs were synthesized by using "an ion exchange process between an anionic resin and a calcium chloride aqueous solution operating at room temperature" [32]. Other commonly used synthesis methods for the fabrication of NPs are the pulsed laser ablation method [33], the microwave-assisted synthesis method [34], and the spray pyrolysis method [35]. Most recently, atomic layer deposition (ALD), molecular layer deposition (MLD), and combinations of the two have been used to produce a wide variety of hybrid organic/inorganic materials with innovative properties [36]. These promising techniques are cyclic vapor-phase deposition processes which have been specially adapted and applied to prevent corrosion damage for silver cultural heritage objects [37].

As a remarkable drawback, an important factor to take into account is that they are different procedures, from both the bottom-up and the top-down, to create nanostructures which can be applied according to its specific needs [22]. Using a suitable (compatible with the stone substrate) nanoparticle solution, it is possible to create a

Characterization Techniques

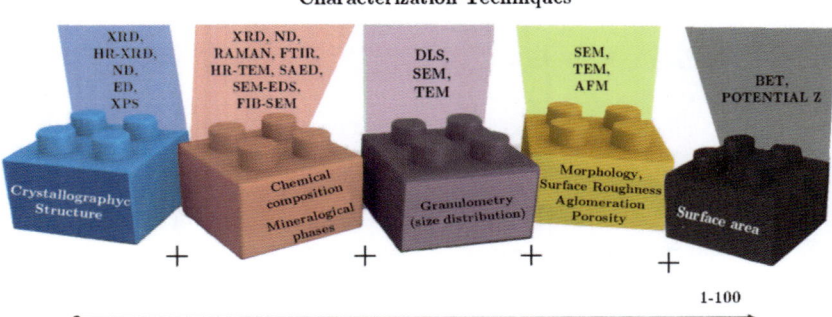

Fig. 6.1 Summary of the analytical techniques most commonly used for the study of inorganic NPs: *XRD* (X-ray diffraction), *HR-XRD* (high-resolution X-ray diffraction), *ND* (neutron diffraction), *ED* (electron diffraction), *XPS* (X-ray photoelectron spectroscopy), *Raman* spectroscopy, *FT-IR* (Fourier transform-infrared spectroscopy), *HR-TEM* (high-resolution transmission electron microscopy), *SAED* (selected area electron diffraction), *SEM-EDX* (scanning electron microscopy-energy-dispersive X-ray spectroscopy), *FIB-SEM* (focus ion beam scanning electron microscopy), *DLS* (dynamic light scattering), *AFM* (atomic force microscopy), *BET* (Brunauer-Emmett-Teller)

consolidating and/or protective product with new properties and characteristics highly compatible with the stone when compared to their bulk counterparts, as shown in the following sections.

6.2.2 Characterization Techniques for Nanomaterials

An extensive number of techniques can be used to evaluate size, shape, polydispersity, composition, purity, and surface properties of nanostructured materials, which have to be studied before their application in stone heritage (Fig. 6.1). Therefore, in this section, some of the most common characterization techniques used to investigate the physicochemical characteristics of nanomaterials designed for their application in the cultural heritage conservation will be briefly discussed.

The crystallography of the nanomaterials is usually studied by diffraction techniques such as X-ray diffraction (XRD), high-resolution X-ray diffraction (HR-XRD), synchrotron radiation X-ray diffraction (SR-XRD), and neutron diffraction (ND). These analytical techniques are very useful tools to investigate, nondestructively, the crystallographic structure, the strains, and the preferred orientations of nanomaterials [38–40]. Specifically, diffraction techniques are also used to measure various structural properties of these nanomaterials, such as defect structure [39, 41], phase composition [42], texture analysis [43], and crystal size [44], allowing many experiments in time resolution. Another fundamental analytical technique for determining the arrangement of atoms in a crystalline solid and the atomic structure determination is electron diffraction (ED) [45]. Moreover, X-ray photoelectron spectroscopy (XPS) is an analytical method with high surface specificity, widely used for elemental analysis and for investigating the electronic structure of nanomaterials [46].

The presence of impurities may significantly impact the effect or even produce unfavorable results in treated stone substrates. In this way, determination of nanomaterial composition, nature, and/or purity can be accomplished through the analysis of their chemical compositions. In general, chemical composition and purity grade are studied by Fourier transform-infrared spectroscopy (FT-IR) and Raman spectroscopy. These spectroscopic techniques have been shown as powerful techniques for the characterization of inorganic nanomaterials and those modified by the adsorption of molecules with different chemical properties [47, 48]. Furthermore, the energy-dispersive X-ray (EDX) analysis is a technique used in conjunction with scanning electron microscopy (SEM). This type of energy provided by X-ray emission is used to identify the elemental composition of samples and to estimate its proportion at different sample areas. Moreover, transmission electron microscopy (TEM) and high-resolution transmission electron microscopy (HR-TEM) are the most commonly used analytical techniques for studying nanoparticles at a spatial resolution down to the level of atomic dimensions (<1 nm) [49]. Thus, imaging, diffraction, and microanalytical information are widely obtained by TEM and HR-TEM. Nevertheless, it is important to take into account the risk of the irradiation damage in the samples caused by the use of electron beams in the transmission electron microscope, which could lead to undesired physical and chemical material property changes or uncontrollable modification of structures [50]. Additional analytical techniques for advanced and precise analysis on different types of nanomaterials are focused ion beam scanning electron microscopy (FIB-SEM) and atomic force microscopy (AFM). These characterization techniques are considered crucial for imaging on the nanometer scale due to their versatility and multifunctionality. While FIB-SEM has shown to be especially useful for studying the three-dimensional morphology of complex material systems [51], AFM allows the visualization and the analysis of individual and groups of nanoparticles, as well as the examination of uncoated and coated surfaces in three dimensions, and can operate in a number of modes, depending on the application and the required information [52, 53].

Among all viable options to measure the particle size of nanomaterials, dynamic light scattering (DLS) and microscopy (SEM and TEM) or a combination of both is important analytical techniques used to determine accurate particle size distribution and polydispersity in a wide range of sizes.

In other words, surface charge potentially affecting "receptor binding and physiological barrier penetration governs the dispersion stability or aggregation of nanomaterials and is generally estimated by zeta potential" [54]. Also, the measurement of the specific surface area (SSA) of nanomaterials is widely determined by the Brunauer-Emmett-Teller (BET) technique [55].

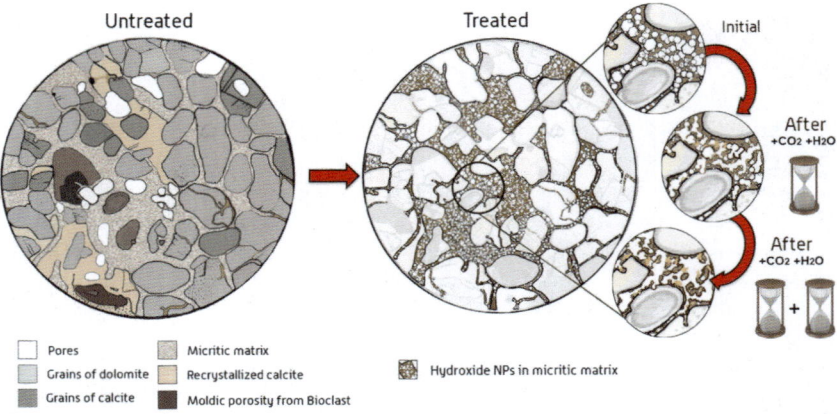

Fig. 6.2 Schematic illustration of a thin film of untreated and treated dolostone by using hydroxide nanoparticles (Reprinted with permission from Sierra-Fernandez et al. [20], Copyright, 2017 Materiales de Construcción. All Rights Reserved)

6.3 Nanomaterials for the Stone Heritage Preservation: Factors Influencing Their Effectiveness

6.3.1 Consolidating Products

The physicochemical compatibility of the inorganic nanomaterials with the stone-built cultural heritage is one of the main advantages of the use of these materials as consolidating products. The reactivity and penetration capacity of a consolidating product within the stone material and, consequently, its effectiveness are potentially increased when its particle size is reduced to the nanoscale. Another significant benefit of nanomaterials is a growth in surface area relative to volume. This greater surface area per unit mass changes or enhances their electronic and optical properties and their chemical reactivity (e.g., carbonation rate) because a greater proportion of atoms are found at the surface compared to those inside.

Significant advances have been made using colloidal nanoparticles based on calcium hydroxide ($Ca(OH)_2$). In this context, to date, the pioneering products developed by Baglioni and his group [27, 56, 57], and the $Ca(OH)_2$ nanoparticles designed by Ziegenbald and co-workers [58], are the most commonly used nanomaterials for stone consolidation. The consolidant action of the hydroxide NPs is based on their transformation into carbonate forming through the action of atmospheric CO_2 in the presence of moisture (Fig. 6.2). In this way, the newly formed carbonate phases act on degraded porous substrates, especially those of carbonatic nature, by binding loose grains and/or modifying their porous structure. However, the effect of relative humidity (RH) has been proven to be a clue in carbonation kinetics [59–61]. Thus, different calcium carbonate polymorphs can be produced depending on the relative humidity, resulting in different physical properties of the treated stony substrate. Therefore, in dry environments (RH values of 33–54%), the carbonation process of the calcium hydroxide NPs is slower, giving rise to the generation of low crystallinity calcium carbonates [62]. However, at RH values of 75–90%, the presence of water acts as an accelerator of the process, and calcium hydroxide is completely converted into calcium carbonate ($CaCO_3$), developing differences in the nucleation of growth of different $Ca(CO)_3$ polymorphs within 7 days [60, 63]. Thus, in addition to the generation of calcite and acicular aragonite, unstable polymorphs like vaterite and crystalline hydrated forms (e.g., monohydrocalcite and ikaite) can be formed [59, 60]. Moreover, under high relative humidity conditions, calcium carbonate polymorphs' stability can be affected, generating structural defects that may give rise to dislocations. Differences in the crystalline habit, crystal size, aggregation, and type of atomic-scale structural defects may be expected to affect the polymorphic properties of $CaCO_3$ and, in particular, carbonation kinetics under specific experimental conditions [64] (Fig. 6.3). Thus, the generation of these different calcium carbonate polymorphs can affect the integrity of the stone substrate, contributing to the loss of cohesion [62]. In this way, questions related to if their physical and chemical characteristics are compatible with the petrophysical properties of the stone to treat or if local environmental conditions to which they are exposed [63, 64] are appropriate for the carbonation process are considered critical.

Furthermore, the NPs' concentration and influence of the dispersions on the effectiveness of hydroxide NPs as consolidants for weathered stone materials should be considered for a successful consolidation treatment. Gomez-Villalba et al. [65] studied the effect of different concentrations of calcium hydroxide NPs obtained by colloidal synthesis (CaLoSiL®) applied on dolostone samples (Redueña, Madrid, Spain). This study showed that the application of calcium hydroxide NPs in high concentrations (15 and 25%) limited their effectiveness as a consolidating product due to their deposition over the stone substrate, producing changes in the color and the brightness of the treated dolostones. Otherwise, the application of the same product in low concentrations can avoid these secondary effects, resulting in an effective method to improve the durability of stone materials [66]. However, low concentrations may not prevent microbial colonization of biodeteriorative organisms.

Moreover, the most suitable application procedure for each case is another important point to study. Therefore, previous research works were centered on the investigation of the absorption and drying kinetics of magnesium and calcium hydroxide nanoparticle dispersions (average sizes from ~70 to ~260 nm) in pure ethanol, applied by different methods and followed by neutron radiography (NR) over time. These studies revealed a maximum particle penetration of 0.55 cm after 100 min of treatment into the dolostone substrates (Laspra dolostone, Asturias, Spain) treated by brushing. However, during this drying phase, radiographs show the accumulation of nanomaterial just underneath the surface (Fig. 6.4), limiting the penetration depth of the NPs into the dolostone treated by brushing. Recently, Borsoi et al. [67] studied the effectiveness of $Ca(OH)_2$ nanoparticle dispersions on two lime-based substrates (Maastricht limestone and lime-based mortar) by different

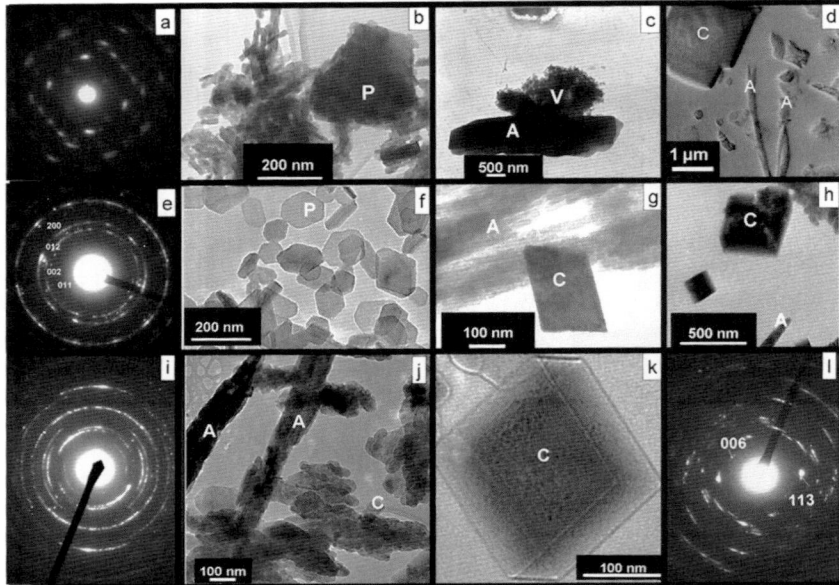

Fig. 6.3 Low magnification TEM images and SAED patterns showing the evolution of the spontaneous carbonation process in two alcoholic colloidal solutions of hexagonal platelike nanoportlandite (solution A (**a**, **b**, **c**, **d**, **i**) and solution B (**f**, **g**, **h**, **j**, **k**, **l**)), at 75% RH, for 0 days (**b** and **f**), 14 days (**c** and **g**), 35 days (**d** and **h**), and 2 years (**j** and **k**) of exposure. TEM images show the initial portlandite ($Ca(OH)_2$) (**b** and **f**) and nucleation of $CaCO_3$ polymorphs during the carbonation. Simultaneous presence of calcite and aragonite after 14 days (**g**) and 35 days (**h**) in solution B. Metastable vaterite crystals associated with aragonite after 14 days in solution A (**c**). Small calcite crystals associated with acicular aragonite crystals in both solutions after 35 days (**d**, **h**). In solution B, well-crystallized calcite crystals after 2 years (**k**) always associated with aragonite (**j**). SAED patterns confirm the initial portlandite in solution A (**a**) and solution B (**e**) and the respective $CaCO_3$ polymorphs for solution A, 14 days (**i**), and solution B, 35 days (**l**). P, portlandite; V, vaterite; A, aragonite; C, calcite. Solution A (Nanorestore) and solution B (CaLoSiL) both dispersed in isopropyl alcohol (Reprinted with permission from Gomez-Villalba et al. [64], Copyright, American Chemical Society. All Rights Reserved)

application methods: full and partial saturation capillarity and partial saturation by nebulization. Their results showed how the addition of a small amount of water (5% by volume) in the $Ca(OH)_2$ dispersion can guarantee moderate kinetic stability so that $Ca(OH)_2$ NPs can be properly absorbed, favoring their penetration into the stone substrates. According to the methodological application, the authors determined that an effective in-depth treatment of Maastricht limestone can be obtained both with NPs dispersed in pure ethanol and a mixture of ethanol (95% by volume) and water (5% by volume) when they are applied by capillarity until full impregnation. However, their results also showed that the use of calcium hydroxide NPs dispersed in pure ethanol induced a partial back migration of the nanoparticles during drying. In addition, the use of NPs (with particle size range: 50–600 nm)

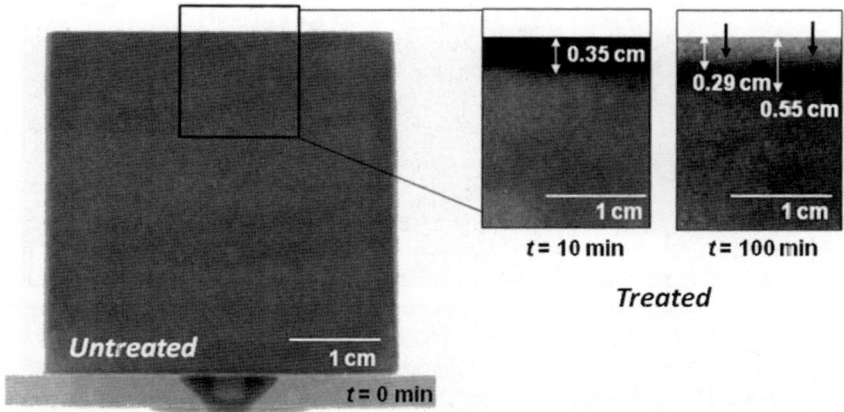

Fig. 6.4 Radiographs of the untreated and treated selected area of Laspra dolostone at different times. The accumulation is visible beneath the drying surface, after 100 min of treatment

dispersed in pure ethanol was recommended when the treatment is carried out through multiple applications by nebulization. The study also suggested the surface impregnation by nebulization as the most suitable application methodology for the treatment of lime-based mortars. The preparation of calcium hydroxide dispersions in mixing solvents, such as butanol with percentages of ethanol, could be considered in the case of the fine porous substrate treatment to improve the penetration depth of this type of nanomaterials [68]. In this sense, while solvents with high boiling points (e.g., butanol or water) would perform better in stone substrates with large pores, solvents with lower boiling points (e.g., ethanol or isopropanol) would reduce the $Ca(OH)_2$ nanoparticle migration back to the stone surface during solvent drying [68].

On the other hand, some routines have been studied that accelerate the carbonation process of the consolidating treatments based on hydroxide NPs, from the role of solvent [39] to the environmental conditions, such as the use of a fermentation system based on water and a yeast-sugar solution [69]. This simple method generates a microclimate saturated with CO_2, high humidity, and ethanol which are able to speed up the carbonation process of $Ca(OH)_2$ NPs.

The kinetic stability of the hydroxide NPs in alcohol dispersions is also another important point to study. Rodriguez-Navarro et al. [39] evaluated the role of alcohol in the kinetic stability and the carbonation behavior of three different calcium hydroxide NP dispersions, which differ in particle sizes and surface areas. The study was focused on $Ca(OH)_2$-alcohol interaction and its effect on $CaCO_3$ polymorph selection and carbonation kinetics, determining that the contact of the particles with alcohol (e.g., ethanol or isopropanol) for long periods of time (>2 months) results in their replacement by calcium alkoxides [39]. The authors suggested that a high conversion of $Ca(OH)_2$ into calcium alkoxides could be desirable if a preconsolidation treatment marked by a high $CaCO_3$ yield with a fast precipitation is required. Most recently, the physical and chemical features of $Ca(OH)_2$ particles dispersed in ethanol and isopropanol, once subjected to storage for different periods of time at different temperatures (40, 60, and 80 °C), were characterized [70]. It was shown that $Ca(OH)_2$ particles reacted with ethanol and isopropanol and were partially transformed into Ca ethoxide and Ca isopropoxide, respectively [70]. This conversion into Ca alkoxides also reduced the rate of carbonation of $Ca(OH)_2$ (nano)particles and induced the formation of metastable vaterite [70].

Another major challenge in an optimal consolidating action of the designed nanomaterials according to the stone substrate is related to their chemical composition. It is important to consider that according to the compatibility, the proportions of magnesium and calcium in limestones and dolostones often differ widely both within a single rock formation and between formations. Therefore, the main research activity has been focused on the synthesis of $Mg(OH)_2$ and mixed solutions of $Mg(OH)_2$ and $Ca(OH)_2$ NPs in search of better compatibility, effectiveness, and durability of the treatment with the calcium-magnesium carbonate substrates. The reason is that the effectiveness of the use of calcium hydroxide NPs is severely reduced in dolostone substrates. The incorporation of the $Ca(OH)_2$ NPs and their subsequent carbonation on $CaCO_3$ entail a change in the dolomite crystals of dolostones. In this context, the application of calcium hydroxide NPs in dolostone could regenerate nano-calcite recrystallization, leading the dolomite dissolution due to the calcium ion enrichment [59]. Also, an alteration in the nucleation and crystals' growth may produce aesthetic modifications, such as changes in color and brightness [60].

Thus, $Mg(OH)_2$ and $Ca(OH)_2$ NPs have been synthesized by the hydrothermal method and by using the solgel synthesis method, carrying out a detailed study of the main synthesis factors that could influence the properties of the different synthesized NPs [29, 50]. The synthesis of the hydroxide NPs by the solgel method has shown the promise of obtaining highly concentrated dispersions of NPs, which is crucial for the upscale production.

Figure 6.5 shows FE-SEM and TEM micrographs of the magnesium hydroxide and the Mg-Ca hydroxide NPs with different weight ratios (10–90 and 50–50 wt%) obtained via the solgel method. This synthesis method resulted in the formation of highly crystalline and well-defined hexagonal flakes, having an average diameter from ~ 30 to ~ 60 nm. The crystal phases and crystallinity were also determined by X-ray diffraction. Thus, Fig. 6.5g shows the X-ray diffraction patterns of the different hydroxide samples, exhibiting the typical diffraction peaks of the $Mg(OH)_2$ and $Ca(OH)_2$ structures and confirming thus the purity and the chemical composition of the different obtained types of hydroxide NPs.

An example of adapting the consolidant composition to the stone substrate is illustrated in Fig. 6.6. In this case, the selection of the nanoparticles as consolidant product in dolostone and limestone substrates was carried out according to the chemical composition and the petrophysical properties of the different lithotypes. Therefore, different types of hydroxide NPs were selected and applied as consolidating products in two types of calcareous stones, widely used in the cultural

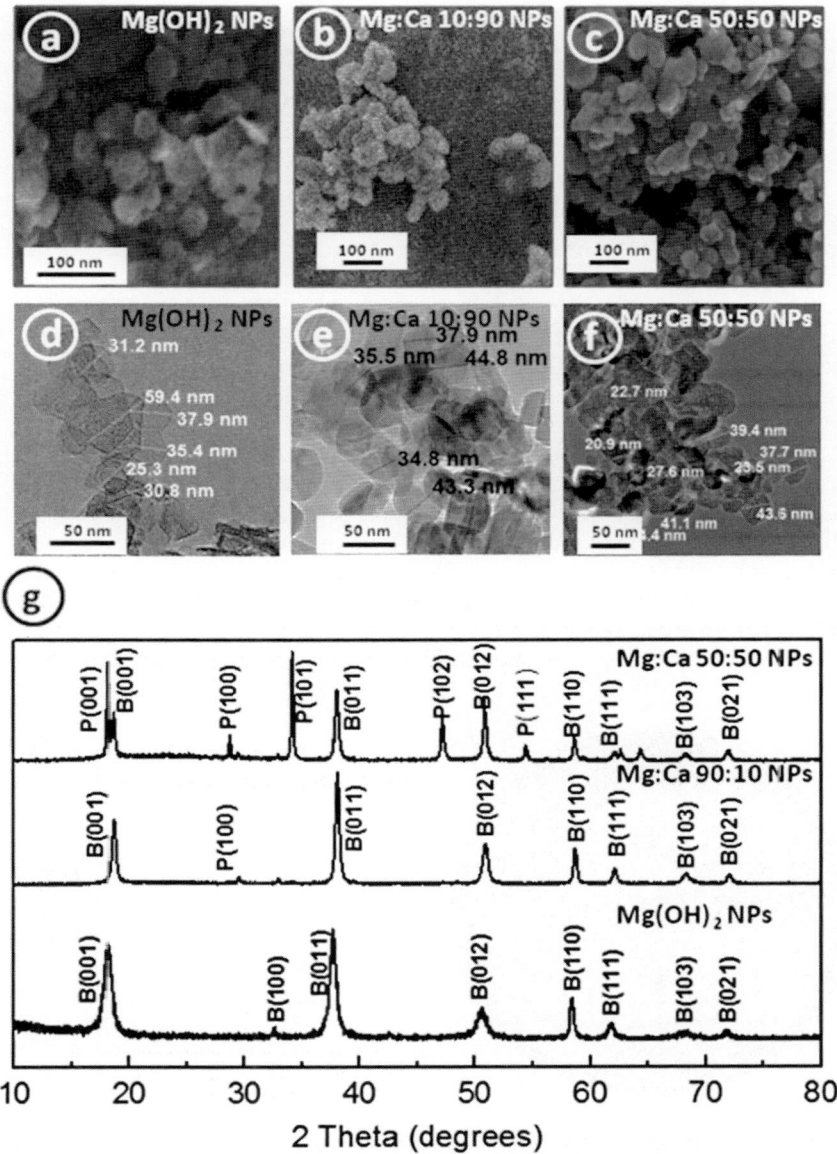

Fig. 6.5 (**a–c**) Low magnification of FE-SEM micrographs, (**d–f**) TEM micrographs of Mg(OH)$_2$, and the Mg-Ca hydroxide NPs with different weight ratios (10–90 wt% and 50–50 wt%, respectively); (**g**) XRD patterns of Mg(OH)$_2$ and the Mg-Ca hydroxide NPs with weight ratios of 10–90% and 50–50%. P, portlandite; B, brucite

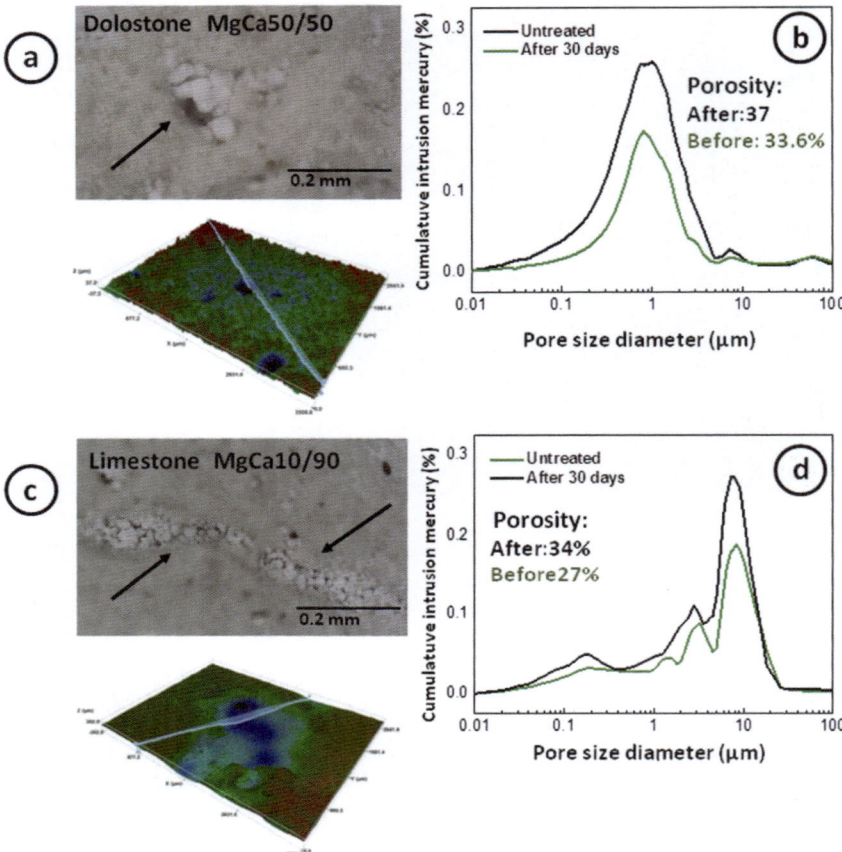

Fig. 6.6 Optical microscopy images and topographic images obtained by confocal microscopy of Laspra dolostone treated by Mg-Ca hydroxide nanoparticles (50/50 wt%) and Conchuela limestone treated by Mg-Ca hydroxide nanoparticles (10–90 wt%) (**a** and **c**, respectively). Pore size distribution (PSD) obtained by mercury intrusion porosimetry (MIP) of Laspra dolostone untreated and treated by Mg-Ca hydroxide nanoparticles (50/50 wt%) and Conchuela limestone untreated and treated by Mg-Ca hydroxide nanoparticles (10–90 wt%) (**b** and **d**, respectively)

heritage of Spain (Laspra dolostone) and Mexico (Conchuela limestone). Whereas the Laspra dolostone is a fossiliferous dolomicrite composed mainly of dolomite, the Conchuela limestone is a biomicritic limestone with a high content of fossil fragments and is mainly composed of calcite [31]. In order to assure an optimal consolidant treatment and according to their physical and chemical characteristics, both types of stone substrates were treated with different types of hydroxide NPs. Laspra dolostone was treated with a solution of $Mg(OH)_2/Ca(OH)_2$ NPs (50:50 wt%) and the Conchuela limestone with a solution of $Mg(OH)_2/Ca(OH)_2$ (10:90 wt%)

dispersed in ethanol in a concentration of 2.5 g/L. Figure 6.6a and c shows the optical micrographs carried out in the treated dolostone and limestone substrates, respectively. These results depicted the $Mg(OH)_2/Ca(OH)_2$ NPs (50–50 wt%) filled the pores and inter-crystalline dolomite grain contacts in the Laspra dolostone. For Conchuela limestone, the presence of the Mg-Ca hydroxides was detected mainly covering the crystal surfaces, favoring the calcite ($CaCO_3$) recrystallization. After the consolidant treatment, a decrease in total porosity was detected by mercury intrusion porosimetry (MIP) in the treated dolostone and limestone (Fig. 6.6b and d, respectively) due to the filling of pores by the consolidating product.

6.3.2 Antifungal Protective Coatings

The biotic factors, especially the growth of fungi on the monumental stone, are among the most active microorganisms in the biodeterioration of stone substrates, representing a crucial threat to monuments worldwide. An important group of microorganisms such as cyanobacteria, microalgae, lichens, fungi, and bacteria [5] can grow in different environments by attaching to the surfaces [71]. When attached to the surface, these microbial cells may develop into biofilms, which can produce the microbial deterioration of stone. The biofilms are thus collections of cells on surfaces that are maintained by electrostatic forces (mainly a combination of van der Waals attraction and chemical bonding) and/or extracellular polymeric substances (EPS) [72]. The EPS are composed mainly of a matrix of polysaccharides, which provide crucial functions such as protection for desiccation, erosion, radiation, as well as storage of organic components and nutrients [73, 74].

The microbial communities can colonize on different areas of stone substrates, becoming a convenient habitat for most rock-dwelling fungi [75, 76]. Therefore, fungi can grow as epilithic, on stone's surface, or endolithic communities, within cracks and pores or actively boring into the substratum [77]. These important biological colonizers of stone heritage may cause aesthetic damage originated by their colored patinas [78] and/or biophysical deterioration, which may be occurring due to the pressure exerted on the surrounding surface material during the growth of hyphal networks (biomass swelling or contraction) through the pore system of stone materials [79]. However, different studies determined that the biochemical actions in stone substrates were more important than mechanical attacks [80, 81]. These actions are marked by the generation and excretion of high concentrations of organic acids by fungi during their metabolic activity [79]. In this way, the fungal species can excrete organic acids (e.g., citric, oxalic), inorganic acids (e.g., sulfuric and nitric acids), H^+, CO_2, and metabolites and can occur in conjunction with biophysical processes [79, 82], resulting in important degradation processes from the pitting to the complete dissolution (Fig. 6.7).

In addition, while the microbial colonization is marked by the climatic conditions and the anthropogenic contamination of atmosphere [67], the degradation and weathering of stone materials are basically determined by its petrographic

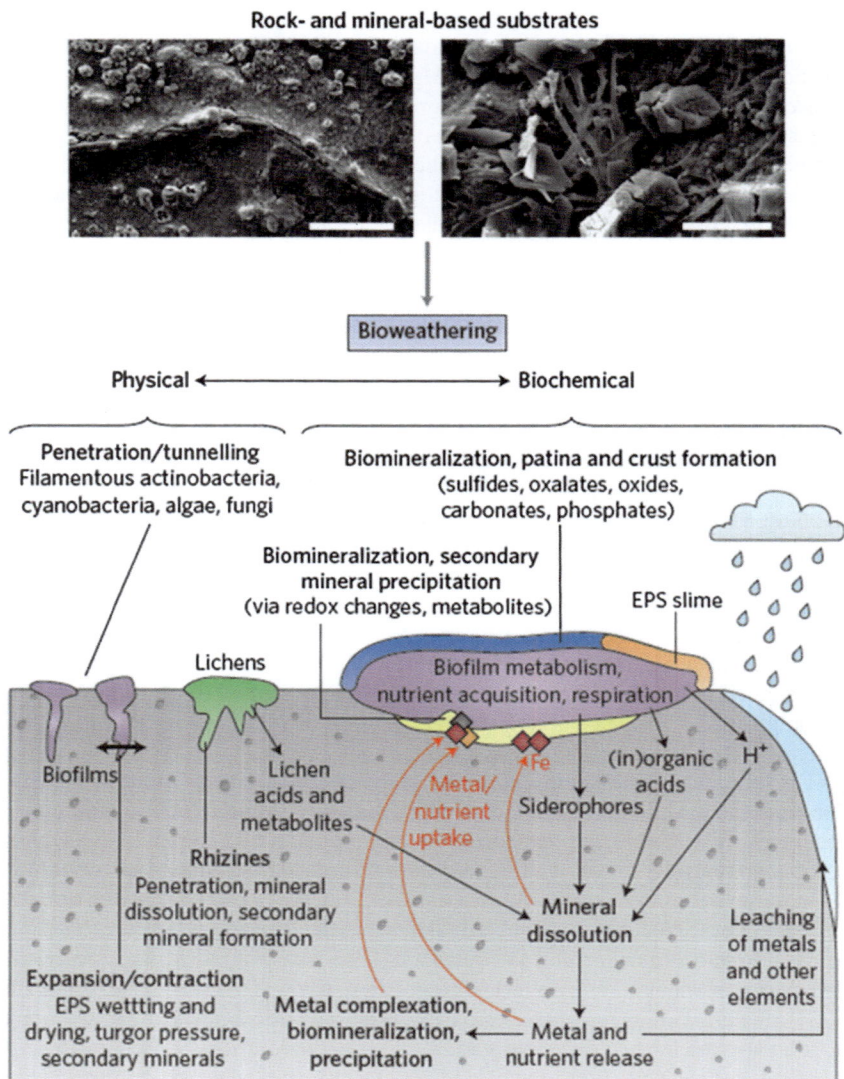

Fig. 6.7 Schematic illustration of the main influences and effects of microorganisms on rock- and mineral-based substrates with the respective SEM images (Reprinted and adapted with kind permission from G.M. Gadd [79], Copyright, Nature Publishing Group. All Rights Reserved)

features (especially the structure and the chemical composition) [30, 76]. Also, the texture and porosity of the material itself are important factors that favor a greater receptivity of the material to be colonized (Fig. 6.8) [76]. The stone microporosity has also been shown as another important factor in the bioreceptivity of stone substrates [31].

In this context, in the last decade, the appearance of nanotechnology has allowed the development of metallic NPs with interesting properties as antifungal agents for stone heritage preservation. To date, it has been reported that metallic NPs, such as Ag [83, 84], Cu [85], and Zn [30, 86, 87], as well as nanostructured metal oxides and combination of both [30, 31], have become attractive alternative sources to combat microbial species that are widely affecting the stone heritage (Fig. 6.9).

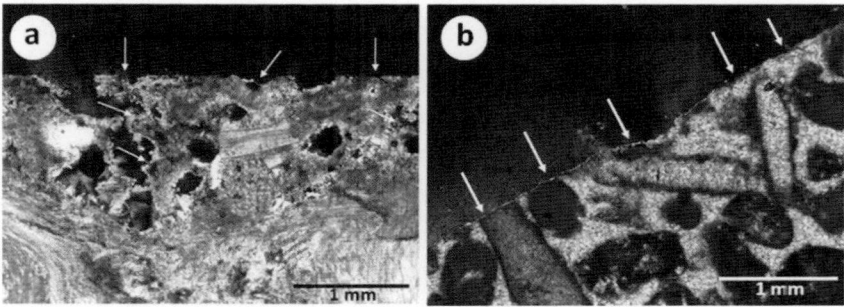

Fig. 6.8 Polarized light optical microscopy (PLOM) images (crossed Nicols) of thin sections from (**a**) Conchuela limestone (Yucatán, Mexico) showing endolithic fungal colonization penetrating down to the stone along cracks and in the internal stone pores and (**b**) the presence of fungal colonization confined to the surface of Laja stone (Yucatán, Mexico). To note the different penetration depths of fungal hyphae attributed to the increased porosity of Conchuela limestone (~27%) in comparison with Laja stone (~7%)

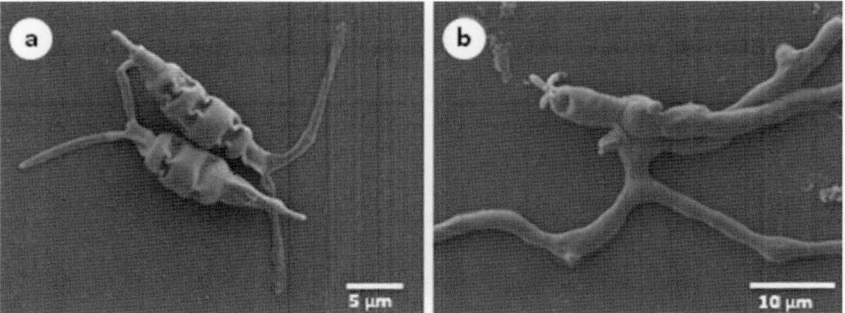

Fig. 6.9 Scanning electron microscopy (SEM) micrographs showing the effect of Zn-doped MgO NPs on conidial germination of *Pestalotiopsis maculans*. (**a**) Conidial cells and appendages (control); (**b**) non-geminated conidia with swollen cell and short appendages after exposition of sublethal concentration of Zn/Mg oxide NPs

In addition to the compatibility of the nanomaterials with the stone to be treated, another important challenging aspect of antimicrobial treatments is the fact that stone artworks are typically affected by a mixed community of microorganisms. These microorganisms may present different levels of susceptibility toward the used chemical compound. Therefore, nanomaterials with a wide action spectrum are of interest for stone protection. Among the different types of oxide NPs that have been applied as antifungal protective agents for stone heritage, MgO [31], ZnO [86], and TiO$_2$ [15, 86] are the most commonly used. These nanostructured materials present important physicochemical properties (mainly their inorganic character, greater surface area, and small particle size), which allow them to exhibit a wide spectrum of antimicrobial activity against different bacterial and fungal species, making them of great interest to the scientific community.

Although, in general, the mechanisms of antifungal activity of metallic NPs have not been completely understood, it is important to understand the main factors influencing the antimicrobial activity of NPs. The study of the antimicrobial mechanisms of nanoparticles constitutes a fascinating and open field for future research. Some studies have revealed that the inhibition of fungal growth is carried out by a combination of one or more mechanisms and it may vary based upon the function and chemistry of nanoparticles [87]. The antimicrobial activity of nanomaterials is generally attributed to three models: oxidative stress induction, metal ion release (disturbance in metal/metal ion homeostasis), and non-oxidative mechanisms [88]. Among these modes, the generation of reactive oxygen species (ROS) and the induction of oxidative stress phenomenon are the main modes of antifungal activity of the metal and metal oxide NPs [89, 90]. These reactive species are mainly superoxide anion (O$_2^-$), hydroxyl radicals (OH$^-$), hydrogen peroxide (H$_2$O$_2$), and organic peroxides, which are produced due to the change in electronic properties and the reduction in particle sizes [91, 92]. The ROS are pernicious to microorganisms, causing damage to almost all organic biomolecules and eventually inducing the cell death [89]. Also, different studies have pointed out that the ROS generation depends on the surface area of the NPs, which can potentially increase with a decrease of the particle sizes [93].

Moreover, the oxidative stress may sometimes happen through disturbing microbial process or oxidizing a vital cellular structure or components without ROS production [92]. Thus, other modes of actions to be considered include damage to cell membranes by electrostatic interaction [94], protein and enzyme dysfunction [95], genotoxicity [96], and photokilling [97]. Furthermore, careful attention should be paid to other physicochemical properties of NPs when studying their antimicrobial mechanisms. Several studies have shown that a large specific surface area and atomic ligand deficiency lead the NP aggregation [98, 99]. An increased aggregation state produces a reduction of surface area that could decrease the interaction between NPs and microorganisms and consequently the inhibition or loss of antifungal properties of NPs [100]. In addition, current research works have shown that particle size [93], shape [101], roughness [88], or surface charge [102] of metal and metal oxide NPs can greatly modify their antimicrobial activity. Additionally, the influence of crystal defect content on the antifungal capacity of nanomaterials should be taken into account. Different studies have highlighted the crucial role of surface defects in the generation of ROS [103, 104]. In this line, recently, Sierra-Fernandez et al. [31] reported the application of Zn-doped MgO NPs as antifungal agents against different fungal species, for dolostone and limestone substrates. Their experimental results have shown that doped Zn greatly improved the antifungal capacities of MgO due to the generation of an increased defect density.

It should be noted that the study of different binding agents to assure an optimal adhesion of NPs being applied over deteriorated stone substrates is an important challenge to reach. In this regard, the use of different systems composed of polymeric matrices and NPs of various concentrations (%w/w) has been the focus of many research works in this field [105–107], in order to obtain coatings with multifunctional properties (hydrophobic, consolidant, and antifungal action) for the stone heritage preservation.

The identification and quantification of microbial communities which are colonizing on stone substrates are crucial steps to implement proper treatments. In this context, different analytical techniques have been used depending on the type of microorganism [108–110]. Moreover, the analysis of affected stone substrates by in situ electron microscopy and fluorescence microscopy, sometimes in conjunction with energy-dispersive X-ray spectroscopy, is commonly used [111–113]. SEM and confocal laser scanning microscopy (CLSM) can also be used to study the biodeterioration of stone substrates and/or determine the effectiveness of antifungal treatments based on nanoparticles in stone specimens [114]. Besides, the combination of TEM and SEM is also widely used in order to identify the microorganism types involved in the biodeterioration of stone and their interaction with the untreated and treated stone surfaces [30, 31, 115]. These powerful analytical techniques present the important advantage of evaluating the effectiveness of nanomaterials as antifungal protective coatings without extracting the microorganisms from the stones and interfering with them [115].

Recently, the analysis of biodeterioration and microbial populations has been developed in parallel with the application of molecular techniques based on phylogenetic descriptions for the study of biodiversity in artworks and historical buildings; these techniques can be used to evaluate the effectiveness after the application of the NPs and the temporal stability. Among the different techniques, RT-PCR (reverse transcription-polymerase chain reaction) is typically used. However, in the field of cultural heritage, preservation is still little due to the constraints originated by the sampling methods, the reproducibility, and/or the physiological niches, among others [116, 117]. Moreover, it is important to consider that these techniques present some risks, such as the inability to detect some organisms, reaching in some cases an underestimation of the colonizing population [116]. Therefore, the complementation of these studies with other analytical techniques, such as traditional microbiological techniques and the monitoring of the physiological activity of the microorganisms on and in materials, is absolutely necessary [118]. Besides, further studies are warranted to optimize protocols concerning sample procedures, extraction, PCR (polymerase chain reaction), and PCR amplifications to determine the best techniques to study the microbial communities present in stone heritage [116].

It is important to note that recent genomic and transcriptomic technology offers new solutions to study and understand the activity of whole microbial communities and also the effect of biocide treatments [118, 119]. The development of new cultivation techniques to describe the biodiversity and behavior of microorganisms colonizing the stone cultural heritage is another important challenge to reach [120].

6.4   Conclusion

The design, development, and application of inorganic NPs for the conservation of built heritage constitute an important advance in the application of nanotechnology for stone heritage preservation. Due to the reduced particle size, these advanced nanomaterials present new and enhanced properties, representing a significant improvement on traditional conservation treatments. The key to success is represented by a multidisciplinary research activity, combining material science, petrophysics, ecology, biology, chemistry, and physics and/or cultural heritage conservation. This synergy of different scientific disciplines opens the possibility to engineer materials that are highly compatible, effective, and durable for stone preservation. However, despite all the efforts done so far, the research is still at the beginning of a fascinating endeavor. It is necessary to study a great number of factors that can influence the effectiveness of innovative nanomaterials as consolidating and antifungal protective treatments for stone works of art. Thereby, although these new nanomaterials have been extensively characterized and applied for cultural heritage preservation, little is known about their behavior, effectiveness, and durability in an outdoor environment. Due to the high complexity of real aged stone surfaces, future research should focus on the analysis of the long-term treatment effectiveness after the application on real cases of stone building surfaces. More studies are needed in order to address the stability behavior of different inorganic nanoparticles being exposed to different thermo-hygrometric conditions, with emphasis on their effect on the crystallization kinetics and the development of structural modifications. These studies should be conducted with the aid of advanced characteristic techniques, in multidisciplinary approaches. The advances in the optimization and application of new analytical nondestructive techniques are expected to strengthen the understanding and consequently the effectiveness of new nanostructured materials specifically designed for stone heritage. Moreover, further studies should be centered on exploring the development of hybrid nanomaterials in order to create multifunctional coatings able to exhibit several properties simultaneously (e.g., self-cleaning and biocidal), simplifying and improving the conservative treatments, without visual impact on the surface. Future research is also necessary to shed light on the human health risks and environmental implications resulting from the use of new nanomaterials for stone cultural heritage preservation.

Acknowledgment These studies were supported by the Community of Madrid under the Geomaterials 2 Programme (S2013/MIT-2914) and Multimat-Challenge (S2013/MIT-2862), the Innovation and Education Ministry (Climortec, BIA2014-53911-R, MAT2016-80875-C3-3-R, and MAT2013-47460-C5-5-P), and the FOMIX-Yuc 2008-108160, CONACYT LAB-2009-01-123913, 188345, and Fronteras de la Ciencia No. 138. A.S-F would like to gratefully acknowledge the financial support of Santander Universidades through "Becas Iberoamérica Jóvenes Profesores e Investigadores, España 2015" Scholarship Program.

References

1. Ruedrich J. Physical weathering of building stones induced by freeze-thaw action: a laboratory long-term study. Environ Earth Sci. 2011;63:1573–86.
2. Ákos T, Licha T, Simon K. Urban and rural limestone weathering; the contribution of dust to black crust formation. Environ Earth Sci. 2011;63:675–93.
3. Steiger M, Charola AE, Sterflinger K. Weathering and deterioration. In: Siegesmund S, Snethlage R, editors. Stone in architecture: properties, durability. Heidelberg: Springer, 2011. p. 227–316.
4. Cardell C, Delalieux F, Roumpopoulos A, Moropoulou A, Auger F, Van Grieken R. Salt-induced decay in calcareous stone monuments and buildings in a marine environment in SW France. Const Build Mater. 2003;17:165–79.
5. Scheerer S, Ortega-Morales O, Gaylarde C. Microbial deterioration of stone monuments-an updated overview. Adv Appl Microbiol. 2009;66:97–139.
6. Jain A, Bhadauria S, Kumar V, Chauhan RS. Biodeterioration of sandstone under the influence of different humidity levels in laboratory conditions. Build Environ. 2009;44:1276–84.
7. Campagna BA, Kumar R, Kumar AV. Biodeterioration of stone in tropical environments: an overview. Madison: The J. Paul Getty Trust; 2000.
8. Doehne E, Price CA. Stone conservation. An overview of current research. 2nd ed. Los Angeles: Getty Publications; 2010.
9. Delgado Rodrigues J. Consolidation of decayed stones. A delicate problem with few practical solutions. Proc Int Semin Hist Constr. 2001:3–14.
10. Dorniecen T, Gorbushina AA, Krumbein WE. Biodecay of cultural heritage as a space/time-related ecological situation-An evaluation of a series of studies. Int Biodeterior Biodegr. 2000;46:261–70.
11. Sterflinger K. Fungi: their role in deterioration of cultural heritage. Fungal Biol Rev. 2010;24:47–55.
12. Murty BS, Shankar P, Raj B, Rath BB, Murday J. Unique properties of nanomaterials. In: Murty BS, Shankar P, Raj B, Rath BB, Murday J, editors. Textbook of nanoscience and nanotechnology. Heidelberg: Springer; 2013. p. 29–65.
13. Guo D, Xie G, Luo J. Mechanical properties of nanoparticles: basics and applications. J Phys D Appl Phys. 2013;47:1–25.
14. Savage T, Rao AM. Thermal properties of nanomaterials and nanocomposites. In: Tritt TM, editor. Thermal conductivity. Physics of solids and liquids. Boston: Springer; 2004. p. 261–84.
15. La Russa MF, Macchia A, Ruffolo SA, De Leo F, Barberio M, Barone P, Crisci GM, Urzi C. Testing the antibacterial activity of doped TiO_2 for preventing biodeterioration of cultural heritage building materials. Int Biodeterior Biodegrad. 2014;96:87–96.
16. Munafò P, Battista G, Quaglirini E. TiO_2-based nanocoatings for preserving architectural stone surfaces: an overview. Const Build Mater. 2015;84:201–18.

17. Colangiuli D, Calia A, Bianco N. Novel multifunctional coatings with multifunctional coat-
    ings with photocatalytic and hydrophobic properties for the preservation of the stone building
    heritage. Const Build Mater. 2015;93:189–96.
18. Natali I, Tomasin P, Becherini F, Bernardi A, Ciantelli C, Favaro M, Favoni O, Forrat Pérez
    VJ, Olteanu ID, Romero Sanchez MD, Vivarelli A, Bonazza A. Innovative consolidating
    products for stone materials: field exposure tests as a valid approach for assessing durability.
    Heritage Sci. 2015;3:6.
19. Graziani G, Sassoni E, Franzoni E. Consolidation of porous carbonate stones by an innova-
    tive phosphate treatment: mechanical strengthening and physical-microstructural compatibil-
    ity in comparison with TEOS-based treatments. Heritage Sci. 2015;3:1.
20. Sierra-Fernandez A, Gomez-Villalba LS, Rabanal ME, Fort R. New nanomaterials for
    applications in conservation and restoration of stony materials: a review. Mater Constr.
    2017;67:107.
21. Buzea C, Pacheco I. Nanomaterials and their classification. In: Kumar Shukla A, editor.
    EMR/ESR/EPR spectroscopy for characterization of nanomaterials. India: Springer; 2017.
    p. 3–45.
22. Daraio C, Jin S. Synthesis and patterning methods for nanostructures useful for biologi-
    cal applications. In: Silva G, Parpura V, editors. Nanotechnology for biology and medicine.
    Fundamental biomedical technologies. New York: Springer; 2012. p. 27–44.
23. Chelazzi D, Poggi G, Jaidar Y, Toccafondi N, Giorgi R, Baglioni P. Hydroxide nanoparticles
    for cultural heritage: consolidation and protection of wall paintings and carbonate materials.
    J Colloid Interface Sci. 2013;392:42–9.
24. Öner D, McCarthy TJ. Ultrahydrophobic surfaces. Effects of topography length scales on
    wettability. Langmuir. 2000;20:7777–82.
25. Fan M, Andrade GFS, Brolo AG. A review on the fabrication of substrates for surface
    enhanced Raman spectroscopy and their applications in analytical chemistry. Anal Chim
    Acta. 2011;693:7–25.
26. Ali G, Kim HJ, Kum JM, Cho SO. Rapid synthesis of TiO_2 nanoparticles by electrochemical
    anodization of a Ti wire. Nanotechnology. 2013;24:185601.
27. Giorgi R, Dei L, Ceccato M, Schettino C, Baglioni P. Nanotechnologies for conservation of
    cultural heritage: paper and canvas deacidification. Langmuir. 2002;18:8198–203.
28. Poggi G, Toccafondi N, Melita LN, Knowles JC, Bozec L, Giorgi R, Baglioni P. Calcium
    hydroxide nanoparticles for the conservation of cultural heritage: new formulations for the
    deacidification of cellulose-based artifacts. Appl Phys A Mater Sci Process. 2014;114:685–93.
29. Sierra-Fernandez A, Gomez-Villalba LS, Milosevic O, Fort R, Rabanal ME. Synthesis and
    morphostructural characterization of nanostructured magnesium hydroxide nanostructured
    magnesium hydroxide obtained by a hydrothermal method. Ceram Int. 2014;40:12285–92.
30. Gómez-Ortiz N, De la Rosa-García S, González-Gómez W, Soria-Castro M, Quintana P,
    Oskam G, Ortega-Morales B. Antifungal coatings based on $Ca(OH)_2$ mixed with ZnO/
    TiO_2 nanomaterials for protection of limestone monuments. ACS Appl Mater Interfaces.
    2013;5:1556–65.
31. Sierra-Fernandez A, De la Rosa-García S, Gomez-Villalba LS, Gómez-Cornelio S, Rabanal
    ME, Fort R, Quintana P. Synthesis, photocatalytic and antifungal properties of MgO, ZnO
    and Zn/Mg Oxide Nanoparticles for the protection of calcareous stone heritage. ACS Appl
    Mater Interfaces. 2017;9:24873–86.
32. Taglieri G, Daniele V, Del Re G, Volpe R. A new and original method to produce $Ca(OH)_2$
    nanoparticles by using and anion exchange resin. Adv Nanopart. 2015;4:17–24.
33. Cueto M, Sanz M, Ouija M, Gámez F, Martínez-Haya B, Castillejo M. Platinum nanopar-
    ticles prepared by laser ablation in aqueous solutions: fabrication and application to laser
    desorption ionization. J Phys Chem C. 2011;45:22217–24.
34. Saoud KM, Ibala I, El Ladki D, Ezzeldeen O, Saeed S. Microwave assisted preparation of
    calcium hydroxide and barium hydroxide nanoparticles and their application for conservation
    of cultural heritage. In: Ionnides M, Magnenat-Thalmann N, Fink E, Zarnic R, Yen AY, Quak

E, editors. Digital heritage. Progress in cultural heritage: documentation, preservation, and protection. Switzerland: Springer; 2014. p. 342–52.

35. Flores G, Carrillo J, Luna JA, Martínez R, Sierra-Fernandez A, Milosevic O, Rabanal ME. Synthesis, characterization and photocatalytic properties of nanostructured ZnO particles obtained by low temperature air-assisted-USP. Adv Powder Tecnol. 2014;25:1435–41.

36. Gregorczyk K, Knez M. Hybrid nanomaterials through molecular and atomic layer deposition: top down, bottom up, and in-between approaches to new materials. Prog Mater Sci. 2016;75:1–37.

37. Marquardt AE, Breitung EM, Drayman-Weisser T, Gates G, Phaneuf RJ. Protecting silver cultural heritage objects with atomic layer deposited corrosion barriers. Heritage Sci. 2015;3:37.

38. Taglieri G, Mondelli C, Daniele V, Pusceddu E, Trapananti A. Synthesis and X-ray diffraction analyses of calcium hydroxide nanoparticles in aqueous suspension. Adv Mater Phys Chem. 2013;3:108–12.

39. Rodriguez-Navarro C, Suzuki A, Ruiz-Agudo E. Alcohol dispersions of calcium hydroxide nanoparticles for stone conservation. Langmuir. 2013;29:11457–70.

40. Narayanan T, Wacklin H, Konovalov O, Lund R. Recent applications of synchrotron radiation and neutrons in the study of soft matter. Crystallogr Rev. 2017;23:160–226.

41. Gomez-Villalba LS, López-Arce P, Alvarez de Buergo M, Fort R. Structural stability of a coloidal solution of Ca(OH)$_2$ nanocrystals exposed to high relative humidity conditions. Appl Phys A Mater Sci Process. 2011;104:1249–54.

42. Ciliberto E, Condorelli GG, La Delfa S, Viscuso E. Nanoparticles of Sr(OH)$_2$: synthesis in homogeneous phase at low temperature and application for cultural heritage artifacts. Appl Phys A Mater Sci Process. 2008;92:37–141.

43. Taglieri V, Daniele V, Macera L, Mondelli C. Nano Ca(OH)$_2$ synthesis using a cost-effective and innovative method: reactivity study. J Am Ceram Soc. 2017;100(12):5766–78.

44. Rodriguez-Navarro C, Ruiz-Agudo E, Ortega Huertas M, Hansen E. Nanostructure and irreversible behavior of Ca(OH)$_2$: implications in cultural heritage conservation. Langmuir. 2005;24:10948–57.

45. Kumar SSR. Transmission electron microscopy characterization of nanomaterials. Berlin: Springer; 2014.

46. Ditaranto N, Van der Werf ID, Picca RA, Sportelli MC, Giannossa LC, Bonerba E, Tantillo G, Sabbatini L. Characterization and behaviour of ZnO-based nanocomposites designed for the control of biodeterioration of patrimonial stonework. New J Chem. 2015;39:6836–43.

47. Licchelli M, Malagoudi M, Weththimuni M, Zanchi C. Nanoparticles for conservation of bio-calcarenite stone. Appl Phys A Mater Sci Process. 2014;114:673–83.

48. De Ferri L, Lottici PP, Lorenzi A, Montenero A, Salvioli-Mariani E. Study of silica nanoparticles-polysiloxane hydrophobic treatments for Stone-based monument protection. J Cult Herit. 2011;12:356–63.

49. Willian DB, Carter CB. The transmission electron microscope. In: Willian DB, Carter CB, editors. Transmission electron microscopy. New York: Springer; 2009. p. 3–22.

50. Gomez-Villalba LS, Sierra-Fernandez A, Rabanal ME, Fort R. TEM-HRTEM study on the dehydration process of nanostructured Mg-Ca hydroxide into Mg-Ca oxide. Ceram Int. 2016;42:9455–66.

51. Bellot-Gurlet L, Dillmann P, Neff D. From archaeological sites to nanoscale: the quest of tailored analytical strategy and modelling. In: Dillmann P, Bellot-Gurlet L, Nenner I, editors. Nanoscience and cultural heritage. Paris: Atlantis Press; 2016. p. 205–30.

52. Manoudis P, Papadopoulou S, Karapanagiotis I, Tsakalof A, Zuburtikudis I, Panayiotou C. Polymer-Silica nanoparticles composite film as protective coatings for stone-based monuments. J Phys Conf Ser. 2007;61:1361.

53. Rodriguez-Navarro C, Jroundi F, Schiro M, Ruiz-Agudo E, González-Muñoz MT. Influence of substrate mineralogy on bacterial mineralization of calcium carbonate: implications for stone conservation. Appl Environ Microbiol. 2012;78:4017–29.

54. Lin PC, Lin S, Wang PC, Sridhar R. Techniques for physicochemical characterization of nanomaterials. Biotechnol Adv. 2014;32:711–26.
55. Xiao Y, Gao F, Fang Y, Tan Y. Dispersions of surface modified calcium hydroxide nanoparticles with enhanced kinetic stability: properties and applications to desalination and consolidation of the Yungang Grottoes. Mater Res Soc Symp Proc. 2015;1656.
56. Giorgi R, Baglioni M, Berti D, Baglioni P. New methodologies for the conservation of cultural heritage: micellar solutions, microemulsions, and hydroxide nanoparticles. Acc Chem Res. 2010;43:695–704.
57. Giorgi R, Ambrosi M, Toccafondi N, Baglioni P. Nanoparticles for cultural heritage conservation: calcium and barium hydroxide nanoparticles for wall painting consolidation. Chem Eur J. 2010;16:9374–82.
58. Ziegenbald G. Colloidal calcium hydroxide: a new material for consolidation and conservation of carbonate stone. In: 11th International Congress on Deterioration and Conservation of Stone III. 2008. p. 1119.
59. López-Arce P, Gomez-Villalba LS, Pinho L, Fernández Valle ME, Álvarez de Buergo M, Fort R. Influence of porosity and relative humidity on consolidation of dolostone with calcium hydroxide nanoparticles: effectiveness assessment with non-destructive techniques. Mater Char. 2010;61:168–84.
60. López-Arce P, Gomez-Villalba LS, Martinez-Ramírez S, Álvarez de Buergo M, Fort R. Influence of relative humidity on the carbonation of calcium hydroxide nanoparticles and the formation of calcium carbonate polymorphs. Powder Technol. 2011;205:263–9.
61. Rodríguez-Navarro C, Elert K, Sevcik R. Amorphous and crystalline calcium carbonate phases during carbonation of nanolimes: implications in heritage conservation. Cryst Eng Comm. 2016;35:6594–607.
62. Gomez-Villalba LS, López-Arce P, Fort R. Nucleation of $CaCO_3$ polymorphs from a coloidal alcoholic suspension of $Ca(OH)_2$ nanocrystals exposed to low humidity conditions. Appl Phys A Mater Sci Process. 2012;106:213–7.
63. Baglioni P, Chelazzi D, Giorgi R. Nanotechnologies in the conservation of cultural heritage: a compendium of materials and techniques. Dordrecht: Springer; 2014.
64. Gomez-Villalba LS, López-Arce P, Álvarez de Buergo M, Fort R. Atomic defects and their relationship to aragonite-calcite transformation in portlandite nanocrystal carbonation. Cryst Growth Des. 2012;12:4844–52.
65. Gomez-Villalba LS, López-Arce P, Álvarez de Buergo M, Zornoza-Indart A, Fort R. Mineralogical and textural considerations in the assessment of aesthetic changes in dolostones by effect of treatments with $Ca(OH)_2$ nanoparticles. In: Rogerio-Candelera MA, Lazzari M, Cano E, editors. Science and technology for the conservation of cultural heritage. London: CRC Press; 2013. p. 235–329.
66. Gomez-Villalba LS, López-Arce P, Zornoza-Indart A, Álvarez de Buergo M, Fort R. Evaluation of a consolidation treatment in dolostones by mean of calcium hydroxide nanoparticles in high relative humidity conditions. Bol Soc Esp Ceram V. 2011;50:85–92.
67. Borsoi G, Lubelli B, Van Hees R, Veiga R, Santos Silva A. Evaluation of the effectiveness and compatibility of nanolime consolidants with improved properties. Constr Build Mater. 2017;142:385–94.
68. Borsoi G, Lubelli B, Van Hees RPJ, Tomasin P. Effect of solvent on nanolime transport within limestone: how to improve in-depth deposition. Colloids Surf A Physicochem Eng Asp. 2016;497:171–81.
69. López-Arce P, Zornoza-Indart A. Carbonation acceleration of calcium hydroxide nanoparticles: induced by yeast fermentation. Apply Phys A. 2015;120:1475–95.
70. Rodriguez-Navarro C, Vettori I, Ruiz-Agudo E. Kinetics and mechanism of calcium hydroxide conversion into calcium alkoxides: implication in heritage conservation using nanolimes. Langmuir. 2013;32:5183–94.
71. Warscheid T, Braams J. Biodeterioration of stone: a review. Int Biodeter Biodegr. 2000;46:343–68.
72. Gadd GM. Geomicology: biogeochemical transformations of rocks, minerals, metals and radionuclides by fungi, bioweathering and bioremediation. Mycol Res. 2007;11:3–49.

73. Perry IVTD, McNamara CJ, Mitchell R. Biodeterioration of stone. In: Sackler NAS Colloquium. Scientific examination of art: modern techniques in conservation and analysis. Washington, DC: National Academy of Sciences; 2005. p. 72–84.
74. May-Crespo J, Ortega-Morales BO, Camacho-Chab JC, Quintana P, Alvarado-Gil JJ, Gonzalez-García G, Reyes-Estebanez M, Chan-Bacab MJ. Photoacoustic monitoring of water transport process in calcareous stone coated with biopolymers. Appl Phys A Mater Sci Process. 2016;122:1060–70.
75. Gorbushina AA. Life on the rocks. Environ Microbiol. 2007;9:1613–31.
76. Cámara B, De los Ríos A, García del Cura MA, Galván A, Ascaso C. Dolostone bioreceptivity to fungal colonization. Mater Constr. 2008;58:113–24.
77. Hirsch P, Eckhardt FEW, Palmer RJ Jr. Fungi active in weathering of rock and stone monuments. Can J Bot. 1995;73:1384–90.
78. Gorbushina AA, Krumbein WE, Hamann CH, Panina L, Soukharjevski S. Role of black fungi in color change and biodeterioration of antique marbles. Geomicrobiol J. 1993;11:205–21.
79. Gadd GM. Geomicrobiology of the built environment. Nat Microbiol. 2017;2:16275.
80. Sterflinger K, Krumbein WE. Dematiaceous fungi as a major agent for biopitting on Mediterranean marbles and limestones. Geomicrobiol J. 1997;14:219–30.
81. Fomina M, Burford EP, Hillier S, Kierans M, Gadd G. Rock-building fungi. Geomicrobiol J. 2010;27:624–9.
82. Urzi C, García-Valles MT, Vendrell M, Pernice A. Biomineralization processes of the rock surfaces in field and in laboratory. Geomicrobiol J. 1999;16:39–54.
83. Bellissima F, Bonini M, Giorgi R, Baglioni P, Barresi F, Mastromei G, Perito B. Antibacterial activity of silver nanoparticles grafted on stone surface. Environ Sci Pollut R. 2014;21:13278–86.
84. Essa AM, Khallaf MK. Biological nanosilver particles for the protection of archaeological stones against microbial colonization. Int Biodeterior Biodegr. 2014;94:31–7.
85. Ditaranto N, Loperfido S, Van der Werf I, Mangone A, Cioffi N, Sabbatini L. Synthesis and analytical characterisation of copper-based nanocoatings for bioactive stone artworks treatment. Anal Bioanal Chem. 2011;399:473–81.
86. Ruffolo SA, La Russa MF, Malagodi M, Oliviero Rossi C, Palermo AM, Crisci GM. ZnO and ZnTiO₃ nanopowders for antimicrobial stone coating. Appl Phys A: Mater. 2010;100:829–34.
87. Gambino M, Ahmed MAAA, Villa F, Cappitelli F. Zinc oxide nanoparticles hinder fungal biofilm development in an ancient Egyptian tomb. Int Biodeterior Biodegr. 2017;122:92–9.
88. Lemire JA, Harrison JJ, Turner RJ. Antimicrobial activity of metals: mechanisms, molecular targets and applications. Nature Rev Microbiol. 2013;11:371–84.
89. Wang L, Hu C, Shao L. The antimicrobial activity of nanoparticles: present situation and prospects for the future. Int J Nanomedicine. 2017;12:1227–49.
90. Savi GD, Bortoluzzi AJ, Scussel VM. Antifungal properties of zinc-compounds against toxigenic fungi and mycotoxin. Int J Food Sci Technol. 2013;48:1834–40.
91. Vatansever F, De Melo WCMA, Avci P, Vecchio D, Sadasivam M, Gupta A, Chandran R, Karimi M, Parizotto NA, Yin R, Tegos GP, Hamblin MR. Antimicrobial strategies centered around reactive oxygen species-bactericidal antibiotics, photodynamic therapy and beyond. FEMS Microbial Rev. 2013;37:955–89.
92. Manke A, Wang L, Rojanasakul Y. Mechanisms of nanoparticle-induced oxidative stress and toxicity. Biomed Res Int. 2013;2013:942916.
93. Raghurath A, Perumal E. Metal oxide nanoparticles as antimicrobial agents: a promise for the future. Int J Antimicrob Agents. 2017;49:137–52.
94. Yamamoto O. Influence of particle size on the antibacterial activity of zinc oxide. Int J Inor Mater. 2001;3:643–6.
95. Chung YC, Su YP, Chen CC, Jia G, Wang HL, Wu JC, Lin JG. Relationship between antibacterial activity of chitosan and surface characteristics of cell wall. Acta Pharmacol Sin. 2004;25:932–6.
96. Kumar S, Singh M, Halder D, Mitra A. Mechanistic study of antibacterial activity of biologically synthesized silver nanocolloids. Colloids Surf A Physicochem Eng Asp. 2014;449:82–6.

97. Ivask A, Voelcker NH, Seabrook SA, Hor M, Kirby JK, Fenech M, Davis TP, Ke PC. DNA melting and genotoxicity induced by silver nanoparticles and graphene. Chem Res Toxicol. 2015;28:1023–35.
98. Bonetta S, Bonetta S, Motta F, Strini A, Carraro E. Photocatalytic bacterial inactivation by TiO_2-coated surfaces. AMB Express. 2013;3:59.
99. Viswanath B, Patra S, Munichandraiah N, Ravishankar N. Nanoporous Pt with high surface area by reaction-limited aggregation of nanoparticles. Langmuir. 2009;25:3115–21.
100. Usman MS, El Zowalaty ME, Shameli K, Zainuddin N, Salama M, Ibrahim NA. Synthesis, characterization, and antimicrobial properties of copper nanoparticles. Int J Nanomedicine. 2013;3:4467–79.
101. Noeiaghaei T, Dhami N, Mukherjee A. Nanoparticles surface treatments on cemented materials for inhibition of bacterial growth. Constr Build Mater. 2017;150:880–91.
102. Chitra K, Annadurai G. Antibacterial activity of Ph-dependent biosynthesized silver nanoparticles against clinical pathogen. Biomed Res Int. 2014;2014:725165.
103. Lin JQ, Zhang HW, Chen Z, Zheng YG. Penetration of lipid membranes by gold nanoparticles: insights into cellular uptake, cytotoxicity, and their relationship. ACS Nano. 2010;4(9):5421.
104. Lakshmi Prassana V, Vijayaraghavan R. Insight into the mechanism of antibacterial activity of ZnO: surface defects mediated reactive oxygen species even in the dark. Langmuir. 2015;31:9155–62.
105. Stankic S, Suman S, Haque F, Vidic J. Pure and multi metal oxide nanoparticles: synthesis, antibacterial and cytotoxic properties. J Nanobiotechnol. 2016;14:73.
106. Pinna D, Salvadori B, Galeotti M. Monitoring the performance of innovative and traditional biocides mixed with consolidants and water-repellents for the prevention of biological growth on stone. Sci Total Environ. 2012;423:132–41.
107. Aflori M, Simionescu B, Bordiani I, Olaru M. Silsesquioxane-based hybrid nanocomposites with methacrylate units containing titania and/or silver nanoparticles as antibacterial/antifungal coatings for monumental stones. Mat Si Eng B. 2013;178:1339–46.
108. La Russa MF, Ruffolo SA, Rovella N, Belfiore CM, Palermo AM, Guzzi MT, Crisci GM. Multifunctional TiO_2 coatings for cultural heritage. Prog Org Sci. 2012;74:186–91.
109. Jurado V, Fernandez-Cortes A, Cuezva S, Laiz L, Cañaveras JC, Sanchez-Moral S, Saiz-Jimenez C. The fungal colonisation of rock-art caves: experimental evidence. Naturwissenschaften. 2009;96:1027–34.
110. De Los Ríos A, Pérez-Ortega S, Wierzchos J, Ascaso C. Differential effects of biocide treatments on saxicolous communities: case study of the Segovia cathedral cloister (Spain). Int Biodeterior Biodegr. 2012;67:64–72.
111. Mihajlovsji A, Seyer D, Benamara H, Bousta F, Di Martino P. An overview of techniques for the characterization and quantification of microbial colonization on stone monuments. Ann Microbiol. 2015;65:1243–55.
112. De Los Ríos A, Wierzchos J, Sancho LG, Green TGA, Ascaso C. Ecology of endolithic lichens colonizing granite in continental Antarctica. Lichenologist. 2005;37:383–95.
113. De los Ríos A, Pérez-Ortega S, Wierzchos J, Ascaso C. Differential effects of biocide treatments on saxicolous communities: case study of the Segovia cathedral cloister (Spain). Int Biodeter Biodegr. 2012;67:64–72.
114. De los Ríos A, Ascaso C. Contributions of in situ microscopy to current understanding of stone biodeterioration processes. Int Microbiol. 2005;8:181–8.
115. Schlafer S, Meyer RL. Confocal microscopy imaging of the biofilms matrix. J Microbiol Methods. 2017;138:50–9.
116. Ascaso C, Wierzchos J, Souza-Egipsy V, de los Ríos A, Delgado Rodrigues J. In situ evaluation of the biodeteriorating action of microorganisms and the effects of biocides on carbonate rock of the Jeronimos Monastery (Lisbon). Int Biodeterior Biodegr. 2002;49:1–12.
117. Saiz-Jimenez C. Biodeterioration: an overview of the state of the art and assessment of future directions. 2003. Available from: http://webcache.googleusercontent.com/search?q=cache:jxplYfxPpSMJ:www.arcchip.cz/w08/w08_saiz_jimenez.pdf+%22biodeterioration%22,+%22stone%22,+%22mechanism%22&hl=pl&ie=UTF-8.

118. Portillo MC, Saiz-Jimenez C, Gonzalez JM. Molecular characterization of total and meta-bolically active bacterial communities of "white colonizations" in the Altamira Cave, Spain. Res Microbiol. 2009;160:41–7.
119. Sterflinger K, Piñar G. Microbial deterioration of cultural heritage and works of art-tiling at windmills? Appl Microbiol Biotechnol. 2013;97:9637–46.
120. Simon C, Daniel R. Metagenomic analyses: past and future trends. Appl Environ Microbiol. 2011;77 1153–61.

Chapter 7
Nanomaterials for the Consolidation of Stone Artifacts

David Chelazzi, Rachel Camerini, Rodorico Giorgi, and Piero Baglioni

7.1 Inorganic Nanomaterials

Building and artistic stone materials are typically exposed to various weathering and degradation agents including physical erosion, chemical corrosion, bio-pollution, and even detrimental materials applied in past restoration interventions. As a result, the mechanical strength of stone is lost progressively from the surface inwards, requiring the use of appropriate consolidants in order to recover the cohesion of the artifacts' layers.

Organic materials such as polymeric coatings and adhesives have been widely used in the conservation practice since the 1960s, and are still employed by restorers owing to their short-term performances (e.g., high adhesive power, saturation of colors, easy application, low costs). However, it is well known that the use of synthetic polymers such as acrylates, polyvinyl acetate, and epoxy resins alters the original physicochemical properties of porous stone materials, producing detrimental effects on a time scale that ranges from 5–10 to 50 years [1]. In particular, the decreased porosity and permeability of stone following treatment with these coatings causes the nucleation and growth of salts at the interface between the polymer layer and the stone surface. Consequently, the growing salt crystals induce high pressure in the pores of stone, directly proportional to the size of the crystals and to the concentration of salts and inversely proportional to the pore radius [2, 3]. This eventually leads to the flaking and disaggregation of the stone surface layers and the disruption of the polymer coating (Fig. 7.1). Moreover, the polymer coatings can undergo degradation resulting in discoloration and change in solubility (due to both cross-linking and chain scissions) [5], which may hinder their removal with the

D. Chelazzi • R. Camerini • R. Giorgi (✉) • P. Baglioni (✉)
Department of Chemistry Ugo Schiff and CSGI, University of Florence,
Via della Lastruccia 3, Sesto Fiorentino, Florence, Italy
e-mail: giorgi@csgi.unifi.it; baglioni@csgi.unifi.it

© Springer International Publishing AG 2018
M. Hosseini, I. Karapanagiotis (eds.), *Advanced Materials for the Conservation of Stone*, https://doi.org/10.1007/978-3-319-72260-3_7

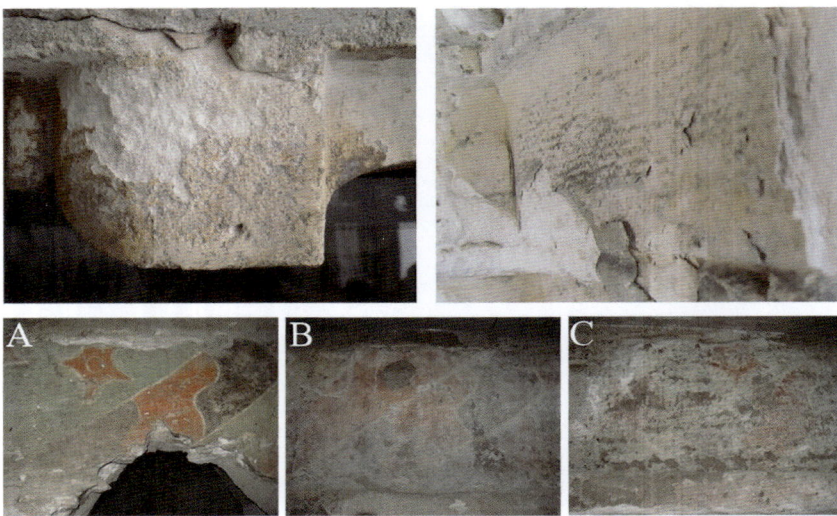

Fig. 7.1 (Top) Typical degradation of stone. Exfoliation and detachment of the outermost layers are manifest. Reprinted with permission from reference number [4], Copyright, Elsevier Inc. (Bottom) Progressing degradative steps (from **a** to **c**) of a wall painting in Cholula (Mexico) treated with a combination of acrylic and vinyl polymers. Polymers led to the complete disruption of the painting in a few years after the application. Reproduced with permission from: "Removal of acrylic coatings from works of art by means of nanofluids: understanding the mechanism at the nanoscale", Michele Baglioni, Doris Rengstl, Debora Berti, Massimo Bonini, Rodorico Giorgi and Piero Baglioni, Nanoscale, 2010, 2, 1723–1732, Copyright, the Royal Society of Chemistry

original solvents used for their application. Finally, many of these solvents are in fact toxic.

While soft nanomaterials such as microemulsions and gels have been developed to safely remove detrimentally aged coatings from the surface of works of art [1], hard nanomaterials have been explored since the end of the 1990s as an alternative to polymers. Namely, this chapter is focused on nanomaterials for the consolidation of stone, without discussing consolidants made of bulk materials or systems used solely for the protection of stone (e.g., self-cleaning).

In particular, dispersions of calcium hydroxide ($Ca(OH)_2$) nanoparticles in alcohols were first developed and applied for the consolidation of wall paintings and mortars, owing to their high compatibility with the original works of art. These particles are usually hexagonal platelets of portlandite, with diameters ranging from 100 to 400 nm and a thickness of 2–40 nm, depending on the adopted synthetic route. The dispersions have a particle concentration that typically ranges from 5 to 25 g/L (up to 50 g/L) and appear as a "milky" fluid that can be applied by brushing, spraying, or immersion. Upon the evaporation of the solvent, the particles deposit in the pore of the artifact and react with carbon dioxide in the presence of water (i.e., faster reaction when relative humidity (RH) is 75–90%, slower when RH is

30–50%), turning into calcium carbonate ($CaCO_3$) via a dissolution–precipitation process [6]. The resulting calcium carbonate network acts as a binder, bridging the pores' walls and improving the mechanical properties of the substrate. Small-angle neutron scattering (SANS) studies have confirmed that carbonation of $Ca(OH)_2$ leads to the formation of bigger aggregates in the stone porous matrix and reduces sharpness (originally formed by cracks during weathering of stone) at the interfaces [7]. The permeability to water vapor is not lost in the process, as opposed to what happens typically when polymer coatings are applied onto the surface [8]. Because $CaCO_3$ is the original binder found in many murals (e.g., fresco paintings) and the main constituent of carbonate stone (e.g., marble, limestone), the physicochemical properties of the treated artifacts are not altered, which minimizes the risk of undesired effects in the long term.

In fact, calcium hydroxide has already been used in the conservation practice, but as an aqueous solution (limewater). Owing to the low solubility of $Ca(OH)_2$ in water (1.7 g/L at 20 °C), it is necessary to use numerous applications in order to achieve some consolidation effect. This is not practical, as uploading large amounts of water in stone porous matrices can favor freeze-thaw cycles (leading to mechanical stress), the solubilization (and subsequent migration and recrystallization) of salts, and the growth of microorganisms. Using dispersions of $Ca(OH)_2$ rather than solutions allows this limit to be overcome. However, aqueous dispersions of calcium hydroxide particles are not stable, as adsorbed water molecules favor the stacking of particles through hydrogen bonds, leading to sedimentation. This prevents the penetration of the particles which accumulate on the treated surfaces, producing white glazes.

On the other hand, synthesizing $Ca(OH)_2$ as solid nanoparticles allows its stable dispersion in low-toxic solvents such as short-chain alcohols (e.g., ethanol, propanol) without the need of any stabilizer. The reduced size of the particles favors their penetration through the porous matrix of stone and increases their reactivity to CO_2. Moreover, the application of solid particles in the nonaqueous phase avoids the risk of altering alkali-sensitive pigments and components as opposed to the application of free, mobile OH^- ions in aqueous solutions.

In the last decades, numerous synthetic processes have been proposed to obtain $Ca(OH)_2$ nanoparticles specifically designed for the consolidation of cultural heritage objects. Historically, the research work carried out in the framework of colloids and materials science has systematically expanded and enhanced the practical knowledge that mankind had gathered from early ages on the use of lime. Indeed, lime putty, obtained by adding water excess during hydration of CaO, has been employed from ancient times both in architecture and decorative arts. Rodriguez-Navarro et al. [9, 10] have shown that the traditional process of aging putty by long-term storage under water results in the formation of sub-micrometer tabular-shaped platelets of $Ca(OH)_2$ which undergo nonreversible aggregation upon drying of the putty.

The first application reported in the literature regarding the use of $Ca(OH)_2$ particles for the consolidation of artifacts is a work by Giorgi et al. [11] where slaked lime particles (3–4 μm) were dispersed by stirring in 1-propanol and effectively used to provide re-cohesion to the weakened painted layers of murals in Florence, Italy. The stability of the particles is favored by the physisorption of alcohol molecules on the particles' surface, leading to the formation of a hydrophobic layer that hinders face-to-face sticking of the platelets [12, 13]. Baglioni et al. [14] patented this method at the end of the 1990s.

Successive upgrades involved the preparation of lime particles of smaller size, which can be achieved through different synthetic processes, as largely reported in the literature. Particles in the range of tens to a few hundreds of nanometers are more effective than micron-sized particles, allowing deeper penetration and higher stability of the dispersions, and are at lower risk of producing white glaze on the treated surfaces. In 2001, Ambrosi et al. [12, 15] succeeded in preparing a dispersion of $Ca(OH)_2$ platelets in the 100–250 nm range using a homogeneous phase synthesis in supersaturating conditions (aqueous medium). These particles (dispersed in alcohol after the synthesis) were used to provide surface consolidation to historical calcareous stone ("Pietra di Nanto") on the external walls of the St. Margherita Abbey in Vigonza (Padua, Italy). Another treatment was carried out on the bricks of the St. Prisca in Aventino church (Rome), which needed a protective surface layer of 50–100 nm. Following these pioneering works, Croveri et al. [16] tested $Ca(OH)_2$ nanoparticles in 2-propanol for the consolidation of biocalcareous stone (*Globigerina*) from Malta. The particles were obtained via a homogeneous phase synthesis in diols developed by Salvadori and Dei [17]. The rationale of this approach is that in nonaqueous media it is possible to reach temperatures higher than 100 °C, which promote the formation of nano-sized particles. By tuning the synthetic process (i.e., type of solvent, molar ratio of reactants, aging time of precipitated particles), it was possible to obtain either small spherical particles (30–60 nm) or hexagonal platelets (50–150 nm) which were then peptized in 2-propanol to remove the diols adsorbed on the platelets' surface. These particles were used on *Globigerina* stone, obtaining the reinforcement of surface layers while maintaining transpiration (as checked with water vapor permeability tests). This aspect is critical for stone artifacts exposed to marine aerosol, in order to minimize degradation promoted by salt crystallization. The same synthetic route was used to prepare dispersions that were applied to calcareous sandstone (*Gallina* stone, from a historical building in Mantova, Italy) and limestone (*Alberese* stone, from the Santa Maria Church in Impruneta, Florence) as seen in Fig. 7.2 [18]. The treatment was carried out by immersion, followed by quick washing to remove excess consolidant and to prevent glazing. After complete carbonation of the particles in 3 weeks, reaggregation of the powdering surface layers was observed.

The penetration of the particles through these low-porosity matrices was in the order of 80 μm as measured by energy dispersive X-ray (EDX) using magnesium hydroxide nanoparticles as markers mixed with the calcium hydroxide dispersion. The specific superficial area of stone was only slightly reduced as confirmed by water absorption measurements. Despite the encouraging results obtained, the synthesis in diols was later deemed time-consuming, and successive applications used calcium hydroxide nanoparticles prepared through homogenous phase synthesis in aqueous medium, later dispersed in alcohol. The method was patented by Baglioni

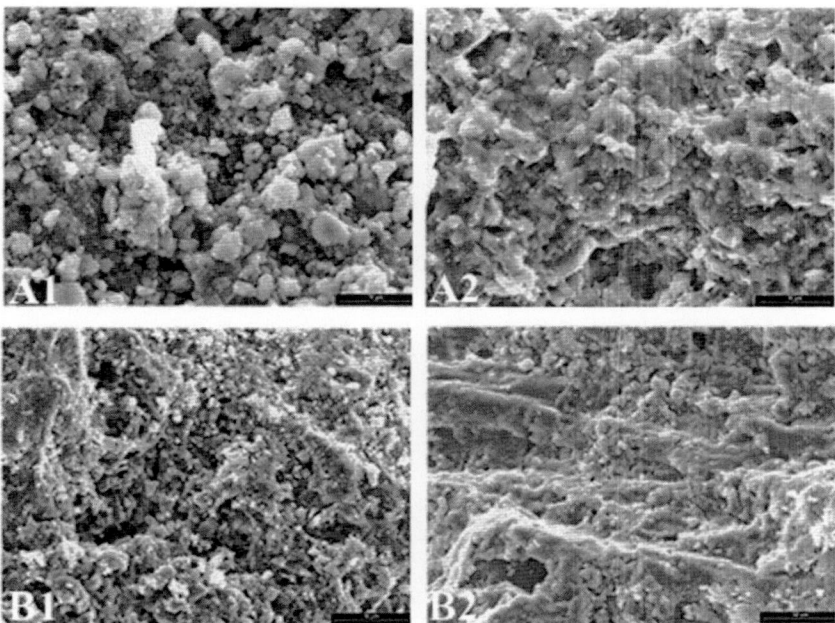

Fig. 7.2 Scanning electron microscopy (SEM) images of Alberese and Gallina stone, respectively, untreated (**A1**, **B1**) and treated with $Ca(OH)_2$ nanoparticles (**A2**, **B2**). The treated stone samples exhibit increased compactness. Bar = 20 micron. Reprinted with permission from reference number [18], Copyright, Elsevier Masson SAS

et al. [19], and dispersions of nanoparticles prepared with this process were also used in non-European settings such as the archaeological Maya site of Calakmul (Mexico), where both wall paintings and limestone were in need of consolidation after being recovered [20]. This synthetic route was refined and upgraded through the years. Besides bottom-up synthetic processes (i.e., where nanoparticles are built atom by atom), top-down approaches were also explored (i.e., reducing materials from bulk size to the nanoscale using energy). For instance, Giorgi et al. [21] carried out a thermomechanical process where the unreacted calcium oxide core of lime particles was hydrated at high temperature and pressure, with the following volume expansion resulting in the fragmentation of the particles. The size distribution and polydispersity of the particles can be tuned by playing on both temperature and pressure. For example, highly crystalline hexagonal portlandite particles of 150–300 nm can be obtained and stably dispersed in 2-propanol without any need of purification or additional treatments, favoring the production of larger batches as opposed to the bottom-up route [21]. Dispersions prepared with this synthetic route were employed for the consolidation of severely degraded dolomite stone ("Pietra d'Angera") of historical buildings (fifteenth and sixteenth centuries) in Milan and Pavia. Because dolomite stone is composed of calcium and magnesium carbonate,

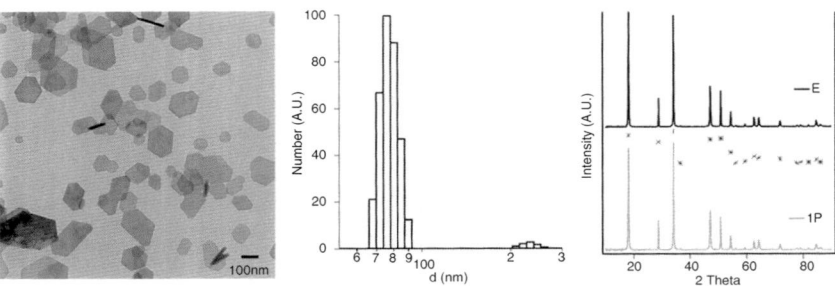

Fig. 7.3 Transmission electron microscopy (TEM) image (left), size distribution obtained by dynamic light scattering (DLS) (center), and X-ray powder diffraction (XRPD) (right) of $Ca(OH)_2$ nanoparticles obtained via a solvothermal synthesis and dispersed in ethanol. Reproduced with permission from: "Calcium hydroxide nanoparticles for the conservation of cultural heritage: new formulations for the deacidification of cellulose-based artifacts", Appl. Phys A (2014),114:685–693, Copyright Springer-Verlag Berlin Heidelberg 2013

it was decided to mix $Ca(OH)_2$ nanoparticles prepared by thermomechanical treatment of lime with $Mg(OH)_2$ nanoparticles obtained through a homogeneous phase reaction in excess of Mg^{2+} [4]. Monitoring of the treated stone indicated that the application of the nanoparticles led to durable preservation.

Besides the results reported above, mostly achieved at CSGI (Florence, Italy), a successive upgrade regarded the use of a variation of a solvothermal alkoxide route for the synthesis of $Ca(OH)_2$ nanoparticles (based on the widely used solvothermal syntheses of oxide nanoparticles), whose main advantage consists in the reproducible and feasible upscaling of the process, with great benefits in terms of costs [22] (Fig. 7.3). These research and assessment activities resulted in the development of two formulations of calcium hydroxide nanoparticles in short-chain alcohols (Nanorestore®, Nanorestore Plus®), both available on the market [23].

Based on this background, other research groups have also contributed to the development and assessment of inorganic nanomaterials for the consolidation of different types of stone.

Lopez-Arce et al. [24] investigated the effectiveness of the Nanorestore® product for the consolidation of dolostone (5–10% calcite, 90–95% dolomite) from Redueña, a Cretaceous geologic formation from the north of Madrid. The stone samples were impregnated with the dispersion, and then the effect of controlled temperature and RH on the consolidation process was monitored using an inclusive set of diagnostic techniques including nuclear magnetic resonance (NMR) imaging and relaxometry, ultrasound velocity, capillarity, water absorption under vacuum, optical surface roughness, and environmental scanning electron microscopy (ESEM) coupled with EDS (Fig. 7.4). The best results were obtained at RH 75% where portlandite transformed into vaterite, monohydrocalcite, and calcite. The newly formed crystals had dimensions that range from the nano- to the micron-size, which results in the improvement of the hydric properties of stone. Consolidation of the treated samples was confirmed by the decrease in porosity, increase in ultrasound velocity

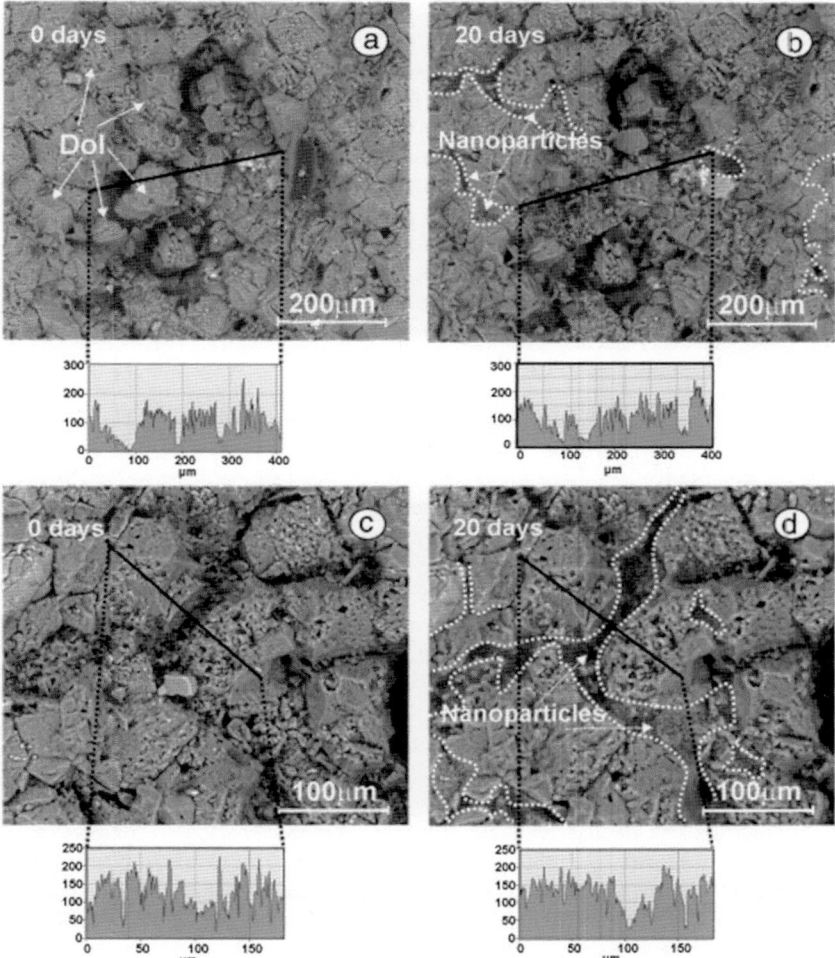

Fig. 7.4 Environmental scanning electron microscopy (ESEM) images of stone specimens before and after consolidation at 75% RH, and the optical line scan profiles obtained after processing of images with DigitalMicrograph software™. (**a**) Pore in specimen submitted to freezing–thawing cycles before consolidation, (**b**) same pore in the specimen 20 days after consolidation, (**c**) contacts between dolomite crystals (Dol), and (**d**) same contacts between dolomite crystals 20 days after consolidation. Reprinted from reference number [24], Copyright, Elsevier Inc

(particularly for higher porosity substrates), and the shift of the relaxometry distribution curve toward shorter times (indicating the filling of big pores and cracks).

Moreover, no alteration of the stone color was detected after treatment. As expected, consolidation at RH 33% resulted in slower carbonation rates and the formation of smaller particles that agglomerate up to micron-size after ca. 3 weeks. In these conditions, calcite recrystallization was observed, but the process also led to the dissolution and fracturing of the dolomite crystals. Therefore, it was advised that the treatment should be carried out at higher RH. As mentioned above, mixed $Ca(OH)_2$ and $Mg(OH)_2$ nanoparticle dispersions have been used on dolomite stone to increase the compatibility of the treatment with no reported issues, which opens perspectives for more research work on this application.

Daniele et al. [25, 26] applied $Ca(OH)_2$ nanoparticle dispersions to natural lithotypes, including *Estoril* and *Pietra Serena* (a gray sandstone). The particles were prepared either via the original homogeneous phase synthesis by Ambrosi et al. [15] or through variations of this method. The effectiveness of the consolidation treatment was confirmed by capillarity, imbibition, and Scotch Tape tests. The particles penetrated up to 30 μm in the case of *Estoril* and up to 1 mm through *Pietra Serena*, which has a larger average pore size. Following consolidation, scanning electron microscopy (SEM) showed the partial filling of macropores in *Pietra Serena* stone. The authors also investigated the effect of particles' concentration and residual water content in the 2-propanol nanolime dispersions. It was observed that the efficiency of the carbonation process increased from 1 to 70% by varying the water/2-propanol ratios from 0.01 to 1, respectively. Surface consolidation was also demonstrated on basalt.

Borsoi et al. [27] applied $Ca(OH)_2$ nanoparticles prepared via solvothermal reaction and dispersed in ethanol to limestone and assessed consolidation by drilling resistance measurement system (DRMS), open porosity and water absorption measurements, as well as through the evaluation of chromatic alteration. Results showed that multiple applications were possible on de-cohered limestone (i.e., with coarse pores) without inducing white glazing, and consolidation was observed up to ca. 16 mm in depth with only moderate alteration of the total porosity and moisture transport in treated stone. The authors also conceived a conceptual model to correlate the drying rate and kinetic stability of nanoparticles in different solvents to the porosity of limestone, measured by mercury intrusion porosimetry (MIP) [28]. Essentially, by comparing the absorption and drying kinetics of different solvents in limestone with the stability of particles in each solvent, predictions were made on the penetration of the particles through coarse or fine porous matrices. For instance, the high stability of ethanol and 2-propanol dispersions and the fast drying rate of these solvents were expected to facilitate back migration of the particles to the drying surface. According to the model, solvents with a higher boiling point (e.g., butanol, water) are expected to produce deeper penetration of the particles in coarse porous stone, while solvents with lower boiling point (e.g., ethanol, 2-propanol) should grant deeper penetration of the particles in fine porous limestone. Experimental data confirmed that butanol dispersions could penetrate up to 20–25 mm through Maastricht limestone without back migration of the particles to the drying front, while ethanol dispersions showed fairly good penetration through fine porous Migné limestone. Further investigation will be required to understand how environmental conditions, the application method (e.g., spraying, impregnation, etc.), and the amount of consolidant deposited affect the transport and penetration of the particles.

Ziegenbalg [29] developed a heterogeneous synthesis of narrow-dispersed calcium hydroxide nanoparticles via a solvothermal method, and the formulation has been produced since October 2006 under the commercial name of CaLoSiL® (by IBZ-Salzchemie GmbH & Co. KG, Germany). This product has been used for consolidating different substrates including murals, mortars, and various types of stone (e.g., limestone, sandstone, and tuff) with positive results. Literature reports several case studies and recommendations on the best protocols to maximize the consolidation effect [29–32]. Recently, Lanzon et al. [33] showed that surface consolidation could be obtained also on adobe bricks using $Ca(OH)_2$ dispersions in 2-propanol, following an approach where consecutive applications of diluted dispersions are carried out, rather than fewer treatments with concentrated suspensions.

Rodriguez-Navarro et al. [34] showed that effective consolidation with $Ca(OH)_2$ nanoparticles could also be achieved on non-calcareous stone using a commercial product (Nanorestore®) or particles made either from aged lime putty or carbide lime putty. The carbide lime was prepared using a patented purification process, and the final product exhibited physicochemical properties similar to aged slaked lime putty [35]. In this case, the particles made from putty gave the best consolidation results, possibly owing to the polydispersity of this system (featuring both nano- and micron-sized particles) that could match the presence of fine and coarse pores in the stone matrix [21]. The effective consolidation of sandstone by calcium hydroxide nanoparticles (Fig. 7.4) was ascribed to the reaction of $Ca(OH)_2$ with SiO_2 grains, leading to the formation of gel-like calcium silicate domains (either amorphous or poorly crystalline) that link calcium carbonate (from carbonated lime) and silicate minerals (Fig. 7.5).

Insights on the carbonation kinetics of $Ca(OH)_2$ nanoparticles and on the evolution of the calcium carbonate phases are also provided in literature [6, 36]. Essentially, the particles react with the short-chain alcohols to pseudomorphically form calcium alkoxides, following a 3D diffusion controlled deceleratory advancement via an interface-coupled dissolution–precipitation process. The transformation shows Arrhenius behavior. During carbonation, amorphous calcium carbonate (ACC) is pseudomorphically formed from $Ca(OH)_2$. The process follows the Ostwald's step rule, with the sequence ACC → vaterite → aragonite → calcite, and the kinetic data were fitted with a deceleratory (pseudo)first-order model with no induction time, where the rate limiting step is the amount of unreacted $Ca(OH)_2$. The formation of vaterite and aragonite is likely to be favored because the hydrolysis of alkoxides releases alcohol, leading to the formation of a hydroalcoholic solution on the portlandite surface during carbonation. Vaterite grows after ACC via a nonclassical nanoparticle-mediated process, starting with heterogeneous nucleation of vaterite nanoparticles that aggregate into nearly iso-oriented structures. Vaterite and aragonite transform into calcite via a dissolution–precipitation mechanism. Monitoring the process via XRD showed that the formation of the solely crystalline $CaCO_3$ phases (vaterite + calcite + aragonite) was fitted to an Avrami–Erofeev model, with an induction time. In fact, Carretti et al. [8] and Baglioni et al. [23] had

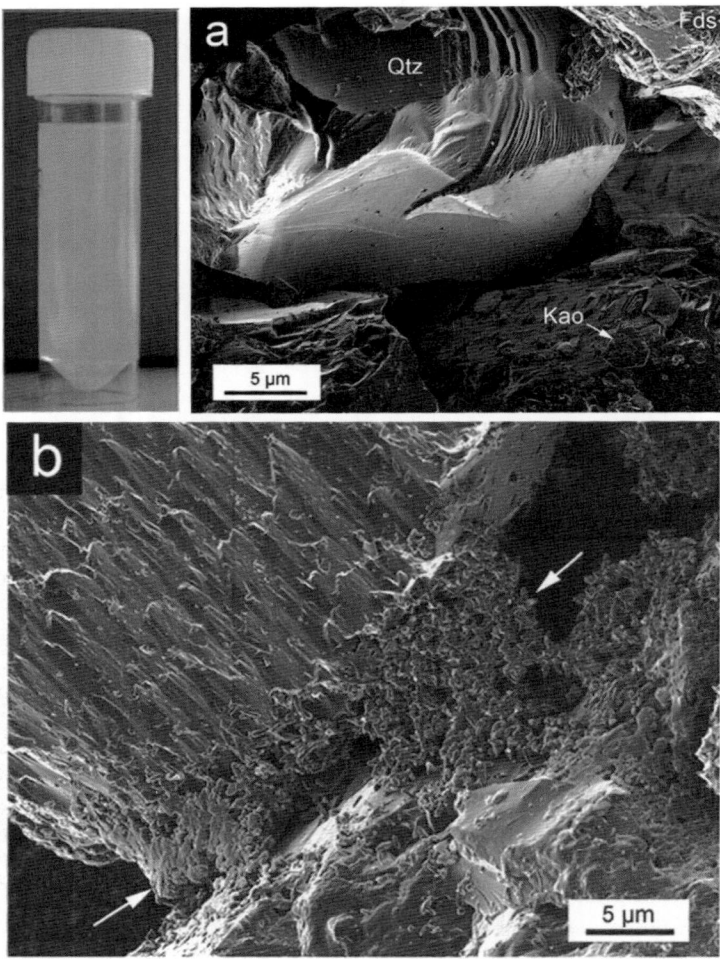

Fig. 7.5 (Top left) A dispersion of Ca(OH)$_2$ nanoparticles in alcohol. (Top right, bottom) Field emission electron scanning microscopy (FESEM) photomicrographs of untreated and treated sandstone: (**a**) control (untreated) sample showing quartz (Qtz) and feldspar (Fds) grains with sparse kaolinite (Kao) platelike crystals and (**b**) newly formed CaCO$_3$ cement (arrows) deposited at the contact between silicate grains after treatment with nanoparticles from aged slaked lime putty. Reprinted with permission from reference number [34], Copyright, American Chemical Society

previously used an Avrami–Erofeev model to fit the time evolution of calcite from portlandite nanoparticles.

The use of different earth-alkaline metals other than calcium was also explored. Giorgi et al. [21] synthesized Ba(OH)$_2$ nanoparticles and applied them as mixed to

$Ca(OH)_2$ dispersions for the consolidation of wall paintings in the presence of sulfate salts, where $Ba(OH)_2$ induced the formation of stable $BaSO_4$ (otherwise, soluble sulfates can hinder the carbonation of calcium hydroxide). Licchelli et al. [37] assessed the application of $Ca(OH)_2$ and $Sr(OH)_2$ nanoparticles to stone. This approach had been previously proposed by Ciliberto et al. [38] based on the rationale that strontium hydroxide, besides undergoing carbonation, can react with gypsum (a common pollutant in stone and mortars) to form $SrSO_4$, which as an insoluble salt does not produce detrimental solubilization–crystallization cycles. Overall, these authors showed that samples of Lecce stone (bio-calcarenite) treated with these formulations exhibit improved resistance to immersion in sulfate solution followed by heating and cooling cycles, while the permeability to water vapor is only slightly altered.

Besides hydroxides, colloidal silica is a largely explored material that has been used for the consolidation of stone. The literature on synthetic routes is vast. For instance, a review by Hyde et al. [39] and recently Facio et al. [40] proposed a facile and low-cost route to obtain mesoporous monolithic silica xerogels via an inverse surfactant micelle mechanism. In the framework of stone consolidation, research work targeted the synthesis of advanced systems to overcome the limitations of traditional products such as tetraethoxysilane (TEOS). TEOS is widely employed, as it is able to penetrate into the stone matrix where it polymerizes through a sol–gel process, forming siloxane chains that bind grains. However, it is well known that alkoxysilanes have the tendency to crack during drying, owing to the high capillary pressure in the newly formed gel network, with the pressure being inversely proportional to the gel's pore radius.

Mosquera et al. [41] developed and patented a sol–gel route for synthesizing silica nanomaterials without volatile organic compounds (VOCs) by mixing a silica oligomer with a nonionic surfactant under ultrasonic agitation. Besides being a sol–gel transition catalyst, the surfactant also allows controlling the structure of the pores in the gel network. Overall, this results in reducing the capillary pressure during drying so as to obtain a crack-free monolithic network. This methodology was later patented and applied to the conservation of granite stonework on the Santa Maria del Campo Church (A Coruña, Spain), a monument representative of the Romanesque style of northwestern Spain (Fig. 7.6) [42, 43].

In that case, severe deterioration of granite had led to granular disintegration and surface detachments, and past restorations carried out in the 1960s with wax had worsened the problem by blocking the surface pores and hindering water vapor permeability. The new surfactant-synthesized nanoconsolidant was tested and compared to two commercial consolidants (ESTEL 1000, an ethyl silicate, and Paraloid B82, an acrylic resin). The nanoconsolidant proved to have the best suitability to granite stone as it reduced the proportion of pores with very small diameters (<0.1 μm). The water transport capacity of the stone was thus decreased, restricting the movement of soluble salts in the stone and consequently reducing the damage from salt migration and crystallization. On the other hand, the acrylic resin accumulated on the stone surface, dramatically reducing the porosity and enhancing the

Fig. 7.6 (Top) View of the main façade of the Santa Maria del Campo Church (A Coruña, Spain), in which areas affected by black crust derived from bee-wax application (from past restorations) can be observed. (Bottom) Evaluation of the effectiveness of the nanosilica consolidant applied on the south façade of the church. (**A**) Picture of tape application during the peeling test carried out on a deteriorated capital. (**B**) Detail of color measurement on the two ashlars of the façade. Reprinted with permission from reference number [42], Copyright, Elsevier Ltd

damage from salt crystallization. As expected, the new product did not show cracking, as opposed to the commercial ethyl silicate.

Other groups also reported on the application of colloidal silica to different types of artistic and historical substrates. Some examples include the consolidation of marble (from the Pisa Tower), limestone, gypsum plasters, and murals [44–46]. Falchi et al. [47] studied the application of nanosilica aqueous dispersions to Lecce stone, focusing on the distribution of the consolidant within the substrate. The particles had a maximum size of 55 nm, which were consistently smaller than the pores of this type of stone (0.5 and 5 μm), but tended to form a xerogel on the stone surface, penetrating up to 2 mm into the matrix. Pretreatment of the stone with ethanol reduced surface tension, improving penetration of the particles. Other authors have assessed the application of colloidal silica to sandstone on the basis of the physicochemical compatibility between the consolidant and the stone matrix. For instance, nanosilica was used on sandstone reliefs at the archaeological site of Tajin (Mexico) where the particles were mixed with ground stone to repair losses [48]. Zomoza-Indart and Lopez-Arce [49] investigated the influence of RH on the consolidating effectiveness of SiO_2 colloidal nanoparticles applied on historical siliceous–carbonate stone. The system behaved similarly to a silica gel upon exposure to lower and higher RH owing to reversible adsorption and desorption of water. The treated stone showed increased surface hardness, while the interior exhibited higher ultrasound velocity and drilling resistance. Moreover, nanosilica has been shown to provide improved mechanical properties to concrete as it accelerates the dissolution of the C_3S phase, favoring the formation of the calcium–silicate–hydrate binder [50, 51].

Finally, besides $Ca(OH)_2$ and SiO_2 nanoparticles, it is worth mentioning a consolidation approach based on the use of calcium alkoxides, proposed by Favaro et al. [52] and Ossola et al. [53] where alkoxides in solution can impregnate porous substrates and form $CaCO_3$ (with nanostructuration) in situ by reacting with moisture and CO_2. These products provided consolidation to marble where they randomly distributed on wider pores and cracks below the stone surface [54].

7.2 Composite Nanomaterials

Composite materials typically contain components with different chemical properties separated by an interface. The use of composite nanomaterials (i.e., where at least one component exhibits nanostructuration) has been widely investigated in numerous applications, including the treatment of historical stone. As for the previous section, only systems aimed at the consolidation of stone (i.e., improving mechanical properties) will be focused on, without discussing products solely designed for the protection of surfaces (hydrophobic layers, self-cleaning coatings, etc.).

Several groups focused on the synthesis of organic–inorganic hybrid nanomaterials for stone consolidation, aiming to overcome the aforementioned limitations of traditional ethyl silicate products. Mosquera et al. [55] prepared a new material mixing TEOS with hydroxyl-terminated PDMS (polydimethylsiloxane) and n-octylamine (as a surfactant). While n-octylamine coarsens the gel's pores and reduces the surface tension of the solvent medium, PDMS integrates in the silica skeleton, forming aggregates that link silica particles and provides flexibility to the material. A 3D model (Fig. 7.7) was developed to describe both the hybrid silica-PDMS structure and the silica structure without PDMS, based on AFM images [56].

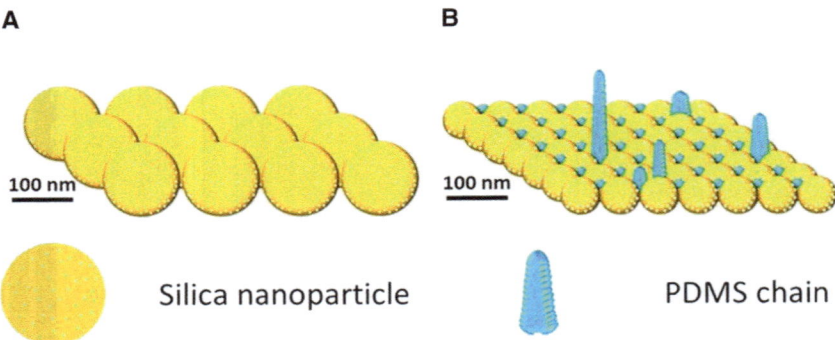

Fig. 7.7 3D structure model for (**A**) silica nanomaterial and (**B**) silica–PDMS nanomaterial surfactant synthesized via a sol–gel route. Reprinted with permission from reference number [56], Copyright, American Chemical Society

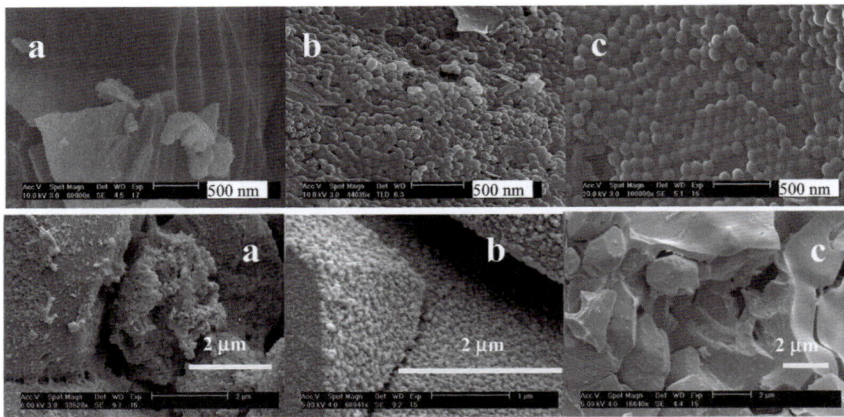

Fig. 7.8 (Top row) Scanning electron microscopy (SEM) micrographs of gels obtained from commercial alkoxysilanes (**a**) and from two different hybrid materials (**b**, **c**) prepared mixing TEOS with hydroxyl-terminated PDMS (polydimethylsiloxane) and *n*-octylamine. (**a**) Pores are not observed at the magnification used. (**b**) A gel network composed of uniform polymeric balls is observed. The presence of *n*-octylamine promotes the coarsening of the gel network. (Bottom row) SEM images of biocalcareous stone treated with two hybrid materials (**a**, **b**) and with a commercial alkoxysilane (**c**). The stone treated with the hybrids exhibits a mesoporous coating of polymer balls, with that can bridge the quartz grains of stone. The commercial alkoxysilane forms a dense coating (**c**), which is affected by cracking. Reprinted with permission from reference [55], Copyright, American Chemical Society

Overall this prevents cracking, leading to the formation of a homogeneous organic–inorganic hybrid xerogel that can bridge the quartz grains (Fig. 7.8).

The new material increases the mechanical resistance of stone, and the removal of the surfactant is simply carried out by drying under regular laboratory conditions.

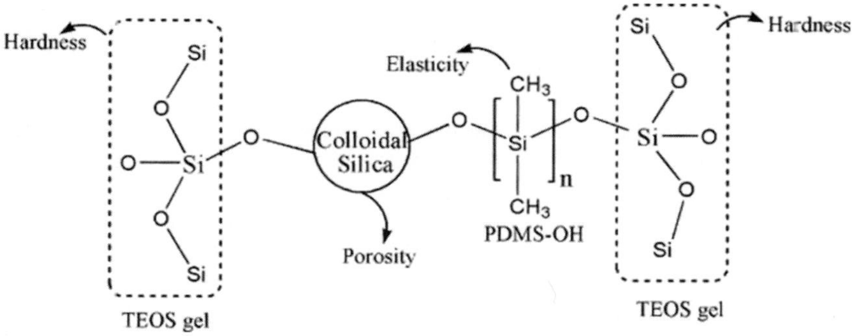

Fig. 7.9 Idealized structure of a TEOS/colloidal silica/PDMS-OH hybrid consolidant. Reprinted with permission from reference number [60], Copyright, John Wiley & Sons, Ltd

Moreover, it was shown that the hybrid had two distinctive scale roughnesses: one corresponding to the silica polymer (low-scale roughness) and one to PDMS aggregates (high-scale roughness). The dual-scale roughness reduced the hysteresis between advancing and receding contact angles. This, together with the fact that PDMS also reduces the surface energy of the material, promoted water repellence. In fact, the treated stone exhibited some stain resistance when exposed to water-containing substances such as cola, vinegar, red wine, and coffee [57].

Zarraga et al. [58] and Salazar-Hernandez et al. [59] had previously studied the effect of hydroxyl-terminated PDMS (PDMS-OH) on TEOS-PDMS-OH formulations using di-n-butyltin dilaurate (DBTL), a well-known neutral polycondensation catalyst for the formation of siloxane bonds. Elastification of the rigid SiO_2 gels is obtained by adding linear segments of PDMA-OH (5%) to a TEOS formulation catalyzed with DBTL, whose viscosity is left comparable to that of commercial products [59, 60]. These hybrid systems were further characterized and investigated as stone consolidants. Salazar-Hernandez et al. [60] proposed the use of a hybrid based on TEOS, colloidal silica (obtained by a variation of the Stöber method), and PDMS-OH for the consolidation of tuff from historical monuments. The colloidal silica particles form micro- and mesoporous structures in the TEOS network, depending on the relative percentage of these two components. The expected cross-linking of the PDMS-OH fragments with agglomerates of silica particles was demonstrated via Si MAS-NMR (magic angle spinning nuclear magnetic resonance), and an idealized structure of the hybrid material was proposed, where PDMS-OH form bridges between the TEOS gel and the colloidal silica aggregates (Fig. 7.9). The PDMS-OH links provide elasticity to the TEOS network, and the colloidal silica forms micro- and meso-porosity as explained above. Overall, this reduces the capillary tension during the drying of the hybrid system. The treated stone exhibited increased hardness and resistance to salt crystallization as compared to commercial formulations. The latter feature is probably due to the fact that the gel formed by the hybrid formulation is porous; thus the application does not dramatically reduce the water vapor permeability of stone. Moreover, the percentage of PDMS-OH affects the hydrophobicity of the formulation.

Alternative approaches have also been proposed by Son et al. [61] who obtained a crack-free gel from a TEOS-based consolidant containing functional (3-glycidoxypropyl)trimethoxysilane (GPTMS) and nano-sized polyhedral oligomeric silsesquioxane (POSS). These components, having flexible segments, reduce the capillary force during drying of the gel. Treatment with the hybrid led to increased ultrasound velocity and shore hardness of granite stone, with better performances than the commercial product tested (i.e., Wacker OH). Moreover, the hybrid coating increased the hydrophobicity of the surface which could be useful for water-barrier applications. Kim et al. [62] added GPTMS and silica nanoparticles of different sizes to a TEOS-based consolidant to create a hybrid that had a similar gelation time as commercial products but developed a crack-free gel. Application to sandstone samples increased contact angle and reduced water adsorption [62]. Liu and Liu [63] added hydroxyl-terminated polydimethylsiloxane (PDMS-OH) and a poly(ethylene oxide)–poly(phenylene oxide)–poly(−ethylene oxide) (PEO–PPO–PEO, F127) surfactant to TEOS to obtain a hybrid for the consolidation of the Chongqing Dazu sandstone sculptures in China. The composite was prepared via a sol–gel method and PDMS-OH formed Si-O-Si bonds with TEOS oligomers. The coating produced micro- and nanoscopic protrusions on the surface of the treated stone with a two-length-scale hierarchical structure. Overall, the surface morphology and roughness of the stone are altered, leading to improved hydrophobicity and preventing the formation of cracks on the surface. The compressive strength of the stone increased by 20%, due to the strong interfacial interactions between TEOS oligomers and PDMS-OH. The 12% decrease in water vapor permeability of the stone upon treatment was deemed a slight alteration (a 50% decrease being indicated as an acceptable threshold, as reported by the literature [64]). The hybrid coating also exhibited enhanced resistance to ultraviolet aging, indicating that this methodology had potential for the treatment of decayed stone.

Besides nanosilica, other inorganic materials have also been considered to form hybrids for stone consolidation. Miliani et al. [65] proposed hybrids obtained with particle-modified silica consolidants (PMCs) filled with titania, alumina, and silica particles. The particle-modified consolidants showed improved performances as compared to unfilled ethyl silicate formulations in terms of protection of sandstone from salt crystallization damage. Moreover, all the PMCs exhibited a reduction in the silica network shrinkage and had a thermal expansion coefficient that was closer (yet still larger) to that of stone, reducing the risk of fracturing during thermal aging cycles. Ksinopoulou et al. [66] prepared a nanocomposite by adding colloidal silica and oxide titania (TiO_2) particles to a TEOS matrix. Both SiO_2 and TiO_2 particles act synergistically to form a gel with coarser pores, which decreases the capillary pressure and improves the permeability of the network. Verganelaki et al. [67] selected silica–calcium oxalate hybrids based on the fact that silica phases have been found to naturally associate with calcium oxalate in well-preserved layers of monuments following some conservation treatments. In fact, stable composites of these two materials are known in the industrial processing of sugar. Nano-calcium oxalate was synthesized by reacting calcium hydroxide with oxalic acid, and the nanoparticles were then added to TEOS; *n*-octylamine was also added to grant the formation of a crack-free gel. Excess of the reactants can catalyze the sol–gel process or form $CaCO_3$ in the stone matrix. The hybrid system was applied on bioclastic limestone where it penetrated more than 1 cm, improving tensile strength with acceptable alteration of color and water vapor permeability.

The use of synthetic polymers derived from traditional restoration practice has also been explored to prepare organic–inorganic composites. For instance, hybrid silica–epoxy polymers have been developed for stone consolidation. In fact, epoxy resins are widely used as highly effective adhesives, and hybrids have been studied to improve on limitations or provide enhanced properties. Cardiano et al. [68] investigated the reaction of the epoxy derivatives 2-(3,4-epoxycyclohexyl)ethyltrimethoxysilane (ECET) and (3-glycidyloxypropyl)methyldiethoxysilane (GLYMS) with the primary amine (3-aminopropyl)triethoxysilane (ATS), studying the obtained solid through differential thermogravimetry, ^{13}C NMR, and Fourier transform (FT) Raman and infrared spectroscopy. When GLYMS reacted with ATS, the resulting materials may be considered as silica–epoxy hybrids where organic polymer chains covalently link to inorganic domains. Porosity, water absorption properties, and dynamic contact angle measurements showed that these compounds created a barrier against water penetration on high-porosity stone, while they had less of an effect on low-porosity samples. Tulliani et al. [69] prepared organic–inorganic materials by increasingly adding TEOS to polysiloxane epoxy formulations. The rationale was that TEOS reduces the viscosity of the formulation without solvent addition, whereas neat epoxy resin solutions are highly viscous and may need the addition of significant amounts of solvents (often toxic) to allow penetration through substrates. Moreover, the inclusion of TEOS in an organic matrix prevents cracking, and the mechanical properties of the film are improved. SEM investigation of the hybrid films showed that the inorganic domains generated in situ are dispersed within the organic network, the silica aggregates having dimensions of 60–80 nm. The films are transparent to visible light, with minimal color alteration of the treated surfaces (plaster). "Scotch tape" testing proved that the formulations were effective in consolidating the upper surface of the plaster, while capillary absorption measurements confirmed that acceptably good water vapor transmission was maintained after treatment. D'Arienzo et al. [70] prepared a composite system starting from Fluormet CP, a commercial blend of acrylic and vinylidenfluoride polymers widely used in stone restoration to provide readhesion of detached parts. The hybrid was obtained dispersing organically modified montmorillonite nanoparticles (Cloisite 30B) in a Fluormet solution upon sonication. Because the presence of the organoclay nanoparticles improves compatibility with silicate stone, the system was applied on tuff (which has high content of both zeolites and amorphous silicates), and assessment was carried out measuring water capillary absorption, permeability to water vapor, abrasion resistance, mechanical properties, color, and morphology (via SEM). The results showed that even a 1–4% content of nanoparticles in the polymer matrix had promising effects for the conservation of tuff, with the best consolidating effect occurring for an addition of 2% wt. of Cloisite 30B as

Fig. 7.10 The Nuestra Señora del Pilar y Santiago de Cocóspera mission at Sonora, México. The site represents a typical case where traditional restoration techniques proved detrimental to the preservation of the *adobe* bricks, requiring new consolidation materials to be developed. Reprinted with permission from reference number [71], Copyright, Elsevier Ltd

verified by uniaxial compression and abrasion tests. Namely, treatment with the nanocomposite led to the formation of a continuous layer in the inner walls of the stone pores (up to 3 mm depth), which linked disaggregated parts. The chromatic alteration was deemed as almost negligible, and the hydrophobicity of the treated surface increased as compared to treatment with neat Fluormet. It must be noted that the permeability of stone was strongly reduced, even if total pore occlusion did not occur. Both the durability and reversibility of the treatment are to be further assessed in forthcoming studies.

Finally, a future perspective involves the possible use of composites for the preservation of the so-called "earthen" building materials, which have been extensively used since ancient times and maintain a central role in sustainable architecture owing to their availability, low manufacturing cost, thermal–acoustic properties, and recyclability. Namely, *adobe* is one of the most popular construction techniques (Fig. 7.10), where tightly compacted mixes of water and soil (sand, silt, and clay) are molded as bricks and dried by prolonged sun exposure.

Additives like vegetable fibers and lime are commonly introduced to improve mechanical properties and reduce shrinkage and cracking. Owing to the susceptibility of clay to moisture, water is the main degradation agent of *adobe*, causing cracks and granular disintegration. Earth consolidation is a widely discussed topic, and many strategies have been proposed, but a compatible and long-lasting solution has not been developed yet. Scarce results have been obtained with sodium and potassium silicates despite their ability to reproduce an inorganic structure in the consolidated soil, as consolidation effects have a specific dependence on the mineralogical composition of the substrate [72]. Other conventional methods include natural organic resins and oils. The nopal mucilage (resin suspended in cactus juice) is commonly used in Latin America [71, 73], even though frequent treatments are needed due to its rapid degradation. Synthetic products such as alkoxysilanes, acrylic resins, vinylic polymers, and polyurethanes are also employed for their compatibility (in the case of silanes) and water repellency but often produce little and superficial hardening. All these traditional methods show limitations and drawbacks, with new consolidants being actively researched. Remarkably, recent work focused on reducing the swelling capacity of clays by turning them into non-expandable binding materials such as calcium silicate hydrate (CSH) using alkaline activators like $Ca(OH)_2$ [74, 75]. In addition, the formation of $CaCO_3$ by carbonation of hydroxide contributes to strengthening the structure. In fact, lime has been used as a stabilizer since ancient times, and evidence of the cementing phases can be found in antique samples [33]. In this perspective, hybrid (nanocomposite) formulations might be a valid solution to a problem whose complexity lies in the composite nature of the target material. The combination of three components such as colloidal silica, cellulose derivatives, and nanolime appears as a valid strategy to obtain consolidants that are highly compatible both with the earth matrix and additives while achieving higher strengthening by the in situ formation of a cementing phase like CSH.

7.3 Conclusion

Starting from the late 1990s, colloid and materials science have provided a wide range of solutions for the consolidation of stone, which significantly improved on traditional consolidants (e.g. polymeric coatings and adhesives). Besides providing effective strengthening, a key point of the innovative materials is their compatibility with the original artifacts, meaning that the physicochemical properties of the works of art are not altered by the consolidation intervention. Synthetic routes and practical assessment of the new systems have been explored systematically, leading to the development of both inorganic (e.g., $Ca(OH)_2$ nanoparticles, colloidal silica) and hybrid (organic–inorganic) materials. While some aspects must be investigated further (e.g., the penetration of particles through low-porosity matrices and the consolidation effects on some lithotypes), several methodologies and products are now available to curators and restorers as valid alternatives to the traditional materials used in the conservation practice.

References

1. Baglioni P, Chelazzi D, Giorgi R. Nanotechnologies in the conservation of cultural heritage. New York: Springer; 2015.
2. Everett DH. The thermodynamics of frost damage to porous solids. Trans Faraday Soc. 1961;57:1541–51.
3. Fitzner B, Snethlage R. Relationship between the influence of salt crystallization and pore distribution. GP News Lett. 1982;3:13–24.
4. Chelazzi D, Poggi G, Jaidar Y, Toccafondi N, Giorgi N, Baglioni P. Hydroxide nanoparticles for cultural heritage: consolidation and protection of wall paintings and carbonate materials. J Colloid Interf Sci. 2013;392:42–9.
5. Favaro M, Mendichi R, Ossola F, Simon S, Tomasin P, Vigato PA. Evaluation of polymers for conservation treatments of outdoor exposed stone monuments. Part II: photo-oxidative and salt-induced weathering of acrylic–silicone mixtures. Polym Degrad Stabil. 2006;91:3083–96.
6. Rodriguez-Navarro C, Elert K, Sevcik R. Amorphous and crystalline calcium carbonate phases during carbonation of nanolimes: implications in heritage conservation. CrystEngComm. 2016;18:6594–607.
7. Bottari C, Crisci GM, Crupi V, Ignazzitto V, La Russa MF, Majolino D, Ricca M, Rossi B, Ruffolo SA, Teixeira J, Venuti V. SANS investigation of the salt-crystallization- and surface-treatment-induced degradation on limestones of historic-artistic interest. Appl Phys A Mater Sci Process. 2016;122:721–30.
8. Carretti E, Chelazzi D, Rocchigiani G, Baglioni P, Poggi G, Dei L. Interactions between nano-structured calcium hydroxide and acrylate copolymers: implications in cultural heritage conservation. Langmuir. 2013;29:9881–90.
9. Rodriguez-Navarro C, Hansen E, Ginell WS. Calcium hydroxide crystal evolution upon aging of lime putty. J Am Ceram Soc. 1998;81:3032–4.
10. Rodriguez-Navarro C, Ruiz-Agudo E, Ortega-Huertas M, Hansen E. Nanostructure and irreversible colloidal behavior of Ca(OH)$_2$: implications in cultural heritage conservation. Langmuir. 2005;21:10948–57.
11. Giorgi R, Dei L, Baglioni P. A new method for consolidating wall paintings based on dispersions of lime in alcohol. Stud Conserv. 2000;45:154–61.
12. Ambrosi M, Dei L, Giorgi R, Neto C, Baglioni P. Stable dispersions of Ca(OH)$_2$ in aliphatic alcohols: properties and application in cultural heritage conservation. Progr Colloid Polym Sci. 2001;118:68–72.
13. Fratini E, Page MG, Giorgi R, Colfen H, Baglioni P, Deme B, Zemb T. Competitive surface adsorption of solvent molecules and compactness of agglomeration in calcium hydroxide nanoparticles. Langmuir. 2007;23:2330–8.
14. Baglioni P, Dei L, Ferroni E, Giorgi R. Sospensioni Stabili di Idrossido di Calcio. 1996. Italian patent FI/96/A/000255, 31 Oct 1996.
15. Ambrosi M, Dei L, Giorgi R, Neto C, Baglioni P. Colloidal particles of Ca(OH)$_2$: properties and applications to restoration of frescoes. Langmuir. 2001;17:4251–5.
16. Croveri P, Dei L, Giorgi R, Salvadori B. Consolidation of Globigerina limestone (Malta) by means of inorganic treatments: preliminary results. In: Kwiatkowski D, Lofvendahl R, editors. Proceedings of the 10th international congress on deterioration and conservation of stone, vol. 1. Stockholm: ICOMOS Sweden; 2004. p. 463.
17. Salvadori B, Dei L. Synthesis of Ca(OH)$_2$ nanoparticles from diols. Langmuir. 2001;17:2371–4.
18. Dei L, Slavadori B. Nanotechnology in cultural heritage conservation: nanometric slaked lime saves architectonic and artistic surfaces from decay. J Cult Herit. 2006;7:110–5.
19. Baglioni P, Dei L, Fratoni L, Lo Nostro P, Moroni M. Process for the preparation of nano-and micro-particles of group II and transition metals oxides and hydroxides, the nano-and micro-particles thus obtained and their use in the ceramic, textile and paper industries. 2005. Patent US20050175530 (A1) (Priority date: 2002-03-28); 2005.

20. Baglioni P, Carrasco Vargas R, Chelazzi D, Gonzalez MC, Desprat A, Giorgi R. The Maya site of Calakmul: in situ preservation of wall paintings and limestone using nanotechnology. In: Saunders D, Townsend JH, editors. Proceedings of the IIC congress 2006, Munich – the object in context: crossing conservation boundaries. London: James and James; 2006. p. 162.
21. Giorgi R, Ambrosi M, Toccafondi N, Baglioni P. Nanoparticles for cultural heritage conservation: calcium and barium hydroxide nanoparticles for wall painting consolidation. Chem Eur J. 2010;16:695–704.
22. Poggi G, Toccafondi N, Chelazzi D, Canton P, Giorgi R, Baglioni P. Calcium hydroxide nanoparticles from solvothermal reaction for the deacidification of degraded waterlogged wood. J Colloid Interface Sci. 2016;473:1–8.
23. Baglioni P, Chelazzi D, Giorgi R, Carretti E, Toccafondi N, Jaidar Y. Commercial Ca(OH)₂ nanoparticles for the consolidation of immovable works of art. Appl Phys A Mater Sci Process. 2014;114:723–32.
24. Lopez-Arce P, Gomez-Villalba LS, Pinho L, Fernandez-Valle ME, Alvarez de Buergo M, Fort R. Influence of porosity and relative humidity on consolidation of dolostone with calcium hydroxide nanoparticles: effectiveness assessment with non-destructive techniques. Mater Charact. 2010;61:168–84.
25. Daniele V, Taglieri G, Quaresima R. The nanolimes in cultural heritage conservation: characterization and analysis of the carbonation process. J Cult Herit. 2008;9:294–301.
26. Daniele V, Taglieri G. Nanolime suspensions applied on natural lithotypes: the influence of concentration and residual water content on carbonation process and on treatment effectiveness. J Cult Herit. 2010;11:102–6.
27. Borsoi G, Lubelli B, van Hess R, Veiga R, Silva AS. Evaluation of the effectiveness and compatibility of nanolime consolidants with improved properties. Constr Build Mater. 2017;142:385–94.
28. Borsoi G, Lubelli B, van Hess R, Veiga R, Silva AS, Colla L, Fedele L, Tomasin P. Effect of solvent on nanolime transport within limestone: how to improve in-depth deposition. Physicochem Eng Aspects. 2016;497:171–81.
29. Ziegenbalg G. Colloidal calcium hydroxide: a new material for consolidation and conservation of carbonatic stones. In: Lukaszewicz JW, Niemcewicz P, editors. Proceedings of the 11th international congress on deterioration and conservation of stone. Torun: Nicolaus Copernicus University; 2008. p. 1109.
30. D'Armada P, Hirst E. Nano-lime for consolidation of plaster and stone. J Architect Conserv. 2012;1:63–80.
31. Dahene A, Herm C. Calcium hydroxide nanosols for the consolidation of porous building materials – results from EU-STONECORE. Herit Sci. 2013;1:11–20.
32. Borsoi G, Lubelli B, van Hees R, Veiga R, Silva AS. Understanding the transport of nanolime consolidants within Maastricht limestone. J Cult Herit. 2016;18:242–9.
33. Lanzon M, Madrid JA, Martinez-Arredondo A, Monaco S. Use of diluted Ca(OH)₂ suspensions and their transformation into nanostructured CaCO₃ coatings: A case study in strengthening heritage materials (stucco, adobe and stone). Appl Surf Sci. 2017;424(Part 1):20–27. In Press, Corrected Proof. Available online 1 March 2017. https://doi.org/10.1016/j.apsusc.2017.02.248.
34. Rodriguez-Navarro C, Suzuki A, Ruiz-Agudo E. Alcohol dispersions of calcium hydroxide nanoparticles for stone conservation. Langmuir. 2013;29:11457–70.
35. Bermejo Sotillo MA, Rodriguez-Navarro C, Ruiz-Agudo E, Elert K. CO₂-capturing binder, production method thereof based on the selection, purification and optimization of carbide lime, and agglomerates having and environmental activity. Eur. Pat. EP2500328; presented May 4th 2010, published September 19th 2012.
36. Rodriguez-Navarro C, Vettori I, Encarnacion R-A. Kinetics and mechanism of calcium hydroxide conversion into calcium alkoxides: implications in heritage conservation using nanolimes. Langmuir. 2016;32:5183–94.
37. Licchelli M, Malagodi M, Weththimuni M, Zanchi C. Nanoparticles for conservation of biocalcarenite stone. Appl Phys A Mater Sci Process. 2014;114:673–83.

38. Ciliberto E, Condorelli GG, La Delfa S, Viscuso E. Nanoparticles of Sr(OH)$_2$: synthesis in homogeneous phase at low temperature and application for cultural heritage artefacts. Appl Phys A Mater Sci Process. 2008;92:137–41.
39. Hyde EDER, Seyfaee A, Neville F, Moreno-Atanasio R. Colloidal silica particle synthesis and future industrial manufacturing pathways: a review. Ind Eng Chem Res. 2016;55:8891–913.
40. Facio DS, Luna M, Mosquera MJ. Facile preparation of mesoporous silica monoliths by an inverse micelle mechanism. Micropor Mesopor Mat. 2017;247:166–76.
41. Mosquera MJ, de los Santos DM, Montes A, Valdez-Castro L. New nanomaterials for consolidating stone. Langmuir. 2008;24:2772–8.
42. De Rosario I, Elhaddad F, Pan A, Benavides R, Rivas T, Mosquera MJ. Effectiveness of a novel consolidant on granite: laboratory and in situ results. Constr Build Mater. 2015;76:140–9.
43. Mosquera MJ, Illescas JF, Facio DS. Product for protecting and restoring rocks and other construction materials. Spanish patent no. P201200152. 2012. Priority data: February 16, 2012.
44. Cericol, La Torre di Pisa: una pietra miliare nella storia del restauro, grazie alle tecnologie di Cericol. Colorobbia Consulting, Florence, Italy. 2017. http://www.cericol.com/. Accessed Oct 2017.
45. Calia A, Masieri M, Bald IG, Mazzotta C. The evaluation of nanosilica performance for consolidation treatment of a highly porous calcarenite. In: C. University, editor. 12th International Congress on the Deterioration and Conservation of Stone. New York; 2012. p. 2–11.
46. Camaiti M, Dellantonio G, Pittertschatscher M. Restauro dello stemma affrescato del Cardinale Bernardo Cles press oil Castello del Buonconsiglio a Trento: nuove soluzioni per il consolidamento di intonaci dipinti staccati dal support murario. In: Ricerce A, editor. Atti del XXIV Convegno di studi Scienza e Beni Culturali su Restaurare i restauri – Metodi, Compatibilità, Cantieri, Marghera (Venezia); 2008. pp. 231–241.
47. Falchi L, Balliana E, Izzo FC, Agostinetto L, Zendri E. Distribution of nanosilica in Lecce stone. Science at Ca' Foscari. 2013;1:49. https://doi.org/10.7361/scicf-441.
48. Grimaldi DM, Nora AP, Porter JH. The preservation of sandstone reliefs at the archaeological site of Tajin, Mexico, using colloidal silica. In: C. University, editor. 12th International congress on the deterioration and conservation of stone. New York; 2012. p. 12–22.
49. Zornoza-Indart A, Lopez-Arce P. Silica nanoparticles (SiO$_2$): influence of relative humidity in stone consolidation. J Cult Herit. 2016;18:258–70.
50. Jo B, Kim C, Tae G, Park J. Characteristics of cement mortar with nano-SiO$_2$ particles. Constr Build Mater. 2007;21:1351–5.
51. Bjornstrom J, Martinelli A, Matic A, Borjesson L, Panas I. Accelerating effects of colloidal nanosilica for beneficial calcium-silicate-hydrate formation in cement. Chem Phys Lett. 2004;392:242–8.
52. Favaro M, Tomasin P, Ossola F, Vigato PA. A novel approach to consolidation of historical limestone: the calcium alkoxides. Appl Organomet Chem. 2008;22:698–704.
53. Ossola F, Tomasin P, De Zorzi C, El Habra N, Chiurato M, Favaro M. New calcium alkoxides for consolidation of carbonate rocks. Influence of precursors' characteristics on morphology, crystalline phase and consolidation effects. New J Chem. 2012;36:2618–24.
54. Natali I, Tomasin P, Becherini F, Bernardi A, Ciantelli C, Favaro M, Favoni O, Forrat Perez VJ, Oltenau ID, Romero Sanchez MD, Vivarelli A, Bonazza A. Innovative consolidating products for stone materials: field exposure tests as a valid approach for assessing durability. Herit Sci. 2015;3:6–19.
55. Mosquera MJ, de los Santos DM, Rivas T. Surfactant-synthesized ormosils with application to stone restoration. Langmuir. 2010;26:6737–45.
56. Illescas JF, Mosquera MJ. Producing surfactant-synthesized nanomaterials in situ on a building substrate, without volatile organic compounds. ACS Appl Mater Interfaces. 2012;4:4259–69.
57. Illescas JF, Mosquera MJ. Surfactant-synthesized PDMS/silica nanomaterials improve robustness and stain resistance of carbonate stone. J Phys Chem C. 2011;115:14624–34.

58. Zarraga R, Cervantes J, Salazar-Hernandez C, Wheeler G. Effect of the addition of hydroxyl-terminated polydimethylsiloxane to TEOS-based stone consolidants. J Cult Herit. 2010;11:138–44.
59. Salazar-Hernandez C, Zarraga R, Alonso S, Sugita A, Calixto S, Cervantes J. Effect of solvent type on polycodensation of TEOS catalyzed by DBTL as used for stone consolidation. S Sol-Gel Sci Technol. 2009;49:301–10.
60. Salazar-Hernandez C, Alquiza MJP, Salgado P, Cervantes J. TEOS-colloidal silica-PDMS-OH hybrid formulation used for stone consolidation. Appl Organometal Chem. 2010;24:481–8.
61. Son S, Won J, Kim JJ, Jang YD, Kang YS, Kim SD. Organic-inorganic hybrid compounds containing polyhedral oligomeric silsesquioxane for conservation of stone heritage. ACS Appl Mater Interf. 2009;2:393–401.
62. Kim EK, Won J, Do JY, Kim SD, Kang YS. Effects of silica nanoparticle and GPTMS addition on TEOS-based stone consolidants. J Cult Herit. 2009;10:214–21.
63. Liu Y, Liu J. Fabrication of TEOS/PDMS/F127 hybrid coating materials for conservation of historic stone sculptures. Appl Phys A Mater Sci Process. 2016;122:743–50.
64. Delgado Rodrigues J, Grossi A. Indicators and ratings for the compatibility assessment of conservation actions. J Cult Herit. 2007;8:32–43.
65. Miliani C, Velo-Simpson ML, Scherer GW. Particle-modified consolidants: a study on the effect of particles on sol-gel properties and consolidation effectiveness. J Cult Herit. 2007;8:1–6.
66. Ksinopoulou E, Bakolas A, Moropoulou A. Modifying Si-based consolidants through the addition of colloidal nano-particles. Appl Phys A Mater Sci Process. 2016;122:267–77.
67. Verganelaki A, Kilikoglou V, Karatasios I, Maravelaki-Kalaitzaki P. A biomimetic approach to strengthen and protect construction materials with a novel calcium-oxalate-silica nanocomposite. Constr Build Mater. 2014;62:8–17.
68. Cardiano P, Ponerio RC, Sergi S, Lo Schiavo S, Piraino P. Epoxy-silica polymers as stone conservation materials. Polymer. 2005;46:1857–64.
69. Tulliani JM, Formia A, Sangermano M. Organic-inorganic material or the consolidation of plaster. J Cult Herit. 2011;12:364–71.
70. D'Arienzo L, Scarfato P, Incarnato L. New polymeric nanocomposites for improving the protective and consolidating efficiency of tuff stone. J Cult Herit. 2008;9:253–60.
71. Martınez-Camacho F, Vazquez-Negrete J, Lima E, Lara VH, Bosch P. Texture of nopal treated adobe: restoring Nuestra senora del Pilar mission. J Archaeol Sci. 2008;35:1125–33.
72. Warren J. Conservation of earth structures. Series in Conservation and Museology. Oxford: Elsevier Butterworth-Heinemann; 1999.
73. Kita Y. The functions of vegetable mucilage in lime and earth mortars – a review, 3rd Historic Mortars Conference, Glasgow; 2013.
74. Elert K, Pardo ES, Rodriguez-Navarro C. Alkaline activation as an alternative method for the consolidation of earthen architecture. J Cult Herit. 2015;16:461–9.
75. Oti JE, Kinuthia JM, Bai J. Compressive strength and microstructural analysis of unfired clay masonry bricks. Eng Geol. 2009;109:230–40.

Chapter 8
Testing Efficiency of Stone Conservation Treatments

Miloš Drdácký

8.1 Introduction

In the field of conservation, new consolidation products or intervention technologies are sometimes introduced before their impact on historical substances has been tested and verified. This happens because evaluating the impact of new agents on cultural heritage is not an easy task, which usually does not aspire to study all possible effects or to attain absolute standard values because relative data are mostly sufficient for decision-making. The effects on real materials and under real conditions of application and action should be examined; nonetheless, in practice investigations are usually carried out in laboratories.

There are no comprehensive standard procedures for assessing consolidation effects on stones, especially when small-sized specimens are to be tested. Tabasso and Simon [1] published a critical review of testing methods for the assessment of stone properties before and after conservation treatment; in it, the reader can find useful information on various standard as well as non-standard techniques. The testing of stones and evaluation of conservation interventions, including recommendations for compatibility criteria, have been studied further mainly in Germany [2, 3], where many innovative ideas have originated.

This chapter presents examples of test methodologies which have not been standardized but were specifically designed or are useful for dealing with practical tasks. The selected tests were intended to provide results which would best enable the characterization of the consolidation effects from the point of view of maximum efficiency, with no or negligible harmful consequences for the given rock.

M. Drdácký (✉)
Institute of Theoretical and Applied Mechanics of the Czech Academy of Sciences,
Centre of Excellence Telč, Telč, Czech Republic
e-mail: drdacky@itam.cas.cz

© Springer International Publishing AG 2018
M. Hosseini, I. Karapanagiotis (eds.), *Advanced Materials for the Conservation of Stone*, https://doi.org/10.1007/978-3-319-72260-3_8

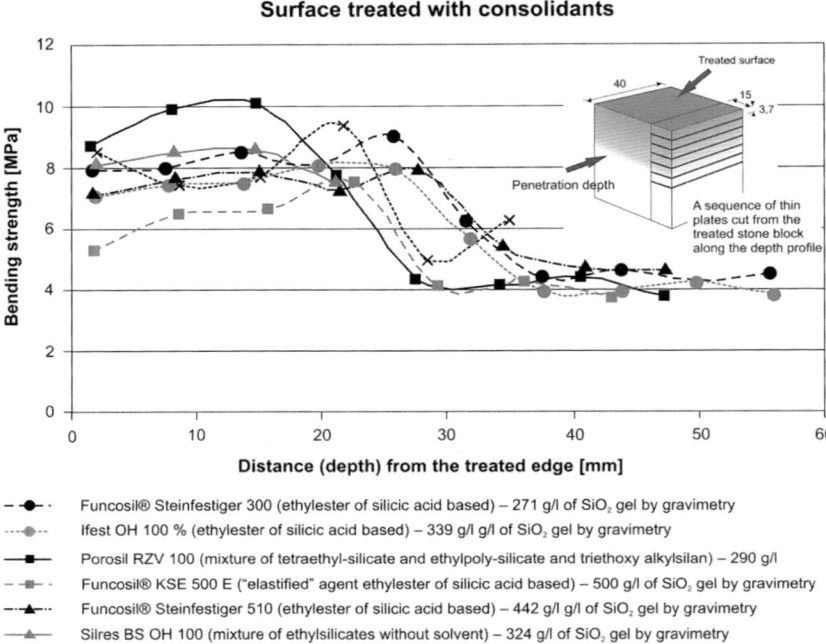

Fig. 8.1 Influence of different consolidants on the bending strength of sandstone. On the right side of the picture, the values measured oscillate around the natural strength of the rock. (Modified from Drdácký and Slížková [4])

Bending strength and Young's modulus of elasticity are typically tested on small-sized specimens—rectangular plates or circular discs. The advantage of the first approach is that several physical characteristics of identical test specimens are tested gradually, which reduces errors originating from material non-homogeneities [4]. Plates with nominal dimensions of 15 × 40 to 50 × 4 mm are used for testing first water uptake, thermal dilation, and then hygric or hydric expansion and flexural strength. The broken halves after the bending test can be used for vapour permeability tests and porosity measurements. The plates are advantageously cut from blocks previously treated on one surface with the tested consolidation agents, which makes it possible to observe changes in the material characteristics along a depth profile perpendicular to the treated surface (Fig. 8.1). Cutting steps of about 5–6 mm are practical and give a plate thickness of 4 mm with a cutting loss of about 1–2 mm. When testing consolidation effects, dry cutting is preferred to avoid possible influences due to intensive water washing. The specimens are typically tested in three-point bending; this is accompanied by measurement of central deflection under the load mounted in a special rig after 24 h of drying at 60 °C and 12 h of conditioning in a climate chamber (40 °C/40%) (Fig. 8.2b). This procedure has been published [4] along with satisfactory results. The identified increase of strength on the treated

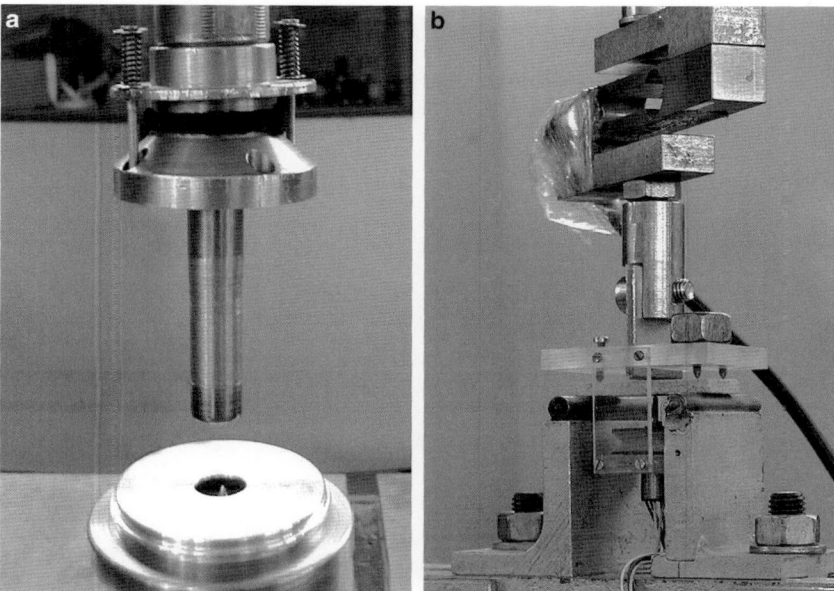

Fig. 8.2 Arrangement of flexural tests: (**a**) biaxial test on stone discs, (**b**) uniaxial test on rectangular plates

surface is clearly apparent in Fig. 8.1 as is the drop in the value of the virgin material at depths where the consolidation agent did not penetrate.

Biaxial flexure tests were mainly developed for ceramic and glass testing, for which several testing arrangements are available and have even been standardized: piston-on-three-ball, piston-on-ring, ball-on-ring and ring-on-ring [5–11]. There are few studies focused on the relationship between the bending strength attained using rectangular plates in three- or four-point bending and that acquired using circular discs subjected to biaxial loading. The effect is naturally influenced by the homogeneity and isotropy of the material tested as well as by the size of the test specimens. Biaxial testing of stone in the mode of a ring-to-ring arrangement for conservation efficacy assessment was recommended by Wittmann and Prim [12]. The method is very effective and rather fast for strength testing, especially if the variant with ball-to-ball support is applied [13]. However, the circular discs are not practical for determining other material characteristics which are decisive for application in the conservation of porous materials. Here, the results of a comparison of uniaxial and biaxial tests on rather isotropic sandstone specimens are presented. As other natural materials, the fine-grained quartz sandstone used exhibits irregularities which influence the scatter of results.

Comparative biaxial bending tests were carried out using plates with a diameter of 55 mm and a thickness of about 4 mm in a special fixture (Fig. 8.2a). The supporting rig had the form of a circular ring with a diameter of 40 mm, on which the specimen tested was centrally placed. The position was controlled with an alignment cap which had a central hole for defining the position of the loading

Fig. 8.3 Testing rig for biaxial bending—schematic sketch (bottom right), sequence of composition (upper pictures, left supporting ring, right disc alignment ring, stone disc in the alignment ring, load—hollow piston—alignment ring)

ring (a cylindrical hollow piston) (Fig. 8.3). The supporting rig carries an LVDT sensor for the measurement of the central deflection of the plate in the course of loading. This is not an ideal situation because the measurement is biased by the surface deformation of the stone disc in contact with the support. A uniaxial flexural test arrangement involving a special fixture placed on the upper surface of the rectangular plate (Fig. 8.2b) makes it possible to eliminate this error.

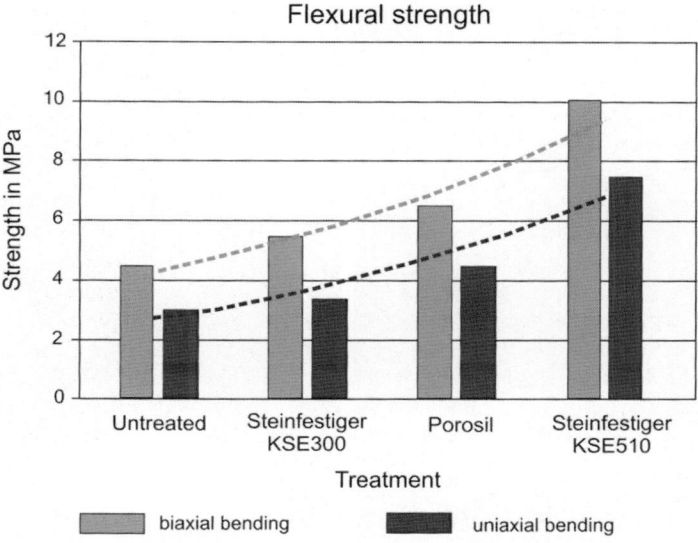

Fig. 8.4 Comparison of average flexural strengths attained using rectangular plates subjected to three-point bending and the biaxial flexural strength from the tests using circular discs

The measured strength in biaxial bending is calculated according to [12] using the following equation:

$$\sigma = \frac{3F}{4\pi\left(h^2\right)} \times \left\{ 2\times(1+v)\times\ln\left(\frac{a}{b}\right) + \frac{(1-v)\times\left(a^2-b^2\right)}{a^2} \right\} \tag{8.1}$$

where F is the maximum force reached during the test, h is the thickness of the plate, v is the Poisson number, b is the loading ring radius and a is the supporting ring radius. Figure 8.4 shows a comparison of the flexural strength attained on rectangular plates subjected to three-point bending and the biaxial flexural strength from the tests using circular discs. It may be observed that both methods give identical tendencies in the determination of the material characteristics of different qualities of stone, here achieved by means of impregnation with different ethyl silicate-based consolidants. They were selected from the agents available on the market to provide significantly different change of mechanical characteristics of the tested stone after treatment in order to demonstrate the capacities of the presented testing procedures. The results show that both methods are applicable for assessing the efficiency of various strengthening agents, with almost the same results. Furthermore, the results correspond to the results of tests on ceramic discs of similar size [8].

8.2 Testing Facilities and Procedures for In Situ Determining Representative Material Characteristics

In contrast to the above laboratory investigations, there are limited testing facilities and procedures for determining representative material characteristics in situ. Again, mechanical characteristics are considered sufficiently significant for consolidation or strengthening treatment efficiency. Surface cohesion characteristics may be determined well by a standardized peeling test if a repetitive procedure is applied [14]. Material characteristics along the depth are measurable using drilling resistance [2]. For quality control of stone masonry conservation, an innovative ultrasonic device has been developed jointly with Rolf Krompholz (GEOTRON-ELEKTRONIK) within the Stonecore project [15] for ultrasonic (US) measurements in a depth profile (Fig. 8.5). It consists in two probes: an US transmitter and a receiver. They are inserted into holes of 20 mm in diameter drilled into the investigated surface layer to a distance of up to 100 mm from their centres and a depth

Fig. 8.5 Double-probe ultrasonic device for in-depth parallel to surface measurements

of up to 60 mm. Certain design features (a flat base and adjustable rods in the transmitter and the receiver) allow for reproducible insertion of the device in the prepared holes in order to acquire a reliable series of measurements for the investigation of changes in material characteristics. The device is portable and fully compatible with ultrasonic equipment for laboratory measurements; therefore, the operator can rely on previously learned skills in data evaluation. The device is robust and well-engineered for difficult outdoor measurement conditions. It makes it possible, for the first time, to measure the properties of materials in situ with an acceptable impact. The main advantage over the standard drilling technique is that it is possible to follow the conservation impact during the intervention process because changes can be measured at an identical place and over the same volume of material after individual impregnation steps. This enables restorers to terminate or continue impregnation repetitions in response to the measured impact—the consolidation depth and strengthening effects. Naturally, the local invasion is much larger and necessitates special repair after testing, which is not a problem with large-dimensioned structures such as stone façades. In such cases, optimum control of conservation and reasonably low material consumption are desirable. Its practical application is illustrated in the following example of the inspection of a quartz sandstone masonry whose surface was treated with the same three types of strengtheners as those described above. The testing wall is presented in Fig. 8.6, where the treated areas as well as the drilled holes are marked. The method is semi-destructive but the holes are quite repairable, and on stone façades the traces are almost invisible. Results in the form of US velocities are shown in Fig. 8.7, in which the effect of consolidation is clearly apparent and the depth and intensity of the agent penetration is evident. Measurement is very fast and reliable, though problems may arise involving hidden material irregularities or gross imperfections on the hole surface preventing reasonably good probe contact. In the example presented, an obstacle of this type complicated measurements in one hole and affected less than 2% of all data acquired.

8.3   Conclusion

Testing the impact of stone consolidation products before their widespread introduction into conservation practice is one of the most important duties of stone consolidant producers. Nevertheless, it is recommended that professional restorers check the impact of market products on real stones. This chapter presented a comparison of two methods for testing the flexural strength of stone samples in laboratory conditions (i.e. biaxial testing of thin discs and uniaxial testing of thin beamlike plates). Both techniques studied using quartz sandstone test specimens yielded results adequately describing the changes in stone after consolidation treatment, and their choice in practice is determined by aspects other than mechanical testing modes, namely, the sample extraction techniques—core drilling or saw cutting—and planning of other types of tests (e.g. dilation behaviour under various moisture and temperature conditions). For in situ testing, a portable

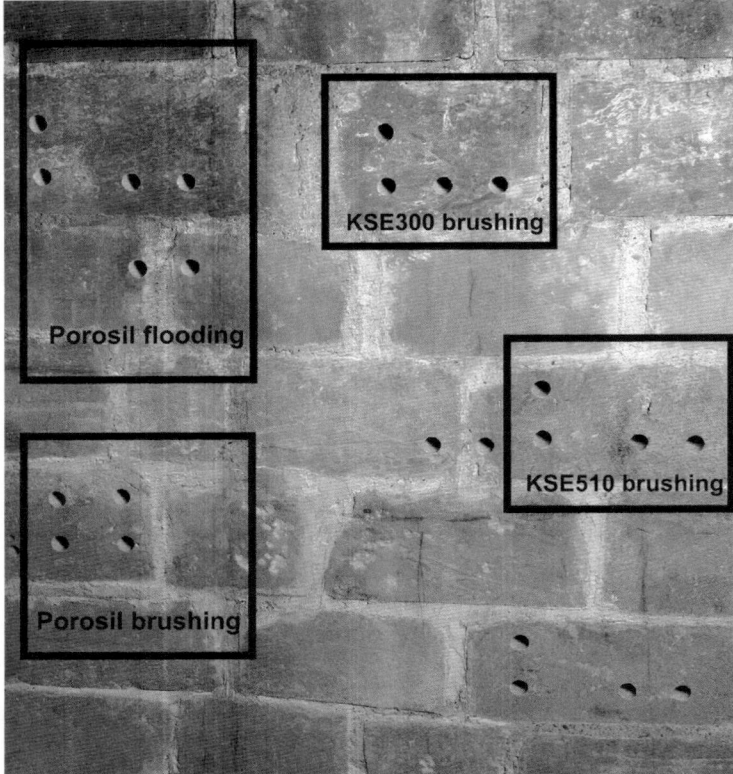

Fig. 8.6 Sandstone wall with treated areas and drilled testing holes

ultrasonic double-hole probe measuring material properties along a depth profile and assessing penetration depth in the near-surface material layer between two drilled holes was presented. This provides restorers with a unique capability to follow the consolidation progress during repetitive impregnation and/or the agent maturing. However, the need to drill 20 mm diameter holes with this technique is more suitable and profitable in the case of façade restoration than in that of small art objects.

Acknowledgement The chapter is based on the results of research supported by the Czech Grant Agency Project G105/12/059 and the kind help of Ivana Frolíková with figure preparation and Luisa Natalia Pena with flexural testing.

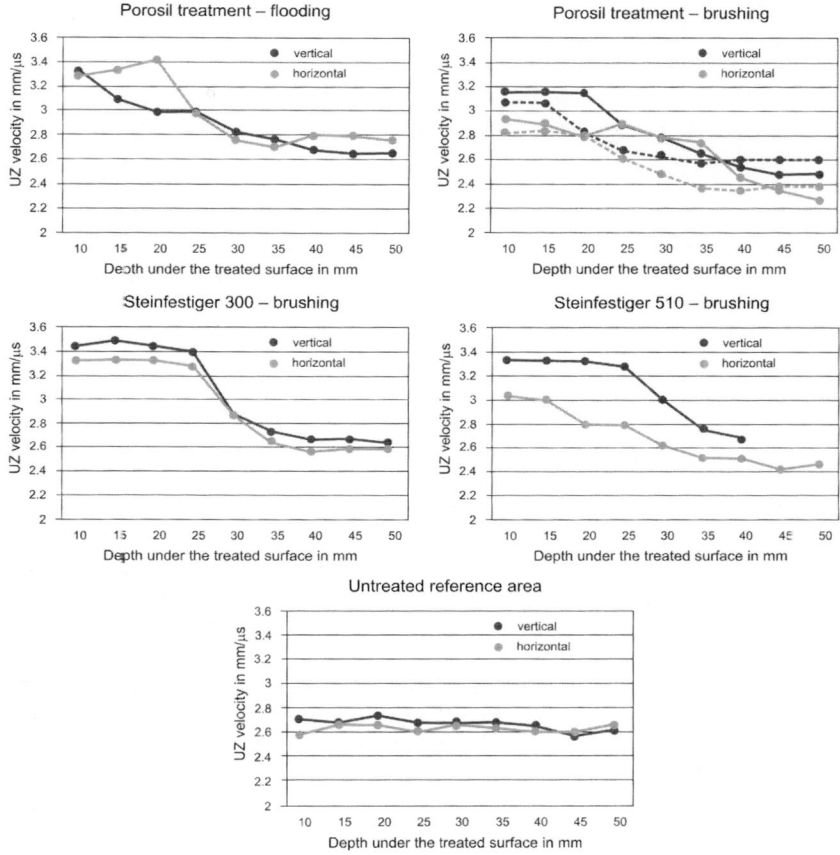

Fig. 8.7 Comparison of ultrasonic velocities in the wall subsurface areas treated with different consolidation agents

References

1. Tabasso Laurenzi M, Simon S. Testing methods and criteria for the selection/evaluation of products for the conservation of porous building materials. Rev Conserv. 2006;7:67–82.
2. Sasse HR, Snethlage R. Evaluation of stone consolidation treatments science and technology for cultural heritage. J Com Natl Sci Tecnol Beni Cult CNR. 1996;5(1):85–92.
3. Snethlage R, Pfanner M. Leitfaden Steinkonservierung. Stuttgart: Fraunhofer IRB Verlag; 2013.
4. Drdácký MF, Slížková Z. Performance of glauconitic sandstone treated with ethylsilicate consolidation agents. In: Łukaszewicz JW, Niemcewicz P, editors. Proceedings of the 11th International congress on deterioration and conservation of stone, vol. 2. Toruń: Nicolaus Copernicus University Press; 2008. p. 1205–12.

5. Giovan MN, Sines G. Biaxial and uniaxial data for statistical comparisons of a ceramic's strength. J Am Ceram Soc. 1979;62:510–5.
6. Fessler H, Fricker DC. A theoretical analysis of the ring-on-ring loading disc test. J Am Ceram Soc. 1984;67:582–8.
7. Chung SM, Yap AU, Chandra SP, et al. Flexural strength of dental composite restoratives: comparison of biaxial and three-point bending test. J Biomed Mater Res B Appl Biomater. 2004;71B:278–83.
8. Xu Y, Han J, Lin H, An L. Comparative study of flexural strength test methods on CAD/CAM Y-TZP dental ceramics. Regen Biomater. 2015;2(4):239–44.
9. Kim J, Kim DJ, Zi G. Improvement of the biaxial flexure test method for concrete. Cem Concr Comp. 2013;37:154–0.
10. Kim J, Yi C, Zi G. Biaxial flexural strength of concrete by two different methods. Mag Concr Res. 2012;64(12):1057–65.
11. Zi G, Oh H, Park S-K. A novel indirect tensile test method to measure the biaxial tensile strength of concretes and other quasibrittle materials. Cem Concr Res. 2008;38(6):751.
12. Wittmann FH, Prim P. Mesures de l'effet consolidant d'un produit de traitment. Mater Constr. 1983;16(94):235–42.
13. Danzer R, Supancic P, Harrer W, Wang Z, Börger A. Biaxial strength testing on mini specimens. In: Gdoutos EE, editor. Fracture of nano and engineering materials and structures. Dordrecht: Springer; 2006. p. 589–90.
14. Drdácký M, Lesák J, Rescic S, Slížková Z, Tiano P, Valach J. Standardization of peeling tests for assessing the cohesion and consolidation characteristics of historic stone surfaces. Mater Struct. 2012;45(4):505–20.
15. Ziegenbalg G, Drdácký M, Dietze C, Schuch D. Nanomaterials in architecture and art conservation. Singapore: Pan Stanford Publishing Pte Ltd.; 2017 (in print).

Chapter 9
Challenges of Alkoxysilane-Based Consolidants for Carbonate Stones: From Neat TEOS to Multipurpose Hybrid Nanomaterials

Bruno Sena da Fonseca, Ana Paula Ferreira Pinto, Susana Piçarra, and Maria de Fátima Montemor

9.1 Stone Consolidation in Built Heritage

Porous carbonate stones are generally easy to carve but are also very durable; important advantages that combined with their worldwide availability, have made this material the selected building stone for many millennia, being still used in a wide range of modern constructions. Most of the porous carbonate stones that have been used as building material are durable; however their chemical and physical properties are not everlasting, and the degradation of stone elements from cultural heritage through physical, chemical, and mechanical processes often results in loss of their value. Typical decay mechanisms in porous carbonate stones result into poor cohesion, which causes the loss of material and thus the reduction of the artistic, historic, and material value due to disappearance of the original geometrical features, such as the highly carved surfaces. Generally, the strength of porous carbonate stones mainly derives from the cohesion provided by "mineral bridges" (e.g., calcite intergrowth) and mechanical indentation (mechanical interlocking effect) [1]. Therefore, the loss of cohesion means the breakdown or weakening of these agglomerating mechanisms that typically occur in zones located near the surface, at a major or minor depth.

B. Sena da Fonseca (✉) • M.F. Montemor
Centro de Química Estrutural, CQE, DEQ, Instituto Superior Técnico,
Universidade de Lisboa, Lisboa, Portugal
e-mail: senadafonseca@gmail.com

A.P. Ferreira Pinto
Department of Civil Engineering, Architecture and Georesources, CERIS, ICIST,
Instituto Superior Técnico, Universidade de Lisboa, Lisboa, Portugal

S. Piçarra
Centro de Química Estrutural, CQE, DEQ, Instituto Superior Técnico,
Universidade de Lisboa, Lisboa, Portugal

Escola Superior de Tecnologia de Setúbal, Centro de Desenvolvimento do Produto e
Transferência de Tecnologia, Instituto Politécnico de Setúbal, Setúbal, Portugal

© Springer International Publishing AG 2018
M. Hosseini, I. Karapanagiotis (eds.), *Advanced Materials for the Conservation of Stone*, https://doi.org/10.1007/978-3-319-72260-3_9

Fig. 9.1 Mechanisms of stone consolidation: cohesion promoted by filling process (**a**) and cohesion promoted by "gluing" process (**b**). Schematic representation of a silica network from alkoxysilane consolidant (**c**)

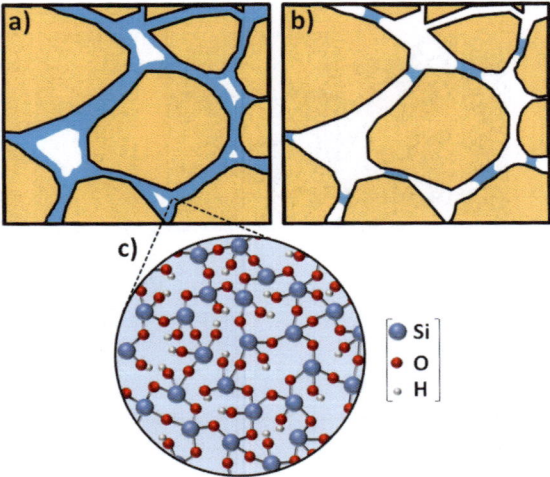

Consolidation actions are intended to restore the cohesion between stone particles, binding them and securing into the underlying sound stone to reduce the rate of stone decay [2, 3] and to avoid an imminent and irreparable loss of material, ultimately preventing the replacement of the damaged object [4]. The solid materials left within the stone pores by most of the stone consolidants, including alkoxysilanes, may act via two different mechanisms [5, 6]: locking the grains by filling the interstices between the loosened grains (Fig. 9.1a) and "gluing" grains together by forming adhesive bridges between adjacent grains (Fig. 9.1b).

Independently of the consolidation mechanism, the enhancement of stone integrity minimizes and/or avoids per se the accelerated loss of material, preserving the current state of the object and its historical, cultural, aesthetic, or social relevance. Although not all the degradation forms can be addressed by consolidation treatments, which should be considered when the decay is in the form of grain disintegration (e.g., powdering, sugaring, sanding) [7], some authors also mention that particular cases involving scaling can be addressed through consolidation treatments [8]. In fact, stone consolidation is a remedial conservation action and should only be carried out when the objects are in a fragile condition, with a strong possibility of being lost in a short time span [9]. In this way, it is important to distinguish requirements for consolidation treatments on severely damaged porous carbonate stone objects from built heritage and requirements for actions intending to enhance mechanical and other properties of sound stone to obtain appropriate building materials for application in new construction. Several authors also emphasize the distinct characteristics between consolidation treatments and protection treatments, since the main aim of a consolidation treatment is not to protect [10, 11].

A consolidation treatment in built heritage only makes sense when integrated in a well-designed conservation plan [4] to ensure long-term survival of the built heritage, being imperatively guided by important conservation principles.

Multidisciplinary approaches respecting the values and authenticity of the objects, and based on the full understanding of the material characteristics to reduce the risks of incompatibility, are required for stone consolidation as in any other conservation actions. Besides, no action shall be undertaken without having confirmed the possible benefits and potential harms on the heritage. If a consolidation treatment or other conservation action is regarded as indispensable, minimum intervention should be adopted, the safety of the original stone object must be preserved, and the treatment should not have negative consequences, including long-term impacts, to avoid undesirable side effects (compatibility principle) [12]. Furthermore, the maintenance of the consolidation action should be adequately proven over a sufficiently long period of time (durability). It is important to stress that the commonly known "artificial aging tests" when used alone access not the durability of the consolidation action but the delayed harmfulness effects that took longer to become perceptible after consolidation [3].

In addition to durability, retreatability is another priority aspect in stone consolidation of built heritage [13]. It is well known that conservation actions are intended to be reversible, so that they can be replaced by others in the future. However it is also recognized that complete reversible actions in stone consolidation are virtually impossible to achieve; hence any consolidation action should not hinder or limit future interventions (principle of retreatability).

9.2   Alkoxysilanes in Stone Conservation

Alkoxysilane-based products, generically (but often inaccurately) called as ethyl silicates, are possibly the most well-established and used stone consolidants. These products have been in the market for more than four decades, and despite the constant chemical modifications to improve their performance or to address environmental and safety issues, the essential constituents responsible for the consolidation action (precursors) have remained almost unchanged over time. Alkoxysilanes containing three or four alkoxy groups are key components in the majority of stone consolidants because of their ability to form a continuous tridimensional network (Fig. 9.1c). As a matter of fact, tetraethoxysilane (TEOS) is the most used alkoxysilane precursor in commercial products, and it has been intensely studied in the stone consolidation field [14].

It is recognized that alkoxysilane-based products are able to answer several of the basic technical requirements in stone consolidation, being cited due to the following advantages: low viscosity and thus good ability to penetrate deeply within the stone pores [15, 16]; chemical stability (thermal, oxidative, and photochemical) [14, 15]; the compounds that result from the chemical reactions involved on the network formation are not deleterious to stone [15, 17]; reduced tendency to form superficial hardened layers due to over-consolidation (Fig. 9.2 shows an example of the drastic consequences of a superficial hardened layer owing to a consolidation treatment) [15, 16]; reduced alterations on visual appearance and water transport

Fig. 9.2 Failure of a
porous carbonate stone as
consequence of a
consolidation treatment
that formed a superficial
hardened layer

properties, which points out the interesting potential of retreatability with a low risk
of incompatibility [15, 16]; versatility, being suitable for treatment of a wide range
of porous materials (e.g., limestone, sandstone, and brick ceramic) [18]; and easy to
apply [18].

Herein, the major differences between the former and current generations of
commercial alkoxysilane-based stone consolidants are briefly discussed. Presently,
alkoxysilane-based consolidants are mainly composed of oligomers containing pre-
condensed units of 4–6 molecules derived from TEOS [19], while the past consolidants were tendentiously composed of monomeric alkoxysilanes [14]. Furthermore,
in the past, harmful organic solvents like toluene, butanone, or mineral spirits were
widely employed to lower viscosity, while presently many of these solvents have
been abandoned from stone consolidant formulations, which are now considered
solvent-free (although alcohols are formed as by-products of the hydrolysis reactions). Nevertheless, some alkoxysilane-based stone consolidants are being used
with cosolvents in their compositions [20].

Currently, two types of commercial products are available: (1) those considered
as pure consolidants, which are almost exclusively composed of TEOS and/or their
oligomers as active components and (2) those having a twofold role, which in addition to the consolidation action also provide hydrophobic effects. Hydrophobicity is
usually achieved by the presence of nonpolar methyl-terminated groups introduced
either in the form of trifunctional alkoxysilanes (e.g., MTMOS) or of polymeric
organosilicon compounds (e.g., PDMS in CTS Estel 1100).

To form a solid material able to act as a consolidant, functional alkoxysilanes
follow the sol-gel process, where the most important chemical reactions taking
place are hydrolysis and condensation. The reactions generate three-dimensional
species with increasing sizes, and gelation occurs when a sufficient portion of

molecules has been condensed to each other. The resulting gel is a substance that contains a continuous network composed of Si—O—Si bonds, enclosing a liquid phase that prevents the solid from collapsing [21]. The evaporation of alcohol and water (by-products of sol-gel reactions) enclosed within the network increases the inorganic polymerization as the sol-gel reactions proceed at lower rates. However, the formation of a suitable Si—O—Si network is not simple because the relative rates between hydrolysis and condensation, as well as the reactions mechanisms, are very sensitive to a wide range of factors (e.g., pH, type of catalyst, solvent nature and its relative amount, water content, temperature, presence of other ions, and type of precursors) [22–25]. Typically commercial consolidants are composed of a precursor, eventually a solvent, and a catalyst, often an organometallic compound. Among other catalysts, dibutyltindilaurate (DBTL) is a well-established organotin neutral catalyst that is present in small amounts in several commercial stone consolidants [14, 26]. The catalysis mechanism of these tin compounds starts with their own hydrolysis, with lauric acid being the reaction side product. The resulting tin-hydroxy compound reacts with the precursor (non-hydrolyzed) to create a tin silox-ane, which will further react with available silanols leading to the formation of Si—O—Si bonds. The resulting product of the last reaction is once again a tin-hydroxyl compound that will continue to react, generating more siloxane bonds [26, 27]. Contrarily to acidic and basic catalysts, there is no detailed information about the silica network formation paths and the real role of the reaction by-products (lauric acid) on the properties of the resulting gel network. It has been proposed that the first step of the sol-gel process (hydrolysis) follows a path similar to that of the basic catalysis [28], leading to the formation of either "linear swollen" or "multiparticle diffusion-limited aggregates" [29]. One of the major drawbacks of DBTL is its toxicity (reproductive and mutagenic), which has forced formulators to replace this catalyst by less toxic (and less effective) ones, but still organotin compounds, whose associated mechanisms are generally similar [20, 27].

In every alkoxysilane-based consolidant, the presence of water is fundamental, as the tin compounds and silanols need to undergo hydrolysis before the next reactions take place. In fact, these catalysts essentially influence the condensation reactions and have an indirect effect in the first step of sol-gel process, thus reducing the amount of water molecules required for the precursors' hydrolysis. However, alkoxysilanes were, and still are, applied "neat" or without water and solvent, with the majority of commercial products using atmospheric moisture or water within the capillary pores [30]. One of the last commercial products that contained water was Brethane® [31, 32] which was mainly composed of a mixture of monomeric methyltrimethoxysilane (MTMOS) as precursor, ethanol, water, and lead naphthenate [32]. The chemistry of this product allowed it to be miscible with the moisture inside stone, had relatively short gel times (2–6 h) and low tendency to crack [32], contrary to those commercially available today.

However, there are some issues constraining the potential efficacy and overall performance of alkoxysilanes in the consolidation of stones, particularly porous carbonate stones (Fig. 9.3). It is known that carbonate media may affect the polymerization of alkoxysilanes [30], while it is also reported that there is a lack of strong

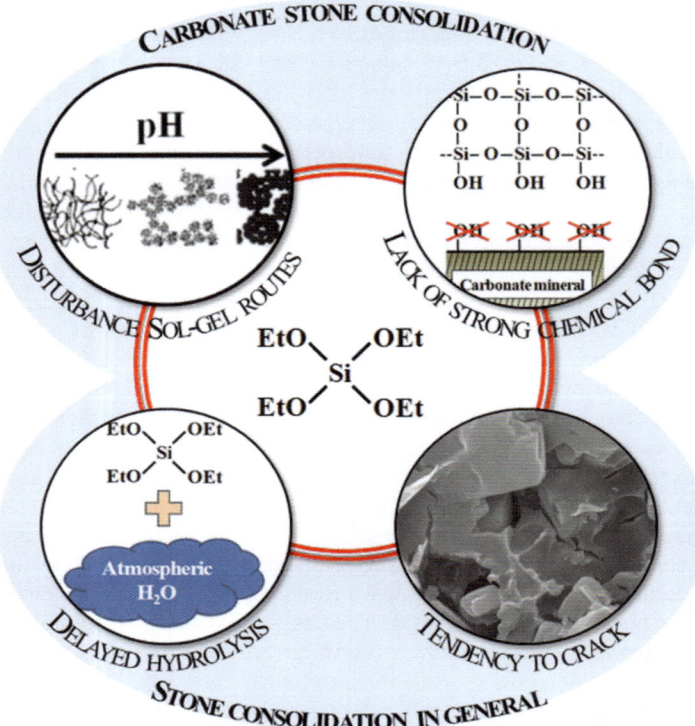

Fig. 9.3 Drawbacks and constrains commonly pointed out to the use of alkoxysilanes as consolidants for carbonate stones

chemical bonding with carbonate lithotypes [16, 19]. In this context, it is important to distinguish the carbonate stones having low porosity containing fissure-shaped narrow voids (marbles) from carbonate stones having high porosity with pore-shaped voids (most limestone varieties) because the challenges for a successful consolidation with alkoxysilane-based products are different in both cases. Some consolidation effects with alkoxysilane-based treatments have been easily achieved in marbles containing fissure-shaped narrow voids due to filling processes. The consolidating material interlocks the loosen grains through a wedging effect [33, 34]. However this mechanism can hardly be effective and durable in carbonate stones which are characterized by pore-shaped voids and are composed of very small grains such as those found in several types of limestone with fine-medium pore sizes present in European historical heritage.

Additionally, there are two other aspects that can be considered as general drawbacks constraining the overall performance of conventional alkoxysilanes: delayed hydrolysis reactions after product application [35, 36] and the gel's tendency to crack during the drying stage [37, 38].

To address some of the identified problems, commercial alternatives have been proposed such as the introduction of elasticized products that are intended to minimize the tendency to crack of ordinary alkoxysilane-based consolidants (e.g., KSE 300 E® and ESTEL 1100®) or the use of specific agents to enhance the adhesion to carbonate substrates (e.g., KSE 300 HV®).

With regard to products especially designed to consolidate porous carbonate stones (KSE 300 HV®), good initial efficacies have been reported in some porous limestones, while no relevant strength increments were obtained in depth for other varieties of limestones [39]. Although certain improvements were obtained in depth, the product (KSE 300 HV®) may create a superficial hardened zone (Fig. 9.4a) that can be responsible for an increased risk of failure [40]. Similarly to the most common alkoxysilane-based consolidants, this product also exhibits a tendency to crack (Fig. 9.4b) but has some potential to effectively improve the strength of certain limestone and to succeed in terms of efficacy, if the superficial hardening can be avoided and its durability ascertained.

Despite the improvements described, several modified commercial alkoxysilanes have not satisfactorily solved all the problems of porous carbonate stone consolidation. Therefore, there is an ongoing search for new and better solutions that can address the development of improved alkoxysilanes and of other alternatives (e.g., nanolimes, solutions based in ammonium oxalate or diammonium hydrogen phosphate, and treatments based in bacterial biomineralization).

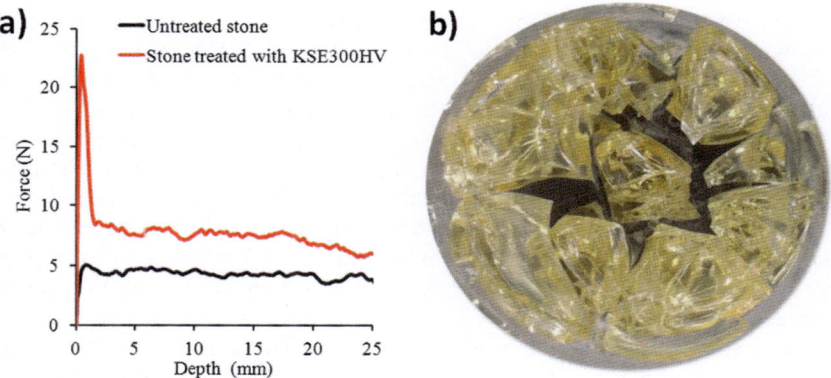

Fig. 9.4 Drilling resistance results of a porous carbonate stone before and after treatment with a commercial alkoxysilane containing adhesion promoters (application by capillary absorption for 3 h) (**a**); cracked xerogel resulting from the polymerization of the same commercial alkoxysilane (**b**)

9.3   Understanding Consolidants to Overcome Challenges

9.3.1   Influence of Carbonate Media on Sol-Gel Processes

Several authors noted an unexpected and excessive evaporation in some tetra-functional alkoxysilane precursors after application into carbonate stones [14, 32]. Less than 4% of solid materials were left within the stone pores of treated limestones (when compared to the initial mass of the original consolidant product), while 25% were obtained when treating sandstones [41]. The phenomenon has been attributed to an excessive evaporation of precursor monomers and by the formation of low molecular species prior to gelation. However, this issue has been a matter of intense debate often involving contradictory perspectives. Nevertheless, it seems intuitive that the excessive evaporation results from modifications of the sol-gel path, but no solid conclusions have been established yet. Danehey et al. [42] studied the sol-gel reactions of a formulation containing MTMOS as precursor in coexistence with quartz and calcite by means of ^{29}Si NMR analysis. The authors concluded that the quartz had little influence on the hydrolysis and condensation routes, while calcite dramatically slowed down the condensation reactions due to surface effects. Goins et al. [43] also adopted a similar procedure that followed the sol-gel reactions of MTMOS formulations in coexistence with carbonate and non-carbonate stones (limestone, marble, and sandstone). Contrary to the explanations provided by Danehey et al. [42], Goins et al. [43] found that the hydrolysis reactions, instead of condensation, were slower under the carbonate environment, while the shortest gelling times obtained could reflect an intensification of the condensation reactions [30, 44].

Despite such conclusions, the excessive evaporation of precursor, lack of efficacy, and the influence of the substrate on the hydrolysis and condensation reactions are still not fully understood. As a matter of fact, by altering the hydrolysis rate, the subsequent condensation reactions also change being stated that, in addition to an excessive evaporation, the carbonate media would change the silica growth route, originating materials with distinct textural and structural properties from the expected ones. This will certainly cause distinct physical and mechanical properties and better or worse possibilities to provide effective consolidation.

The hypothesis of pH change due to carbonate media was raised and briefly discussed as a possible explanation for the changes in the sol-gel process and resulting structures and seems to agree with the distinct morphological features typical from acid (in siliceous varieties) and basic/slightly acid catalysis (in carbonate varieties) [30].

Interestingly, no references were found about excessive evaporation of precursors owing to an apparent slow down/change or inhibition of the sol-gel reactions due to the carbonate environment in new commercial consolidants. In fact, the susceptibility of alkoxysilanes to be influenced by the carbonate media is extremely dependent on their pre-condensation degree [45]. Therefore, the phenomena is significantly minimized in commercial consolidants mainly because pre-polymerized solutions composed of species with higher molecular weight are now used [17]. Another possible contribution to minimize the influence of the carbonate media is the use of organotin catalysts that theoretically manage the condensation mechanisms and make the overall sol-gel process less susceptible to pH fluctuations, although not completely.

Nevertheless, for the development of new and different solutions to consolidate carbonate stones with alkoxysilanes, this is an important issue that needs to be carefully addressed and investigated further. Therefore, one of the first challenges in the development of new alkoxysilane-based stone consolidants to treat carbonate stones is to guarantee that the sol-gel process and polymerization mechanisms will occur within stone pores according to expectations/predictions in order to enable an effective consolidation action. This can be achieved by using pre-condensed formulations and by the careful selection of sol-gel catalysts. However, as the degree of pre-polymerization influences the viscosity and thus its ability to penetrate within porous materials, it is always essential to guarantee that the product has low viscosity and reasonable gelling times for the same reasons. To be effective, stone consolidants need to reach the underlying sound stone, and significant penetration depths are often required. Thus, several authors have purposed minimal penetration depths within 15 and 20 mm as a requirement for stone consolidants [1].

9.3.2 Organo-Functional Alkoxysilanes as Adhesion Promoters

Although there are reported significant strength increments provided by alkoxysilane products without demonstrating strong adhesion to carbonate stones, it is generally assumed that this is an important drawback in the consolidation of porous carbonate stones. Carbonate substrates do not have anchor points on their surfaces that can chemically react with the hydrolyzed species and cannot establishing strong chemical bonds between the consolidating material, composed by a silica network, and the loose carbonate grains (Fig. 9.5), contrary to siliceous substrates [16, 34].

Goins et al. [43] highlighted this phenomena by applying neat MTMOS in polished quartz and calcite crystals' surfaces, as well as in both mineral powders. While the quartz crystals showed fuzzy and etched surfaces and a monolithic powder was obtained, the calcite powder remained unconsolidated, an effect that suggests a lack of efficacy in enhancing the powder cohesion. Without strong chemical bonds, the consolidation action of carbonate stones occurs due to the physical lock of loosen grains. Nevertheless, mixed consolidation mechanisms or exclusively adhesive bridges are often preferable due to the lower influence on the porous structure.

The challenge of improving the adhesion between the silica network and carbonate substrates has been addressed by following two different strategies: a hybrid route (i.e., by combining common precursors with organo-functional alkoxysilanes that can act as coupling agents to improve chemical adhesion) or by using primers before applying conventional alkoxysilanes.

B. Sena da Fonseca et al.

Fig. 9.5 SEM image of a porous limestone treated with a commercial ethyl silicate, evidencing the inexistence of strong adhesion between carbonate minerals and consolidating material

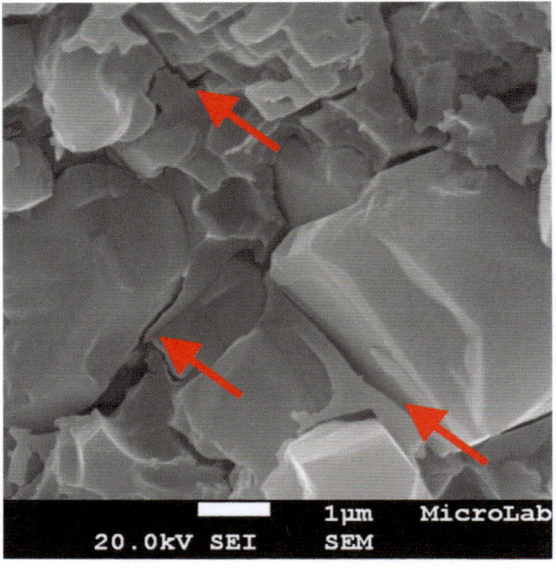

The use of organo-functional alkoxysilanes to consolidate carbonate stones is not recent and dates back to the early 1990s. One of the first attempts to evaluate the consolidation effect of hybrid formulations on carbonate stones was performed by Wheeler et al. [46] who tested the use of four trimethoxysilanes with adhesion promoters (i.e., diamine, amine, acrylic, and epoxy groups) either alone or in MTMOS-based solutions. Despite the nonacceptable color modifications, promising consolidation actions were reported, and significant improvements on the rupture modulus were obtained for limestones treated with amino-trimethoxysilanes. It was verified that coherent films covering calcite crystals were obtained with phosphate and amino alkoxysilanes, revealing that both may have a place in improving the performance of carbonate stone consolidants [43]. More recently these two functional groups were tested further as adhesion promoters, either by incorporation in conventional alkoxysilane formulations or by carbonate substrate functionalization as a primer for subsequent treatments [47]. Specifically, the amino-functional alkoxysilanes (3-aminopropyl) triethoxysilane (APTES) and N1-(3-trimethoxysilylpropyl) diethylene-triamine (SiDETA) were tested alone or in combination with TEOS-based laboratory-developed consolidants. A polyethylene glycol (PEG) containing formulation was also proposed to create a product with low tendency to crack and enhanced adhesion (due to SiDETA) [47]. The incorporation of SiDETA did not improve the drilling resistance of the treated stone, but it caused important changes on the final textural parameters of the consolidating materials as it had a nucleation effect in the TEOS-based product [47]. However, depending on the characteristics of the original formulation, SiDETA incorporation could cause additional strength gains in the carbonate stones (Fig. 9.6a). Nevertheless, the

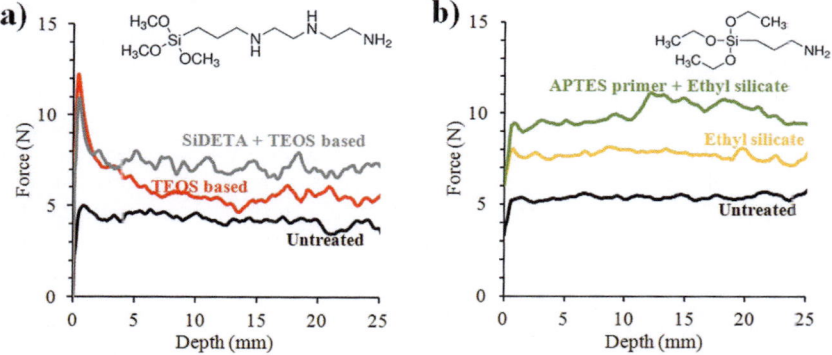

Fig. 9.6 Drilling resistance of untreated and treated porous carbonate stone by capillary absorption (3 h). Treated with a TEOS consolidant developed in laboratory and with the same consolidant modified with SiDETA (**a**) (adapted from [47]); treated with a commercial consolidant with and without APTES as primer (**b**) (adapted from [48])

authors could not conclude whether these additional gains were due to effective chemical adhesion of the silica matrix to the carbonate substrate or to modification on the silica network itself. Therefore, conclusions about the incorporation of amino-functional alkoxysilanes in TEOS-based solutions are not simple since the amino groups also play an important role on the sol-gel processes, which may hinder the use of the modified formulation as stone consolidant (changes on viscosity, precipitation of silica, etc.).

The use of organo-functional alkoxysilanes as primers aims at functionalizing carbonate minerals to create stronger chemical interactions between the substrate and conventional alkoxysilane-based posttreatments. Contrarily to the previous approach, functionalization reduces the risk of unexpected changes on the sol-gel routes of consolidants. This strategy provided mixed results, but illustrates the capabilities to improve further the consolidation action of ethyl silicates [48, 49]. Ferreira Pinto and Delgado Rodrigues [48] found clear improvements on the consolidation action of ethyl silicate-based product by using APTES as primer in porous limestone (Fig. 9.6b). However, Xu et al. [49] functionalized limestone powder with 3-aminopropyl methyl dimethoxysilane (AE-APMDMS) to be post-treated with a TEOS-based consolidant and obtained surprisingly low compressive resistance and superficial hardness, much lower than that obtained in samples exclusively treated with the TEOS-based consolidant.

Although strength improvements can be obtained in certain circumstances by adopting both strategies, the reported results still raise some doubts about the use of coupling agents and its effective role on the performance of alkoxysilane products. Whereas in addition to some strength increments, no practical gains in the overall consolidation performance were experimentally demonstrated.

9.3.3   Delayed Hydrolysis Reactions

Neutral catalysts commonly used in commercial consolidants have a direct influence on the condensation reactions but can indirectly decrease the overall water demand to complete the hydrolysis reactions. However, the presence of external water is still required by most stone consolidants which is usually supplied by atmospheric moisture. Consequently, despite various manufacturers recommend application of their products in environments where relative humidity is greater than 40% [50], some authors detected uncompleted reactions even after several months at fairly higher relative humidity [51, 52]. Therefore, it is unanimous that the sol-gel reactions of these systems require more time to stabilize than that usually recommended in the technical data sheets [35, 53, 54], despite the fact that the transition from sol to gel is relatively fast [37]. Two consequences of this are that the final consolidation action is only reached several months after the application and the treatment confers a hydrophobic effect to the stone surface for several months. The reason behind this phenomenon is related to the residual ethoxy groups from unhydrolyzed precursors that are nonpolar and thus have hydrophobic characteristics before hydrolysis is completed. Therefore, the fluid transport properties of the treated stones are time dependent and rely upon the hydrolysis and condensation rates/degrees even after a solid state has been reached [36].

The hydrophobic behavior, even if provisional for several months, can be an important constraint for other conservation activities [14, 55] and can be a factor to increase the risk of salt crystallization damage when the stone has a source of water and soluble salts behind the treated zone [55]. This effect will be discussed later. Another problem associated to alkoxysilane products relies on the fact that atmospheric water is needed in sol-gel reactions and that heterogeneous reactions may occur when the substrate has some water in the liquid form as these products are immiscible with liquid water [17]. Therefore, for the stone to be treated, it must be as dried as much as possible before the treatments are applied [17].

The challenge of producing formulations with no hydrophobic time span has been addressed through the optimization of application procedures and treatment conditions by supplying additional water to enhance the hydrolysis/condensation reactions and to reduce the duration of the hydrophobic effect. Adding water prior to the conventional consolidant application is not a suitable option, since most of the consolidants are immiscible with it. Consequently some authors have proposed to accelerate the reactions of commercial consolidants by supplying water after consolidant application, either by rinsing treated stones with $EtOH:H_2O$ solutions or by using water-impregnated cellulose pulp poultices [35, 54]. Although none of the options completely reestablished the water absorption of the original stone, both methods proved to be effective in accelerating the hydrolysis reactions. Another approach to avoid the provisional hydrophobic behavior is to develop formulations that already contain the amount of water demanded by the precursors. To overcome the immiscibility between silica precursors and water, several strategies can be followed, with the use of a cosolvent being the simplest [45]. The use of intense

Fig. 9.7 Capillary water absorption of a porous carbonate stone: untreated (black), treated with a laboratorial developed formulation containing water on its composition 2 months after application (blue), treated with a "pure" commercial ethyl silicate consolidant (BS) 2 months after application (red) and 10 months after application (green)

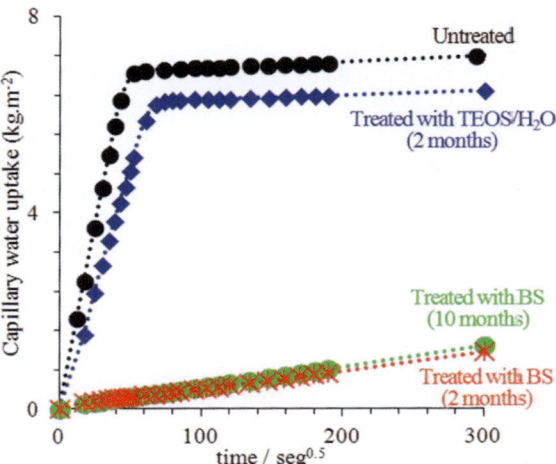

ultrasonication also enables the creation of suitable mixtures when surfactants are employed [56, 57]. In fact, this strategy is being used to produce hydrophobic, water-repellent, or self-cleaning treatments for porous building stones [57, 58]. Furthermore, the use of a solvent was successfully used to develop products with an exclusive consolidation purpose that did not rely on atmospheric moisture to treat porous carbonate stones [45]. Figure 9.7 shows the difference in the capillary water absorption between a porous limestone treated with a commercial pure consolidant based on atmospheric moisture and a consolidant containing water in its composition, with the ratio of $1TEOS/2.1H_2O$ (and ethanol).

9.3.4   Tendency to Crack

The major reported drawback influencing the performance of alkoxysilanes as stone consolidants is their tendency to crack during the drying stages [59]. The practical consequences of cracking within the stone pores have been widely reported in literature, as it leads to decreased mechanical strength of the consolidating material and reduces its overall performance [60]. A direct relation between the poor consolidation efficacy, incompatibility issues, or lack of long-term performance and the product's tendency to crack is, however, hard to prove since any attempt to reduce the cracking will also change the sol-gel routes, polymerization mechanisms, and consequently the structure of the final consolidating material of a given product.

This phenomenon is explained by the forces applied on the silica network during the drying of the liquid phase of the gel (i.e., capillary pressure, osmotic pressure, disjoining pressure, and moisture stress) [61, 62]. Evaporation of water and solvent

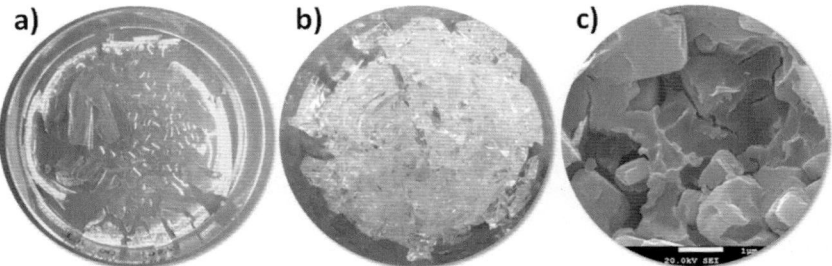

Fig. 9.8 Appearance of the dry residue of a commercial consolidant polymerized within a Petri dish (**a**), inside a vessel (**b**), and within stone pores (SEM) (**c**)

(by-products of sol-gel reactions) tends to expose the solid phase, which is compensated by a spread of the liquid from the interior to the exterior surface. Differential shrinkage occurs when the liquid evaporates from one surface of the gel (as in the majority of the cases) leading to faster shrinkage in the exterior drying surface than in more interior zones, resulting in differential stresses and the formation of cracks [63]. The magnitude of the stress on the silica network is proportional to the capillary stress which is dependent on the pore radius resulting from microstructure heterogeneities, making the cracking to occur more easily when pores have different sizes [61, 63]. Indeed, most of the commercial consolidant origin significantly cracked materials (Fig. 9.8).

The challenge of producing crack-free products was attempted by loading commercial consolidants with nanoparticles of different natures (e.g., Al_2O_3, TiO_2, and SiO_2) to act as fillers [64]. This strategy seems to be the simplest to achieve alkoxysilanes with reduced drying shrinkage and low cracking tendency, due to the nanoparticles' physical properties and the increase of the silica gel matrix porosity, which often has a particulate microstructure (Fig. 9.9a), with nanoparticles acting as nuclei for the precipitation of silica [64].

More complex approaches based on the development of new alkoxysilane-based formulations loaded with SiO_2 or hydroxyapatite nanoparticles have been proposed [65, 66].

In addition to their use in crack-free products, the incorporation of nanoparticles having different chemical and physical properties also allows to control the final properties of the consolidating material such as stiffness, thermal expansion, or color [64].

One important drawback of using conventional alkoxysilanes as consolidants is the absence of compatibility with regard to their contrasting physical properties in relation to carbonate stones, as the consolidating materials are composed of silica. Although their initial consolidation efficacy is not affected, undesirable stresses in the interface between treated and non-treated zones may occur when, for example, the stone is subject to heating caused by solar exposure or when the stone becomes wet [67]. The lack of compatibility is commonly cited as drawback, but important properties such as thermal expansion or linear expansion by water absorption are

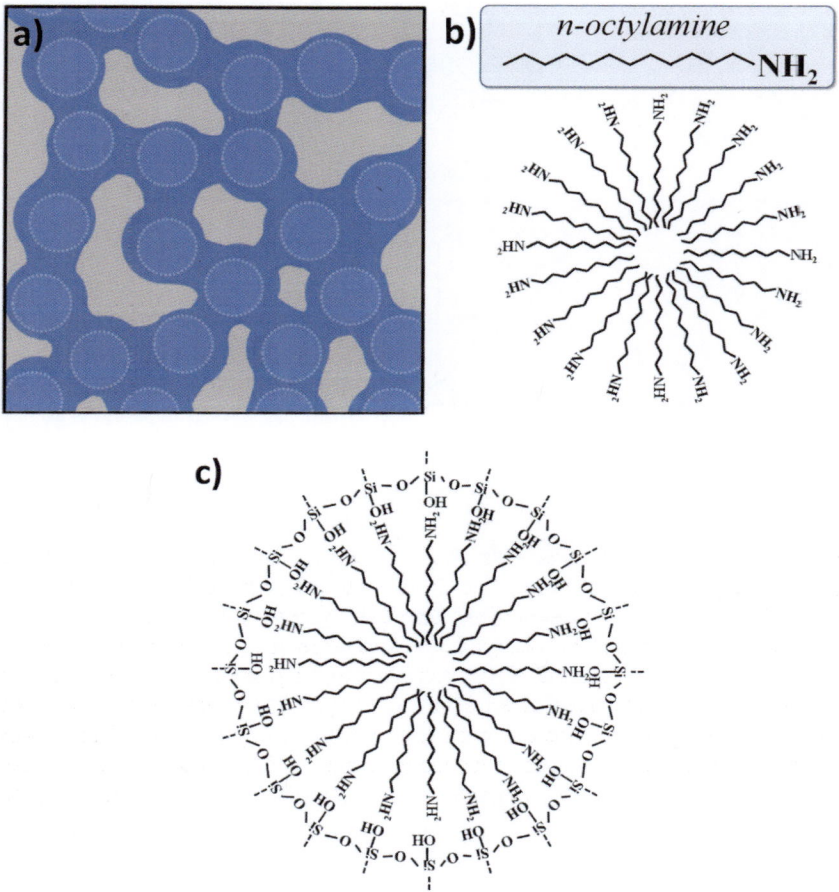

Fig. 9.9 Schematic representation of an alkoxysilane product loaded with nanoparticles, where nanoparticles act as nuclei for silica precipitation (dark blue) (**a**) (adapted from [64]); proposed mechanism for the silica network development by using n-octylamine (adapted from [84]) (**b, c**)

rarely assessed in treated stones. Nevertheless, there are some cases where the conventional alkoxysilanes created dangerous interfaces by increasing both coefficients in porous carbonate stones, but this action seems not to be a general rule for all situations [67]. In any case, the strategies used to prevent gel cracking constitute a good basis to tailor the properties of the consolidating materials according to the properties of the carbonate stones.

The development of alkoxysilane-based consolidants by hybrid routes is another alternative to produce crack-free products. The use of hydroxyl-terminated polydimethylsiloxane (PDMS-OH) [68] is a widespread approach, although suit-

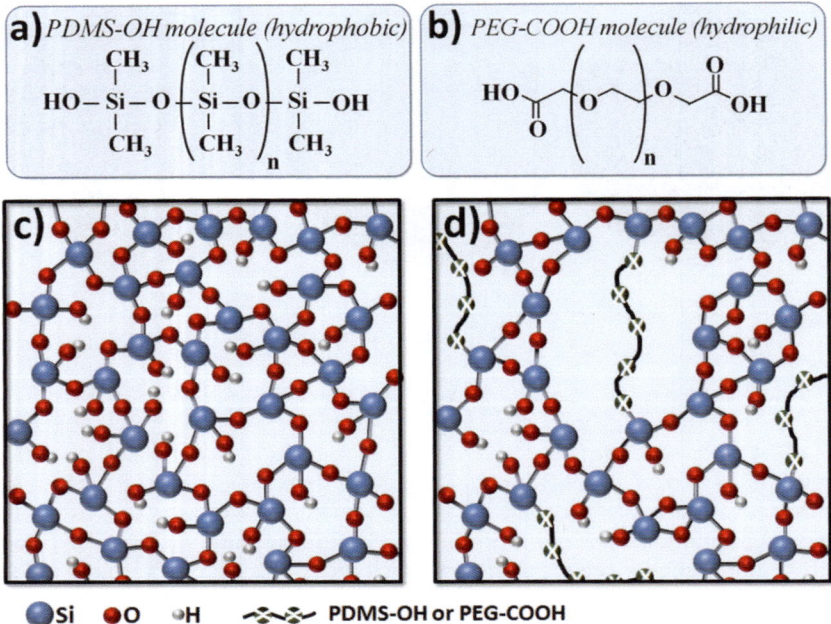

Fig. 9.10 "Chains" proposed by several authors to reduce the cracking tendency of alkoxysilanes: (**a**) PDMS and (**b**) PEG-COOH. Schematic representation of typical silica network (**c**) and modified with "chains" (**d**)

able results can be hardly obtained by simple incorporation of these chains into commercial consolidants. The PDMS-OH is used to manage xerogel rigidity as it can act as "elastic bridges" to produce modified network structures (Fig. 9.10). In fact, by changing the ratio of TEOS/PDMS-OH and the processing conditions, it is possible to produce xerogels with distinct physical and mechanical characteristics, varying from hard brittle solids to rubbery solids [69]. Depending on the reaction parameters, it can contribute to decrease shrinkage and to increase porosity, giving the gel a greater ability to resist to the stresses imposed by capillary pressure [68, 70]. Despite PDMS-OH being primarily used to reduce the tendency to crack, the formulations containing PDMS-OH are also hydrophobic because of the apolar character of the PDMS-OH chains.

The use of agents acting as template for the growth of the silica network (i.e., n-octylamine) was proposed to solve the same problem [37, 71]. N-octylamine tends to form micelles (Fig. 9.9b) that silanols can condense around (Fig. 9.9c). The development of a silica network from the n-octylamine micelles originates pores in the consolidating material, similarly to those formed when using nanoparticles. This approach causes the coarsening of the gel pores, which reduces the pressure on the silica matrix during the drying stage. In addition, the roughness caused by n-octylamine also reduces the hysteresis of the stone surface, providing water-repellent properties to treated stones [72].

The combined use of PDMS-OH and n-octylamine [70, 73, 74] produced non-negligible strength increments to silica-containing stones [56] and carbonate stones [72]. Additionally, exceptionally high water contact angles on the stone surfaces were obtained due to the combined effect of PDMS-OH's low surface energy and surface roughness derived from n-octylamine incorporation [73, 74]. These multi-purpose hybrid nanomaterials prevent the direct absorption of water, which is the main abiotic factor of stone degradation but can also provide staining resistance to the treated stone (e.g., graffiti) [72]. Supplementary functionalities were added in these already multipurpose solutions, namely, self-cleaning properties achieved by the photocatalytic activity of TiO_2 that can be incorporated in the form of nanoparticles [75, 76] or in the form of titanium alkoxides [77]. Gains in compressive strength and superficial hardness in silicate-containing stones were obtained with the abovementioned strategies [65, 66], while the ability of these solutions to provide protective or self-cleaning properties is featured more so than their performance as porous carbonate stone consolidants [78].

Multiple properties have been evaluated in these cases, which restrict a detailed analysis of all strands involving consolidation treatments in built heritage. It is important to note that satisfactory initial efficacies, assumed with slight increases on the tensile strength or drilling resistance, or significant penetration depths, although fundamental, are not sufficient to fully validate a potential stone consolidant whereupon compatibility and durability issues are also essential.

The multiple functions obtained by using most of the abovementioned strategies are interesting for particular cases where the intervention requires consolidation and hydrophobic or water repellency actions. In fact, it is likely that a product providing superficial and in-depth hydrophobicity/water repellency is more durable than the traditional superficial water-repellent treatments. However, this strategy may have several pitfalls, and a pure consolidant that lacks hydrophobicity may be often required in conservation practice, either due to conflicting activities (e.g., cleaning, pointing, grouting, plastic repair, etc.) [14, 55] or due to the risk of water/soluble salts trapped behind the treated zone [11, 79, 80]. According to several authors [11, 79, 80], consolidation treatments that have hydrophobic characteristics, especially in-depth, have a potential harmful effect if there is a source of water behind the treated zone, which is not uncommon in ancient buildings and monuments. If water is present behind the treated zone (through rising dump, infiltrations in joins, etc.), it tends to evaporate at the exterior surface, being the drying front coincident or almost coincident with the stone exterior surface (depending on the drying conditions). If soluble salts are dissolved in water, they tend to crystallize in the form of efflorescence (at a superficial level) and have a minimal risk of severely damaging the stone, as schematically represented in Fig. 9.11a. However, hydrophobic treatment in depth tends to recede the drying front (Fig. 9.11b) [11, 79], and the deposition of the soluble salts occurs in the transition zone between treated and untreated stone, nearby the receded drying front. The salts tend to accumulate in the transition zone, which brings increased risk of catastrophic damage. The good breathability of

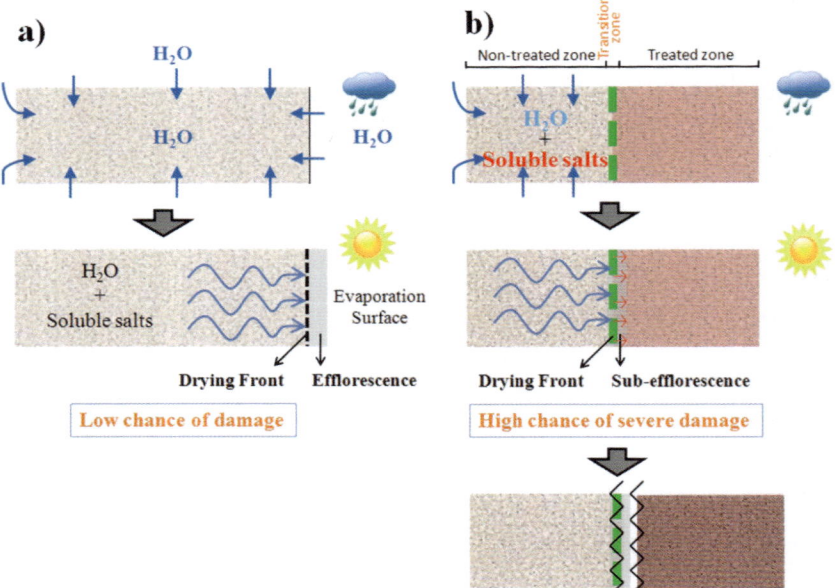

Fig. 9.11 Schematic representation of soluble salts crystallization in an untreated stone (**a**) and in a treated stone with a hydrophobic zone (**b**) (adapted from [79])

the treated zone is important in reducing the risk of damage by water freezing mechanisms, but it may not minimize the risk of damage by salt crystallization (sub-efflorescence). In fact, it can even potentiate such risk by forcing the deposition of salts at the interfacial zone [17].

Effective consolidants that do not constrain the circulation of liquid water within stone pores with reduced propensity to create marked interfaces can be considered an alternative option in these cases. Within this context, the use of non-hydrophobic PEG chains with end groups capable to react with the silica network was tested to consolidate porous carbonate stones (Fig. 9.10b) [45, 81]. Materials with reduced tendency to crack and good consolidation behavior when applied on porous carbonate stones were obtained, while minimal influence on the fluid transport properties of the stones was expected [45, 81]. The use of calcium oxalate nanoparticles, synthesized with calcium hydroxide and oxalic acid, also reduced the tendency of TEOS-based consolidants to crack without causing major effects on water transport properties of porous carbonate stones [82, 83]. This inorganic filler also intends to enhance the compatibility of the consolidating material and to create a mesoporous microstructure while it is also participating in the sol-gel processes.

The validation of these recently developed pure consolidants for the treatment of porous carbonate stones still requires confirmation through further experimentation and testing methods with focus on compatibility and durability issues.

9.4   Conclusion

Conventional alkoxysilane-based materials have recognized drawbacks that hinder successful consolidation treatments in carbonate stones from built heritage. Although commercial solutions specifically dedicated to consolidate carbonate stones with improved properties are available, there are still improvements needed as emphasized by the ongoing research toward more advanced solutions. Drawbacks of the available products are wide, and requirements for consolidation in built heritage are very demanding, as they encompass performance criteria at several levels while fulfilling conservation principles. The complexity and sensibility of sol-gel chemistry allied with the conceptual incompatibility between silica-composed consolidants and carbonate minerals does not allow predicting successful consolidation actions. However, the versatility of alkoxysilanes has enabled the development of multiple solutions with distinct properties and functionalities. It is important to distinguish pure consolidants from multipurpose solutions intended to consolidate and to create hydrophobic, water-repellent, or self-cleaning surfaces, as different concepts and requirements are clearly involved. The cracking tendency of the alkoxysilanes has been widely addressed where strategies relying in hybridization, surfactant-template, or nanoparticle loading are showing relevant improvements. Several of the strategies were initially developed and tested on siliceous stones where the main target was not necessarily conservation, although they were subsequently tested in carbonate stones from built heritage and showed encouraging consolidation actions. The proposed formulations often confer hydrophobic or water repellency to the treated stone, which can be interesting if the objective is to obtain simultaneously a hydrophobic/water-repellent and consolidation treatment, although the risk of incompatibility tends to increase. It has been established that no action should be undertaken without demonstrating that it is indispensable and that the intervention should guarantee the safety and durability of the objects. The temporary hydrophobicity is also commonly pointed as a constraint in commercial alkoxysilane-based consolidants due to its incompatibility with other conservation actions and the potential risk of creating temporary interfaces which can be solved by designing formulations relying in ternary systems (precursor/water/solvent). To reduce the tendency to crack in these formulations, without impairing the hydrophobic characteristics, the use of non-hydrophobic chains or hydrophilic nanoparticles is an interesting option. Further improvements can be performed by addressing the lack of strong chemical bonding by using organo-functional alkoxysilanes, although recent investigations can hardly identify practical advantages of their use. These formulations, intended to be pure consolidants, constitute an interesting alternative to complement multipurpose options for the consolidation of porous carbonate stones. The perturbation of sol-gel routes by the carbonate environment brings additional and often unconsidered challenges to the development of consolidants for carbonate stones. The carbonate media tends to influence the sol-gel reactions, not favoring the formation of solid materials with consolidation ability. Furthermore, the possibility of obtaining materials within the carbonate stone pores that are different from those obtained inside siliceous stones or vessels from the same initial formulation should not be discarded, no matter what catalytic system is employed. Despite the limitations and constraints pointed out, there are multiple proposals to consolidate porous carbonate stones with new alkoxysilane-based formulations that are able to provide consolidation actions at different degrees and at major or minor depths. This may make alkoxysilanes a viable alternative to consolidate carbonate stones in important built heritage again, if their compatibility and durability is properly validated.

Acknowledgments The authors acknowledge Fundação para a Ciência e Tecnologia (FCT) for the financial support, CQE—UID/QUI/00100/2013. The author B. Sena da Fonseca also acknowledges Fundação para a Ciência e Tecnologia (FCT) for the financial support through grant SFRH/BD/96226/2013.

References

1. Sasse HR. Engineering aspects of monument preservation. In: Restoration of buildings and monuments. 2001. p. 197.
2. Clifton JR. Stone consolidating materials: a status report. Washington, DC: U.S. G.P.O.; 1980.
3. Price CA, Doehne E. Stone conservation: an overview of current research. Santa Monica, CA: Getty Conservation Institute; 2011.
4. Ginell W, Wessel D, Searles C. ASTM E2167–01 standard guide for selection and use of stone consolidants. West Conshohocken, PA: ASTM International; 2001.
5. Clifton JR, Frohnsdorff GJC. Stone-consolidating materials: a status report. In: Conservation of historic stone buildings and monuments report, C.o.C.o.H.S.B.a.M.N. Materials. 1982. p. 287–311.
6. Foulks WG, editor. Historic Building Façades: the manual for maintenance and rehabilitation. New York: Wiley; 1997. p. 203.
7. Natali I, et al. Innovative consolidating products for stone materials: field exposure tests as a valid approach for assessing durability. Heritage Science. 2015;3(1):6.
8. Grimmer AE. A glossary of historic masonry deterioration problems and preservation treatments. In: Department of the interior, national park service, preservation assistance division. Washington, DC: For sale by the Supt. of Docs., U.S. G.P.O.; 1984. p. 65.
9. ICOM-CC, Resolution adopted by the ICOM-CC membership at the 15th Triennial Conference, New Delhi, 22–26 Sep 2008. p. 2.
10. Fidler J. Stone consolidants: inorganic treatments. Conserv Bull. 2004;44:33–5.
11. Matteini M. Inorganic treatments for the consolidation and protection of stone artefacts. Conserv Sci Cult Herit. 2008;8:13–27.
12. Cassar M. Education and training needs for the conservation and protection of cultural heritage: Is it a case of one size fits all'? Education and training needs for the conservation and protection of cultural Heritage. 2003:159–63.
13. Hauff G. Durability and retreatability of ethyl silicate treatments. Abstract of International Course on Stone Conservation SC13 – The Getty Conservation Institute. 2013. p. 1.
14. Wheeler G, G.C. Institute. Alkoxysilanes and the consolidation of stone. Los Angeles, CA: Getty Publications; 2005.
15. Ozturk I. Alkoxysilanes consolidation of stone and earthen building materials. University of Pennsylvania. 1992. p. 100.
16. Delgado Rodrigues J. Consolidation of decayed stones. A delicate problem with few practical solutions. In: Proc. Int. Seminar on Historical Construction. Guimarães. 2001. p. 3–14.

17. Scherer GW, Wheeler GS. Silicate consolidants for stone. Key Eng Mater. 2009;391:1–25.
18. De Clercq H, De Zanche S, Biscontin G. TEOS and time: the influence of application schedules on the effectiveness of ethyl silicate based Consolidants/Tetraethoxysilan (TEOS) und die Zeit: der Einfluss unterschiedlicher Anwendungsfolgen auf die Wirksamkeit von Steinfestigern auf der basis von Ethylsilikat. Restoration of Buildings and Monuments. 2007;13(5):305–18.
19. Snethlage R. Stone conservation, Springer Berlin Heidelberg. In: Siegesmund S, Snethlage R, editors. Stone in architecture. Berlin, Heidelberg; 2014. p. 415–550.
20. Tesser E, et al. Study of the stability of siloxane stone strengthening agents. Polym Degrad Stab. 2014;110:232–40.
21. Brinker CJ. Sol-Gel processing of silica. In: Colloidal silica. CRC Press; 2005. p. 615–35.
22. Sakka S. Handbook of sol-gel science and technology. 1. Sol-gel processing. Boston: Kluwer Academic Publishers; 2005.
23. Milea CA, Bogatu C, Duta A. The influence of parameters in silica sol-gel process. Braşov Bulletin of the Transilvania University. 2011;4(53):59–66.
24. Parkhill R. Investigation of water based silica and organically modified silicate sol-gel systems. Oklahoma: University of Oklahoma; 1999. p. 253.
25. Belton D., Deschaume O, Perry CC. An overview of the fundamentals of the chemistry of silica with relevance to biosilicification and technological advances. FEBS J. 2012;279(10):1710–20.
26. Hernández CS, et al. DBTL as neutral catalyst on TEOS/PDMS anticorrosive coating. J Sol-Gel Sci Technol. 2016:1–8.
27. Cervantes J, Zárraga R, Salazar-Hernández C. Organotin catalysts in organosilicon chemistry. Appl Organomet Chem. 2012;26(4):157–63.
28. Schiavon G. Sol-Gel derived nanocomposites synthesis. München: Technischen Universität München; 2000. p. 152.
29. Méndez-Vivar J. The interaction of dibutyltin dilaureate with tetraethyl orthosilicate in sol-gel systems. J Sol-Gel Sci Technol. 2006;38(2):159–66.
30. Goins ES. Alkoxysilane stone consolidants: the effect of the stone substrate upon the polymerization process. London: University College London (University of London); 1995. p. 200.
31. Morris RK, Coldstream N, Turner R. The West front of TINTERN Abbey Church, Monmouthshire. Antiqu J. 2015;95:119–50.
32. Horie CV. Materials for conservation: organic consolidants, adhesives and coatings. Amsterdam: Butterworth-Heinemann; 2010.
33. Ruedrich J, Weiss T, Siegesmund S. Thermal behaviour of weathered and consolidated marbles. Geol Soc Lond, Spec Publ. 2002;205(1):255–71.
34. Charola AE, Wheeler GE, Freund GG. The influence of relative humidity in the polymerization of methyl trimethoxy silane. Stud Conserv. 1984;29(Suppl 1):177–81.
35. Franzoni E, Graziani G, Sassoni E. TEOS-based treatments for stone consolidation: acceleration of hydrolysis–condensation reactions by poulticing. J Sol-Gel Sci Technol. 2015;74(2):398–405.
36. Karatasios I, et al. Modification of water transport properties of porous building stones caused by polymerization of silicon-based consolidation products. In: 16th International Conference on Polymers and Organic Chemistry. Hersonissos, Crete, Greece; 2016.
37. Mosquera MJ, et al. New nanomaterials for consolidating stone. Langmuir. 2008;24(6):2772–8.
38. Mosquera MJ, Pozo J, Esquivias L. Stress during drying of two stone consolidants applied in monumental conservation. J Sol-Gel Sci Technol. 2003;26(1–3):1227–31.
39. Berto T, Godts S, De Clercq H. The effects of commercial ethyl silicate based consolidation products on limestone. In: Science and art: a future for stone. 2016. p. 271–9.
40. Ferreira Pinto AP, Delgado Rodrigues J. Consolidation of carbonate stones: influence of treatment procedures on the strengthening action of consolidants. J Cult Herit. 2012;13(2):154–66.
41. Wheeler G, et al. Toward a better understanding of B72 acrylic resin/methyltrimethoxysilane stone consolidants. In: MRS Proceedings. Cambridge Univ Press; 1990.
42. Danehey C, Wheeler GS, Su SC. The influence of quartz and calcite on the polymerization of methyl-trimetoxysilane. In: Delgado Rodrigues J, Henriques F, editors. Seventh International Congress on the deterioration and conservation of stone. Lisbon: LNEC; 1992. p. 1043–52.

43. Goins ES, Wheeler G, Wypyski MT. Alkoxysilane film formation on quartz and calcite crystal surfaces. In: Proceedings of the Eighth International Congress on deterioration and conservation of stone. Berlin; 1996.
44. Goins ES, et al. The effect of sandstone, limestone, marble and sodium chloride on the polymerisation of MTMOS solutions. In: Proceedings of 8th Congress on deterioration and conservation of stone. Berlin; 1996. p. 1243–54.
45. Sena da Fonseca B, et al. Development of formulations based on TEOS-dicarboxylic acids for consolidation of carbonate stones. New J Chem. 2016;40(9):7493–503.
46. Wheeler G, Fleming S, Ebersole S. Comparative strengthening effect of several consolidants on Wallace Sandstone and Indiana Limestone. In: Proceedings of the 7th International Congress on deterioration and conservation of stone. Lisbon: LNEC; 1992.
47. Sena da Fonseca B, et al. TEOS-based consolidants for carbonate stones: the role of N1-(3-trimethoxysilylpropyl)diethylenetriamine. New J Chem. 2017;41(6):2458–67.
48. Ferreira Pinto AP, Delgado Rodrigues J. Hydroxylating conversion treatment and alkoxysilane coupling agent as pre-treatment for the consolidation of limestones with ethyl silicate. In: Mimoso JDRJM, editor. Proceedings of the International Symposium on Stone Consolidation in Cultural Heritage – Research and Practice. Lisbon: LNEC; 2008. p. 131–40.
49. Xu F, et al. Use of coupling agents for increasing passivants and cohesion ability of consolidant on limestone. Prog Org Coat. 2014;77(11):1613–8.
50. Wacker, Technical data sheet for SILRES® BS OH 100, SILRES, editor. 2014. p. 3.
51. Zornoza-Indart A, et al. Consolidation of a Tunisian bioclastic calcarenite: from conventional ethyl silicate products to nanostructured and nanoparticle based consolidants. Constr Build Mater. 2016;116:188–202.
52. Karatasios I, et al. Modification of water transport properties of porous building stones caused by polymerization of silicon-based consolidation products. In: Pure and Applied Chemistry. 2017.
53. Naidu S, Liu C, Scherer GW. Novel hydroxyapatite-based consolidant and the acceleration of hydrolysis of silicate-based consolidants. MRS Online Proceedings Library Archive. 2014;1656.
54. Naidu S, Liu C, Scherer GW. Hydroxyapatite-based consolidant and the acceleration of hydrolysis of silicate-based consolidants. J Cult Herit. 2015;16(1):94–101.
55. Sassoni E, et al. Consolidation of a porous limestone by means of a new treatment based on hydroxyapatite. In: Proceedings of 12th International Congress on deterioration and conservation of stone. New York City (USA); 2012.
56. Illescas JF, Mosquera MJ. Producing surfactant-synthesized nanomaterials in situ on a building substrate, without volatile organic compounds. ACS Appl Mater Interfaces. 2012;4(8):4259–69.
57. Zarzuela R, et al. CuO/SiO$_2$ Nanocomposites: a multifunctional coating for application on building stone. Mater Des. 2017;114:364–72.
58. Pinho L, et al. A novel TiO$_2$–SiO$_2$ nanocomposite converts a very friable stone into a self-cleaning building material. Appl Surf Sci. 2013;275(0):389–96.
59. Han X, et al. Bridged siloxanes as novel potential hybrid consolidants for ancient Qin terracotta. Prog Org Coat. 2016;101:416–22.
60. Remzova M, et al. Effect of modified ethylsilicate consolidants on the mechanical properties of sandstone. Constr Build Mater. 2016;112:674–81.
61. Scherer GW. Theory of drying. J Am Ceram Soc. 1990;73(1):3–14.
62. Scherer GW, et al. Shrinkage of silica gels aged in TEOS. J Non-Cryst Solids. 1996;202(1):42–52.
63. Brinker CJ, Scherer GW. Sol-gel science: the physics and chemistry of sol-gel processing. Boston: Academic Press; 1990.
64. Miliani C, Velo-Simpson ML, Scherer GW. Particle-modified consolidants: a study on the effect of particles on sol–gel properties and consolidation effectiveness. J Cult Herit. 2007;8(1):1–6.
65. Liu R, et al. Preparation of three-component TEOS-based composites for stone conservation by sol–gel process. J Sol-Gel Sci Technol. 2013;68(1):19–30.

66. Salazar-Hernández C, et al. Conservation of building materials of historic monuments using a hybrid formulation. J Cult Herit. 2015;16(2):185–91.
67. Delgado Rodrigues J, Costa D. Occurrence and behaviour of interfaces in consolidated stones, in Structural studies of historical buildings IV. Volume 1: architectural studies, materials and analysis. Computational Mechanics Publications; 1995. p. 245–52.
68. Zárraga R, et al. Effect of the addition of hydroxyl-terminated polydimethylsiloxane to TEOS-based stone consolidants. J Cult Herit. 2010;11(2):138–44.
69. Mackenzie J, Huang Q, Iwamoto T. Mechanical properties of ormosils. J Sol-Gel Sci Technol. 1996;7(3):151–61.
70. Xu F, Li D. Effect of the addition of hydroxyl-terminated polydimethylsiloxane to TEOS-based stone protective materials. J Sol-Gel Sci Technol. 2013;65:212–9.
71. De Rosario I, et al. Effectiveness of a novel consolidant on granite: laboratory and in situ results. Constr Build Mater. 2015;76:140–9.
72. Illescas JF, Mosquera MJ. Surfactant-synthesized PDMS/silica nanomaterials improve robustness and stain resistance of carbonate stone. J Phys Chem C. 2011;115(30):14624–34.
73. Mosquera MJ, de los Santos DM, Rivas T. Surfactant-synthesized ormosils with application to stone restoration. Langmuir. 2010;26(9):6737–45.
74. Mosquera MJ, et al. New nanomaterials for protecting and consolidating stone. J Nano Res. 2009;8:1–12.
75. Pinho L, Mosquera MJ. Titania-silica nanocomposite photocatalysts with application in stone self-cleaning. J Phys Chem C. 2011;115(46):22851–62.
76. Pinho L, Mosquera MJ. Photocatalytic activity of TiO_2–SiO_2 nanocomposites applied to buildings: influence of particle size and loading. Appl Catal B Environ. 2013;134–135:205–21.
77. Kapridaki C, Maravelaki N-P. TiO_2–SiO_2–PDMS nanocomposites with self-cleaning properties for stone protection and consolidation, vol. 416. London: Geological Society Special Publication; 2015. p. 1.
78. Li D, et al. The effect of adding PDMS-OH and silica nanoparticles on sol–gel properties and effectiveness in stone protection. Appl Surf Sci. 2013;266:368–74.
79. Derluyn H, et al. Salt crystallization in hydrophobic porous materials. In: Hydrophobe V. Aedificatio Publishers; 2008.
80. Scherer GW, Wheeler GS. Silicate consolidants for stone. Key Eng Mater. 2009;391:1–25.
81. Sena da Fonseca B, et al. Polyethylene glycol oligomers as siloxane modificators in consolidation of carbonate stones. Pure Appl Chem. 2016;88(12):1117–28.
82. Kapridaki C, et al. Producing self-cleaning, transparent and hydrophobic SiO_2-crystalline TiO_2 nanocomposites at ambient conditions for stone protection and consolidation. In: Self-cleaning coatings. 2016. p. 105–41.
83. Verganelaki A, et al. Characterization of a newly synthesized calcium oxalate-silica nanocomposite and evaluation of its consolidation effect on limestones. In: Toniolo L, Boriani M, Guidi G, editors. Built heritage: monitoring conservation management. Cham: Springer; 2015. p. 391–402.
84. Prado AGS, Airoldi C. Different neutral surfactant template extraction routes for synthetic hexagonal mesoporous silicas. J Mater Chem. 2002;12(12):3823–6.

Chapter 10
Analytical Investigations and Advanced Materials for Damage Diagnosis and Conservation of Monument's Stucco

Rodica-Mariana Ion

10.1 Introduction

Since ancient times until the modern age, stone has been an important building material, being found both in the building components of edifices and in their interior or exterior decorations. Today, many monuments are affected by stone deterioration, especially by weathering, which means physical disintegration or chemical decomposition initiated by the interaction between the stone and climate, biosphere, or pollution [1, 2]. In order to describe the decay and to measure its extent, it is necessary to understand the causes and mechanisms of decay.

Currently, preserving historical monuments requires taking measures to prevent diminishing, the process of degradation, and permanent loss. One of the consequences of historical monument degradation is due to the complexity of the destructive factors (e.g., microclimate, biological factors, atmospheric pollution, human factors, etc.), which yield to characteristic changes of the binder materials in the building structure [3]. These materials constitute the cohesion elements of the components from the historical monuments, starting with the foundation and masonry to the fine artistic components that adorn architectural surfaces, such as mural painting or stucco decorations.

Nowadays, scientific investigations play an essential role in the material characterization of art objects and artifacts, the recognition of execution techniques and technologies, its authentication and its state of preservation, as well as the causes and mechanisms leading to their deterioration.

R.-M. Ion (✉)
Research center for scientific investigations and conservation/preservation of industrial, cultural and medical heritage – SCI-HERITAG, Bucharest, ICECHIM, Bucharest, Romania

Doctoral School of Materials Engineering, Valahia University, Targoviste, Romania
e-mail: rodica_ion2000@yahoo.co.uk

© Springer International Publishing AG 2018
M. Hosseini, I. Karapanagiotis (eds.), *Advanced Materials for the Conservation of Stone*, https://doi.org/10.1007/978-3-319-72260-3_10

The conservation and restoration of monuments' architecture, including its artistic components, presume all new treatment solutions responding to the international principles in the field (i.e., compatibility, reversibility, adequate aesthetic presentation, resistance to aging, absence of side effects) and the specificity of the constituent materials of our monuments [4].

Under such context, the issue of compatibility, as an essential principle in restoration, presumes the avoidance of any other material of a different nature than the original and whose side effects may be unpredictable. This is the reason for which the mineral materials of old masonry, plasters, or mural paintings should be treated with mineral materials as well.

In this regard, information about the degradation processes (e.g., freeze-thaw, salt crystallization) is extremely important, because it offers selection criteria for materials with high resistance of the binders to the freeze-thaw, mechanical, and aesthetic features, keeping the tonality chromatic [5, 6]. External factors of degradation, methods, and laboratory techniques used for the analysis of stucco and its degradation products with eloquent examples on stucco models, prepared and studied in the laboratory, will be discussed in this chapter.

10.2   Forms of Stucco Degradation

Stucco degradation could involve different forms and should be clearly distinguished, as follows: *weathering* due to natural atmospheric phenomena; *decay* provoked by any chemical or physical modification of the intrinsic stone properties leading to a loss of value or to the impairment of use; *degradation* which involves the decline in condition, quality, or functional capacity; *deterioration* as the process of making or becoming worse or lower in quality and value.

The architectural surfaces of historical monuments suffer various forms of degradation, some of the most frequent being mentioned here [7, 8]. *Removal of masonry plasters or supports* may exist only between the plaster and masonry layer or, at the same time, between multiple layers of plaster (Fig. 10.1a). *Substrate cracks* could be cracks of the support, corresponding to the wall dislocations (centimeters) or contraction cracks (crackles) like a fine network of the cracks that make up a part of the specific aspect of mural painting (Fig. 10.1b). *Gaps* explained as discontinuities of different expansions and depths of surface of the plaster layer or the support of murals. These are caused by migration and crystallization of the walls inside the plaster layer, the phenomenon of freeze-thaw, atmospheric pollution in the presence of humidity and temperature variations, and friability of the mortar (Fig. 10.1b). *The presence of inappropriate foreign plasters and bodies* is due to some restoration interventions with incompatible materials (e.g., cements, synthetic materials, mortars, gypsum or lime with inadequate structure, or improper fitting), repairs, electrical installations, and the installation of objects on the mural surface (Fig. 10.1c). *Deposits of adhering impurities* arise from the following causes: atmospheric pollution, soot (burning candles, candles, etc.), natural or artificial disasters (fires, rains,

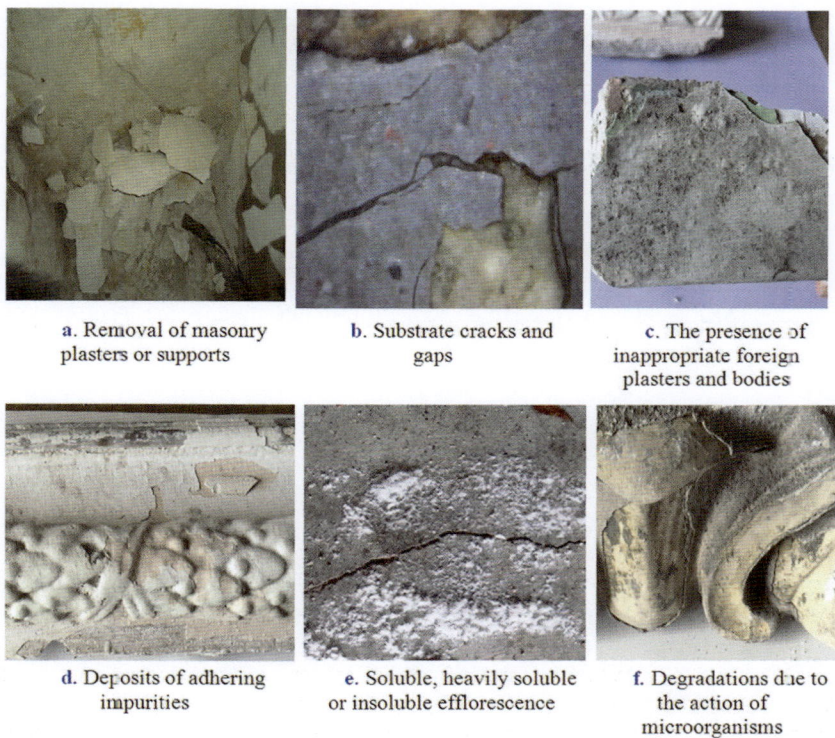

a. Removal of masonry plasters or supports

b. Substrate cracks and gaps

c. The presence of inappropriate foreign plasters and bodies

d. Deposits of adhering impurities

e. Soluble, heavily soluble or insoluble efflorescence

f. Degradations due to the action of microorganisms

Fig. 10.1 Forms of stucco stones degradation. (**a**) Removal of masonry plasters or supports, (**b**) substrate cracks and gaps, (**c**) the presence of inappropriate foreign plasters and bodies, (**d**) deposits of adhering impurities, (**e**) soluble, heavily soluble, or insoluble efflorescence, and (**f**) degradations due to the action of microorganisms

floods), and bird excrement. They may favor some chemical reactions or undergo a biological attack that can irreversibly affect the architectural surface (e.g., the color layer in the case of wall paintings), (Fig. 10.1d). *Soluble, heavily soluble, or insoluble efflorescence* is due to salts (e.g., nitrates, sulfates, chlorides, carbonates) that crystallize under certain conditions of temperature and humidity on architectural surfaces (Fig. 10.1e). *Degradations due to the action of microorganisms* develop in optimal microclimate conditions: humidity, temperature, light, and atmospheric pollution (Fig. 10.1f).

By examining the effects of pollution on individual historic buildings over periods of several hundred years, it is possible to correlate pollution levels with observed damage. Typical pollutants including sulfur oxides, nitrogen oxides, and carbon dioxide are capable of dissolving in water to create an acidic solution and are capable of reacting with calcareous materials. Also, the reaction products of air pollution, such as soluble salts, often remain on sheltered stone surfaces and result in ongoing damage.

When sulfur dioxide levels are higher than 30 $\mu g/m^3$, this becomes a significant contributor to decay [9]. Dissolution was found to be mostly due to gypsum dissolution originating from dry deposition with less contribution from karstic processes due to carbonic acid or from the neutralization of acid rain [10–12].

Beyond these aspects, the main chemical degradation identified in historical monuments is due to some expansive compounds as ettringite and thaumasite (compounds well known in Portland cement chemistry), which could favor a sulfate attack on hardened cements [13].

Ettringite, $3CaO \cdot Al_2O_3 \cdot 3CaSO_4 \cdot 32H_2O$, is a hydrophobic compound that has a negative effect on the resistance structure of the hard material. Thaumasite, $CaCO_3 \cdot CaSiO_3 \cdot CaSO_4 \cdot 15H_2O$, can be formed from ettringite by the incorporation of Si^{4+} ions (octahedral coordination) into its structure by substituting Al^{3+} ions at a high content of sulfate, with an excess of water while in the presence of $CaCO_3 \cdot [CO_2]$ and at a temperature lower than 10 °C [14].

The action of degradation by thaumasite formation involves a dilation mechanism similar to that generated by ettringite or the decomposition of the material through consumption of calcium hydrosilicates, which is responsible for mechanical resistance [15].

10.3   Diagnosing Stucco Damage

In literature there are numerous papers that first describe the situation and history of some particular monuments and the physical properties of the stones involved. Moreover, the aforementioned literature reported various types of characterization techniques which are briefly discussed here.

Mineralogical-petrographic characterization allows for the identification of mineralogical phases present in a structure such as the original materials and degradation products, microstructure, and texture of the minerals [16]. *Chemical characterization* identifies the chemical composition (elemental and molecular) of the used material (substrate), materials of neoformation, and materials intentionally added on to the surfaces [17]. *Physical characterization* takes measurements of crystalline structure, porosity, density, water behavior, and the surface hardness. *Mechanical characterization* studies the mechanical properties of materials, such as the elastic modulus, the tensile, compression and bending strength, and their variations as a function of the material degradation [18, 19], resistance to salt crystallization.

A general list of stucco characterization techniques are X-ray diffraction (XRD), X-ray fluorescence (XRF), Fourier-transform infrared spectroscopy (FTIR), Raman spectroscopy, scanning electron microscope and some energy-dispersive X-ray (SEM-EDX) analyses, transmission electron microscopy (TEM), thermal analysis, and, when is possible, chromatographic techniques as destructive techniques gas chromatography (GC-MS) or high-performance liquid chromatography (HPLC) [20].

The aforementioned characterization techniques should obey the physical integrity of the object's material, should be a fast technique allowing the analysis of a large number of samples, and should be universal and adaptable so that several materials and samples of different shapes and sizes can be analyzed. Finally, the techniques should offer compositional data of the sample and be sensitive and can be performed for both major and trace elements.

Attention is therefore directed to the elaboration of material conservation, in line with the fundamental ethic principles of conservation. The first principal is *reversibility*, where any material applied should be easily removed without causing any damage to the original material. The next is that of *compatibility*, where the materials used in the restorative intervention must be chosen so that they are compatible with the original material on which they will be applied. *Recognizability* is where any intervention should be recognizable; differing from the original material to avoid the falsification, while not disturbing the overall view. The *minimum intervention* principal is where any restorative intervention must be closely related to the needs of each situation, conserving the original material as much as possible and the intervention being minimally invasive. Finally, *durability* is the material's ability to resist the aggressive actions from the environment and maintain essential properties, within adequate safety levels, throughout the lifetime of the material.

10.4 Salts Effects

Along with air pollution, soluble salts represent an important cause of stone decay. Salts cause damage to stone in several ways. The most important is the growth of salt crystals within the pores of a stone, which can generate stresses that are sufficient to overcome the stone's tensile strength or compressive strength and turn the stone into a powder. A prime example is sodium sulfate, one of the most damaging soluble salts, which can exist as the anhydrous salt thenardite (Na_2SO_4) or the decahydrate mirabilite ($Na_2SO_4 \cdot 10H_2O$) [21–23]. Thenardite increases in volume by more than three times during conversion to mirabilite, and it has been argued that this growth in volume was the cause of hydration damage [22].

Some cubes of stucco models with 40 mm sides were taken. They were prepared after the following receipt: gypsum/lime/water = 53:27:20. After dried for 72 h and weighed, the samples were immersed in a salt solution (10% NaCl or 14% Na_2SO_4) for 2 h. After that they were dried at 60 °C and weighed again. Differences in the weights were noted. This procedure of drying, weighing, immersion, and reweighing was repeated five times. Each time, any change in weight was noted and was expressed as a percentage of the original weight.

For the stucco models, the main effects of the salts (NaCl, Na_2SO_4) at different temperatures are discussed herein. The deterioration signs, visible after physical weathering induced by freeze-thaw action, were as follows: sulfate crust, weight increase (1 day), blistering, loss of materials (2 days), fracture with traces of salts (3 days), some efflorescence upper face, and a string of salts in the fracture at 1 month (Fig. 10.2).

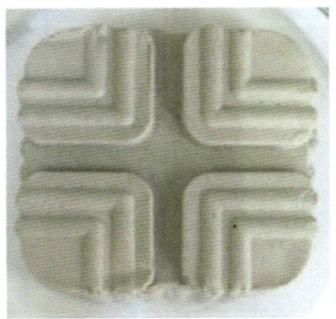

a
Stucco treated with Na$_2$SO$_4$ and kept at 25 °C
(t=0)

b
Stucco treated with Na$_2$SO$_4$ and kept at 25 °C
(t=8 days)

c
Stucco treated with NaCl at 5 °C (t=0)

d
Stucco treated with NaCl at 5 °C (t=8 days)

e
Stucco treated with NaCl at 25 °C (t=8 days)

f
Stucco treated with NaCl at 50 °C (t=8 days)

Fig. 10.2 The aspects of stucco models under different external factors. (**a**) Stucco treated with Na$_2$SO$_4$ and kept at 25 °C ($t = 0$), (**b**) stucco treated with Na$_2$SO$_4$ and kept at 25 °C ($t = 8$ days), (**c**) stucco treated with NaCl at 5 °C ($t = 0$), (**d**) stucco treated with NaCl at 5 °C ($t = 8$ days), (**e**) stucco treated with NaCl at 25 °C ($t = 8$ days), (**f**) stucco treated with NaCl at 50 °C ($t = 8$ days)

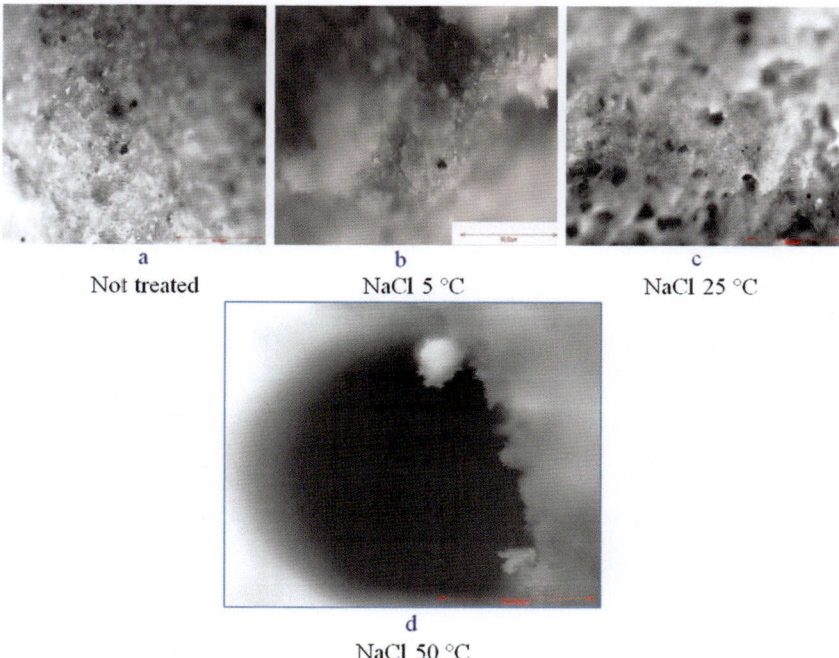

a	b	c
Not treated	NaCl 5 °C	NaCl 25 °C

d
NaCl 50 °C

Fig. 10.3 The microscopy images of stucco damaged by different salts at different temperatures: (**a**) not treated, (**b**) NaCl 5 °C, (**c**) NaCl 25 °C, (**d**) NaCl 50 °C

These tests were repeated at different temperatures (5, 25, 20 °C). By increasing the temperature (25 °C), these effects became stronger and longer. Also, NaCl had a more visible effect than Na_2SO_4. Crystallization of $CaSO_4$ in the pores of stone caused a more pronounced decay of stone due to its weathering, as seen in Fig. 10.2. The amount of salt within the pore space of the stones is a crucial factor of the decay, shown in Fig. 10.3.

These results correlate with porosity data, in the presence of NaCl at different temperatures (Fig. 10.4) where the highest porosity was detected at 25 °C. The higher porosity, the greater the effect of weathering is in different forms including alveolar weathering, granular disintegration, and efflorescence [24]. This is explained by the transformation of gypsum in calcium sulfate hydrate, calcium sulfate (VI) hemihydrate, bassanite, $Ca(OH)_2$, and calcite, visible by FTIR and XRD experiments [18], and is illustrated in Fig. 10.5.

The FTIR bassanite and anhydrite spectra have similar features to gypsum spectra, even though there are many bands distinct for differentiation of these minerals: triplet bands at 1015, 1012, and 1005 cm^{-1} and also 590–675 cm^{-1} for anhydrite, bassanite, and gypsum in FTIR spectra. Also, the bands near 980 and 450 cm^{-1} are specific for the SO_4 groups. The specific H_2O bands are located at anhydrite (3450 cm^{-1}), bassanite (3465 and 3615 cm^{-1}), and gypsum (3250, 3408, 3500, and

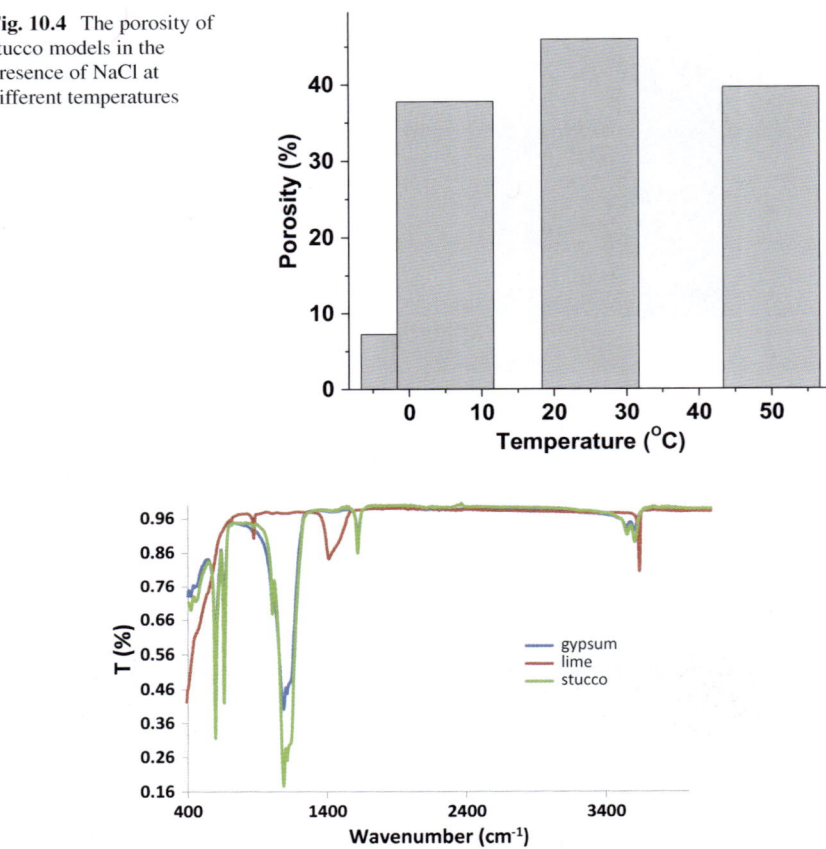

Fig. 10.4 The porosity of stucco models in the presence of NaCl at different temperatures

Fig. 10.5 FTIR spectra of lime, gypsum, and stucco model

3555 cm^{-1}); specific H$_2$O vibrations occur at 1623 cm^{-1} for anhydrite, at 1629 cm^{-1} for bassanite, and at 1629 and 1690 cm^{-1} for gypsum [25].

10.5 Frost Action

The damage generated by frost action is caused by the stresses, induced by ice crystallization, and exceeding the strength of the rock. Freezing-thawing actions can be considered as one of the processes that contribute in the deterioration of stones located in cold regions, characterized by air temperatures below freezing point. The amount of water within the pore space of the stones is a crucial factor of the decay. Soft and porous stone soaks up water and is weathered by alveolar weathering, granular disintegration, and efflorescence [24]. However, the main cause for the

| a | b | c |
| After 24 h immersed in water | After 24 h at -18 °C | After 12 days at -18 °C |

Fig. 10.6 The microscopy images of stucco models supported the freeze-thaw treatment: (**a**) after 24 h immersed in water, (**b**) after 24 h at −18 °C, (**c**) after 12 days at −18 °C

freeze point reduction in natural building stones is the pore space distribution. The smaller the pore, the stronger the water retention is strained. Frost-induced weathering, although often attributed to the expansion of freezing water captured in cracks, is generally independent of water-to-ice expansion [26]. Ice crystal growth weakens the rocks which, in time, break up. Intermolecular forces acting between the mineral surfaces, ice, and water sustain these unfrozen films which transport moisture and generate pressure mineral aggregate surfaces. For the tested stucco specimens. they were kept immersed in water for 24 h. They were then placed in a freezing machine at −18 °C for 12 days. Afterward, the samples were thawed or warmed at atmospheric temperature. This procedure was repeated several times for up to 12 days, and the behavior of stone was carefully observed.

The deterioration signs visible after physical weathering of building stones induced by freeze-thaw action were sulfate crust, loss of materials (2 days), fracture (3 days), and efflorescence upper face (Fig. 10.6a–c).

10.6   Stucco Models Treated with Nanoparticles

Nanoparticles applied to cultural heritage must have the following attributes: stability and sustained photoactivity, biologically and chemically inert, nontoxic, low cost, suitability toward visible or near UV light, high conversion efficiency. high quantum yield, react with wide range of substrates, high adaptability to various environments, and good adsorption in solar spectrum. In this chapter, the current state of knowledge in the application of different types of nanoparticles for the improvement of conservation strategies of the cultural built heritage is discussed.

Frequently applied nanoparticles are $Ca(OH)_2$, $Mg(OH)_2$, and hydroxyapatite. There are both positive and negative properties for this type of application. For example, $Ca(OH)_2$ is characterized by its slow and sometimes incomplete carbonation process which may leave particles of calcium hydroxide on the surface [26, 27], while the penetration depth of the hydroxide is 2 mm or less. Subsequently, this method has been improved by reducing the size of the enhancing particles at nanometric scale to increase the depth of penetration and to avoid whitening on the surface treated [25]. The samples show no remarkable change in color, and the cohesion of the surface was obvious. The use of nanoscale calcium hydroxide in lime mortars may improve fluidity and reduce porosity.

More recently, the mineral hydroxyapatite (HAp, $Ca_{10}(PO_4)_6(OH)_2$) has been experimentally tested as a consolidant for limestones, chalk stone, and marbles [28]. HAp has a much lower solubility than calcite while being the least soluble and the most stable calcium phosphate phases in aqueous solutions at pH values higher than 4.2 [29]. The lattice parameters, which refers to the physical dimension of unit cells in a crystal lattice, have for hydroxyapatite and calcite relatively close values: $a = b = 9.43$ Å and $c = 6.88$ Å for HAp [30] and $a = b = 9.96$ Å and $c = 17.07$ Å for calcite, respectively, (a, b, and c are either a measure of length or angle that defines the size and shape of the unit cell of a crystal lattice).

The problems that encompassed herein were the evaluation of the stability of the coating layer of these nanoparticles on the stucco models during freeze-thaw cycles. Methods were developed to test a material and to clean the model stucco using HAp in a 1% water suspension, with the appearance of stucco before and after treatment shown in Fig. 10.7. No significant evidence of color alteration was observed, which is a promising solution for stucco treatment and preservation [31].

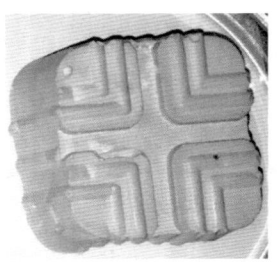 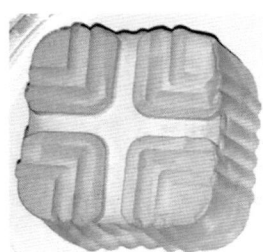 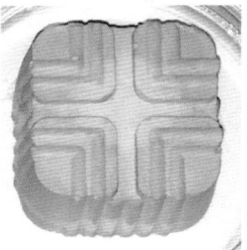

Stucco initial freezed   Stucco treated with Ca(OH)$_2$ Stucco treated wit Mg(OH)$_2$
                                          and freezed   and freezed

Stucco treated with HAp and freezed 1 h   Stucco treated with HAp after 72 h freezing

Fig. 10.7 The aspect of stucco in different conditions: (**a**) stucco initially freezed, (**b**) stucco treated with Ca(OH)$_2$ and freezed, (**c**) stucco treated with Mg(OH)$_2$ and freezed, (**d**) stucco treated with HAp and freezed 1 h, (**e**) stucco treated with HAp after 72 h freezing

Colorimetric measurements (CIELAB) were performed to check color change (ΔH and ΔC) of the sample and were correlated with the overall colorimetric difference between the samples. By measuring the total color difference (ΔE^*), the difference between treated and untreated samples indicated that the color properties are in direct relation to aging (i.e., the brightness decreases the stucco color and the hue is easily changed to red) [25]. The increase of L from negative values to positive ones has shown that there is a temperature variation that induced structural changes but also that there is a slight change of chroma. Chroma (C^*) changed from positive to negative, while the tone (H^*) increased slightly and steadily. All these results are consistent with literature data [32] and demonstrate the structural and aesthetic effects of temperature destruction of stucco, but were not significant in the case of HAp. In that case, ΔE^* had a value of 4.16, lower than 5, which is the imposed condition for a recommended material in this case.

10.7   Conclusion

There is good evidence for a scientific diagnosis of the deterioration stage for stone material types, surface structure, mineral composition, chemical composition, porosity properties, thermal and photochemical properties, and their response to freeze-thaw and salt crystallization. New and revolutionary treatment methods with nanomaterials (hydroxyapatite and derivatives, calcium, and magnesium hydroxides) are able to offer stucco consolidation, making them valuable for different surfaces. The hydroxyapatite shows better properties in comparison with $Ca(OH)_2$ and $Mg(OH)_2$, as follows: chemical stability, porosity properties, thermal and photochemical properties, and their response to freeze-thaw and salt crystallization.

Acknowledgment This paper received financial support from MEN-UEFISCDI through the projects PNII 261/2014 and PN 16.31.02.04.04.

References

1. Brimblecombe P. History of air pollution and deterioration of heritage. In: Zezza F, editor. Weathering and air pollution: Primo corso della Scuola universitaria C.U.M.conservazione dei monumenti, Lago di Garda (Portese), Venezia, Milano, 2–9 September 1991. Bari: Mario Adda Editore; 1992. p. 23–32.
2. Camuffo D, Del Monte M, Sabbioni C, Vittori O. Wetting, deterioration and visual features of stone surfaces in an urban area. Atmos Environ. 1982;16(9):2253–9.
3. Brimblecombe P, Grossi CM. Damage to buildings from future climate and pollution. APT Bull. 2007;38(2–3):13–8.
4. Ashurst J, Dimes FG, editors. Conservation of building and decorative stone. Butterworth-Heinemann series in conservation and museology. Oxford and Woburn, MA: Butterworth-Heinemann; 1998.

5. Fitzner B. Documentation and evaluation of stone damage on monuments. In: Kwiatkowski D, Lofvendahl R, editors. Proceedings of the 10th international congress on deterioration and conservation of stone, Stockholm, June 27–July 2, 2004, vol. 2. Stockholm: ICOMOS; 2004. p. 677–90.
6. Fitzner B, Heinrichs K. Damage diagnosis on stone monuments: weathering forms, damage categories and damage indices. In: Přikryl R, Viles HA, editors. Understanding and managing of stone decay (SWAPNET), vol. 2001. Prague: Charles University, The Karolinum Press; 2002. p. 11–56.
7. Doehne E. Building material decay and salt weathering: a selected bibliography. In: Siegesmund S, Weiss T, Vollbrecht A, editors. Supplement to natural stone, weathering phenomena, conservation strategies and case studies, Geological Society Special Publication 205. London: Geological Society of London; 2003. www.geolsoc.org.uk/webdav/site/GSL/shared/Sup_pubs/2003/SUP18182.rtf. Accessed Aug 2017.
8. Fitzner B, Heinrichs K. Damage diagnosis on stone monuments – weathering forms, damage categories and damage indices. In: Prikryl R, Viles HA, editors. Understanding and managing stone decay, proceeding of the international conference "stone weathering and atmospheric pollution network (SWAPNET 2001)". Prague: The Karolinum Press, Charles University; 2002. p. 11–56.
9. Sharma RK, Gupta HO. Dust pollution at the Taj Mahal: a case study. Conservation of stone and other materials. In: RILEM proceedings, vol. 21. Paris: E&F.N. Spon; 1993. p. 11–8.
10. Baedecker PA, Reddy MM. The erosion of carbonate stone by acid rain: laboratory and field investigations. J Chem Educ. 1993;70(2):104–8.
11. Camaiti M, Bugani S, Bernardi E, Morselli L, Matteini M. Effects of atmospheric NOx on bio-calcarenite coated with different conservation products. Appl Geochem. 2007;22(6):1248–54.
12. Bai Y, Thompson GE, Martinez-Ramirez S. Effects of NO_2 on oxidation mechanisms of atmospheric pollutant SO_2 over Baumberger sandstone. Build Environ. 2006;41(4):486–91.
13. Ion RM, Ion ML, Radu A, Suica Bunghjez IE, Fiearscu RC, Fierascu I, Teodorescu S. Mortar pe bază de nanomateriale pentru conservarea fațadelor construcțiilor. Rom J Mat. 2016;46(4):412–8.
14. Ciliberto E, Ioppolo S, Manuella F. Ettringite and thaumasite: a chemical route for their removal from cementious artefacts. J Cult Herit. 2008;9(1):30–7.
15. Steiger M, Siegesmund S. Special issue on salt decay. Environ Geol. 2007;52(2):185–6.
16. Binda L, Lualdi M, Saisi A. Non-destructive testing techniques applied for diagnostic investigation: Syracuse Cathedral in Sicily, Italy. Int J Archit Herit. 2007;1(4):380–402.
17. Blauer Bohm C. Techniques and tools for conservation investigations. In: Kwiatkowski D, Lofvendahl R, editors. Proceedings of the 10th international congress on deterioration and conservation of stone, Stockholm, June 27–July 2, 2004, vol. 2. Stockholm: ICOMOS Sweden; 2004. p. 549–59.
18. Hamilton A, Hall C, Pel L. Sodium sulfate heptahydrate: direct observation of crystallization in a porous material. J Phys D Appl Phys. 2008;41(21):212002.
19. Ambrosi M, Dei L, Giorgi R, Neto C, Baglioni P. Colloidal particles of $Ca(OH)_2$: properties and applications to restoration of frescoes. Langmuir. 2001;17(14):4251–5.
20. Bracci S, Delgado Rodrigues J, Ferreira Pinto A, Matteini M, Pinna D, Porcinai S, Sacchi B, Salvadori B. Development and evaluation of new treatments for the conservation of outdoor stone monuments. In: Lukaszewicz JW, Niemcewicz P, editors. Proceedings of the 11th international congress on deterioration and conservation of stone, 15–20 September 2008, Toruń, Poland. Toruń: Nicolaus Copernicus University; 2008. p. 811–8.
21. Brajer I, Kalsbeek N. Limewater absorption and calcite crystal formation on a limewater-impregnated secco wall painting. Stud Conserv. 1999;44(3):145–56.
22. Espinosa Marzal RM, Scherer GW. Crystallization of sodium sulfate salts in limestone. Environ Geol. 2008a;56(3–4):605–21.
23. Fidler J. Lime treatments: lime watering and shelter coating of friable historic masonry. APT Bull. 1995;26(4):50–6.

24. Giorgi R, Dei L, Baglioni P. A new method for consolidating wall paintings based on dispersions of lime in alcohol. Stud Conserv. 2000;45(3):154–61.
25. Ion RM, Fierascu RC, Fierascu I, Bunghez IR, Ion ML, Caruțiu-Turcanu D, Teodcrescu S, Raditoiu V. Stone monuments consolidation with nanomaterials. Key Eng Mater. 2015;660:383–8.
26. Hansen E, Doehne E, Fidler J, Larson J, Martin B, Matteini M, et al. A review of selected inorganic consolidants and protective treatments for porous calcareous materials. Rev Conserv. 2003;4:13–25.
27. Giorgi R, Bozzi C, Dei L, Gabbiani C, Ninham BW, Baglioni P. Nanoparticles of $Mg(OH)_2$: synthesis and application to paper conservation. Langmuir. 2005;21:8495–501.
28. Ion RM, Turcanu-Caruțiu D, Fierăscu RC, Fierăscu I, Bunghez IR, Ion ML, Teodorescu S, Vasilievici G, Rădiţoiu V. Caoxite-hydroxyapatite composition as consolidating material for the chalk stone from Basarabi-Murfatlar churches ensemble. Appl Surf Sci. 2015;358:612–8.
29. Ion RM, Turcanu-Carutiu D, Fierascu RC, Fierascu I. Chalk stone restoration with hydroxyapatite–based nanoparticles. The Scientific Bulletin of Valahia University – Materials and Mechanics. 2014;12(9):24–8.
30. Sassoni E, Naidu S, Scherer GW. Preliminary results of the use of hydroxyapatite as a consolidant for carbonate stones. MRS Proceedings: Cambridge University Press; 2011.
31. Sassoni E, Franzoni E. Evaluation of hydroxyapatite effects in marble consolidation and behaviour towards thermal weathering. In: Proceedings of Built Heritage–Monitoring Conservation Management, Milan, Italy. 2013. pp. 1287–95.
32. Charola AE, Munich G, Ceteno SA. Conclusions and recommendations resulting from the colloquium on historic and archaeological mortars: analysis and characterization. US/ICCMOS Scientific J III. 2001;1:15.

Chapter 11
Nanotechnology for the Treatment of Stony Materials' Surface Against Biocoatings

Carlos Alves and Jorge Sanjurjo-Sánchez

11.1 Introduction

Rocks are important materials in old and modern structures due to their strength and durability. However, they can suffer surface alterations due to their interaction with diverse environmental agents. Among them, biological agents are a widespread factor in outdoor situations (rarer indoors) that can contribute to visual modifications, some of which might be considered as having a negative impact on the surface [1–3]. The following discussion will consider biological colonization that imparts undesired effects on stone surfaces (there are situations where that might not be the case). This affects not only historical structures but also contemporary ones. Several treatments have been proposed for hindering biological growths on building materials, including nanoproducts. General reviews of nanoproducts for the built environment and their properties, considering diverse types of effects, as well as the concepts involved, can be found in the literature [3–8].

This chapter is focused on nanoproducts used for protection that inhibit the development of biocolonization. In general, the products discussed here consist of nanoparticles (NPs) of diverse substances. Focus will be placed on studies that consider tests with organisms on natural stone but will also consider results with microorganisms from other porous inorganic materials like concrete, mortars, and bricks. This work's perspective will be that of the end user that needs a solution for treatment of a problem. This chapter will not consider results that were not actually tested on this kind of materials, e.g., studies in vitro tests or on glass or other

C. Alves (✉)
LandS/Lab2PT, Landscape, Heritage and Territory Laboratory (FCT UID/AUR/04509/2013;
FEDER COMPETE POCI-01-0145-FEDER-007528) and Earth Sciences Department,
School of Sciences, University of Minho, Braga, Portugal
e-mail: casaix@cct.uminho.pt

J. Sanjurjo-Sánchez
University Institute of Geology, University of A Coruña, A Coruña, Spain

© Springer International Publishing AG 2018

M. Hosseini, I. Karapanagiotis (eds.), *Advanced Materials for the Conservation of Stone*, https://doi.org/10.1007/978-3-319-72260-3_11

223

materials that in practice have very low-porosity surfaces, such as glazed ceramics, or studies on wood. Studies that consider the degradation of organic products or properties related to biological colonization, e.g., water repellency, but that do not have data on treatment effects on microorganisms will not be considered also. Herein, only applications on existing materials' surfaces will be considered, and hence, studies where nanosubstances are added during preparation of materials as mortars or concrete will be also excluded. This work is primarily concerned with stone surfaces and will not consider applications as paintings for façades.

However, in the section concerning the impacts of NPs, a wider scope will be considered, as results will be included from studies that do not meet the conditions that were set above since they can have relevant data on the impact of the proposed products.

This chapter will address the aforementioned problems. Furthermore, the main methodologies that have been used in dealing with this issue, including their relevance for products' testing and the parameters used to assess the effects of the treatment (including their impacts on materials), are addressed in detail. This will be followed by a presentation of the main results that have been obtained with the use of NPs within the scope defined above. Results will be discussed along three main themes: water testing conditions, organisms, and substrate. Afterward, discussions will be presented on possible impacts of the proposed NPs in the materials and in the environment. Finally, a general overview of the topics covered in the chapter along with a discussion of collected data and the methodologies to tackle the aforementioned issues will be presented.

11.2 Problem Statement

Non-intentional alterations of materials in the built environment are commonly referred as "pathologies," by analogy with medical studies. And, as in medical studies, an understanding of the disease is essential for the correct treatment. The expression "biological colonization" can include microorganisms, plants, and the influence of other organisms (e.g., "animals nesting on and in stone") [9]. However, Fitzner and Heinrichs [10] only consider microorganisms and higher plants under biological colonization (moss is not explicitly mentioned in the classification presented by the authors). The biological colonization considered in this chapter covers what is usually considered in this context, which is similar to what is called biofilms by the "International Council on Monuments and Sites-the International Scientific Committee for Stone" (ICOMOS-ISCS) [9], but without the 2 mm limit for thickness. Albeit most situations will be within that limit, mosses attain a higher thickness, and some lichens can create coatings that are thicker. However, much like algae, fungi, and lichens, moss can develop on stone surfaces. Plants usually grow on macro-voids between the stones or in mortar joints, fractures, or fissures. Macro-animals and plants constitute relative rare occurrences on built stones, and they rarely have a permanently marked impact. Biological colonization herein will cover what Dorn [11] calls "lithobiontic coatings," on epilithic position, and that will include bacteria, cyanobacteria, algae, fungi, lichens, and moss.

The biological colonization considered here (biocoatings) requires the presence of light and water. Biofilm distributions are markedly related to moisture patterns, and biological colonization can be looked upon, even in quotidian situations (e.g., mold stains), as a marker of moisture, and as such is widespread on outdoor surfaces. Biocoatings' distributions can be affected by material characteristics, but they can show frequent variations due to surface orientation in relation to sun exposition or shading effects. They can also present heterogeneous patterns on a given surface due to heterogeneous moisture distribution on that surface. Biofilms can also be found indoors but normally are confined to areas where there are or were moisture infiltrations. These moisture patterns and their characteristics can be important for treatment assessments as they define the conditions that the treatments will face. Following the example of the terminology used for the description of water in joints of rock massifs [12], water conditions on building surfaces can be described as going from damp surfaces due to condensation, through wetting, dripping, sprinkling, flowing, and up to total immersion. These situations correspond to different conditions in terms of chemical or physical leaching, which is an issue of marked relevance for the long-term performance of treatments. Two different temporal regimes can also be considered: intermittent and perennial. Most situations of a built environment will be intermittent, related to weather variations, such as variations in relative humidity and temperature that can promote condensation episodes. However, for biocoatings, the main variations will be related to rain episodes, including occasional flooding due to water-level variations. There could also be situations of a perennial water regime such as some portions of fountains or elements immersed in seawater. But, for a given temporal instance, water regimes could show marked differences due to geometrical characteristics of the built element. These relations are illustrated in Fig. 11.1.

Figure 11.1a shows how on a given surface it is possible to observe biological colonization patterns related to substrate variations. Even on low-porosity materials, such as grayish granite used in pavements, the development of lichens or moss can occur related to surface roughness. This represents the kind of biological colonization that will be considered here. Plants such as the ones observed in spaces between the stones will not be included. It is possible to observe in Fig. 11.1b the effect of the orientation of granite stones (horizontal surfaces are more affected by biocoatings) but also the effect of the openings on the horizontal slabs that create water-flowing zones that favor biological colonization on the vertical wall and that stop at a certain distance from the top. Similar distributions patterns are observed (Fig. 11.1c) in an old granitic wall with darkish strips below the crenels. It can also be seen that the intensity of the biological colonization decreases with distance from the horizontal surface where water accumulates. Depending on exposition, which can result from wall orientation and shading effects, walls can show an extensive development of biocoatings (Fig. 11.1d).

Fig. 11.1 Illustration of factors influencing the development of biological colonization and associated water conditions: (**a**) lichens and moss (on grayish granite pavement stones) (attributed to surface roughness) and vascular plants (and more moss) in the soil joints between the stones; (**b**) showing marked biological colonization on horizontal stone slabs and vertical strips that are attributed to wetter zones related to openings on the horizontal slabs (at the bottom of the wall, it is possible to see some plants growing in crevices between flooring stones set in the "calçada à Portuguesa" style, nearer the base of the wall where they are protected from pedestrians); (**c**) a similar effect observed in an old granitic wall with darkish strips below the crenels (it can also be seen that the intensity of the biological colonization decreases with distance from the crenels; this image illustrates a situation where biocoatings might be considered to be desirable, contributing to an "aged" look); (**d**) extensive biocoatings on a granite wall

11.3 Treatment Methodologies and Tests

As in the case of human diseases related to an ubiquitous pathogen, strategies for treatment require either avoiding the fixation of the pathogen or its destruction. Three main methodologies can be considered in relation to the fight against biocoatings. First, hydrophobicity or water repellency, i.e., attempts to avoid the persistence of water on the material surface (an essential condition for biological colonization development); second, toxicity, where the products will cause the death of the organisms; and third, photochemical reactions, which have some similarities with the previous one in the sense that it include the photocatalytic destruction of matter but that might also include hydrophilic effects, i.e., promoting the presence of a film of water avoiding the fixation of pollutants (the joint effects are commonly referred to as self-cleaning).

It is worth mentioning that there are treatment methodologies and materials that combine more than one effect. For example, it is frequent to combine toxic or photochemical products with binders that have a water repellency effect. However, in this chapter, only the effects attributed to the NPs are discussed. While almost all studies assessed the use of NPs to avoid the development of biological colonization, NPs have also been used as a biocide to eliminate the existing biological colonization.

The main focus will be, as referred above, on tests performed with organisms on natural stone (or other porous inorganic materials). This can be considered, in this context, direct tests. Other tests can be considered as indirect tests, assessing properties that are relevant for the development of biocoatings but that do not include tests on the actual development of biocoatings. This is very common in the case of products for water repellency, where the properties that are assessed deal with hydrophobicity, such as drop angle (the higher the better regarding surface hydrophobicity) and sorptivity. A different perspective, referred to as testing by proxy, concerns the testing of the degradation of organic products that can be considered relevant in assessing the effects of the treatment on organic matter. Tests performed on nonporous materials like glass could also be included among the tests by proxy. While both indirect tests and tests by proxy give information relevant for assessing the effects of the treatments, they have the limitation of not including the actual performance in relation to living organisms, which are able to reproduce and can grow in very aggressive environments. The main issue of testing on nonporous surfaces will be that they do not assess the effect of absorption of the products in the pores and the effects of surface roughness. Some of the collected results may be able to justify these reservations.

Another set of issues in terms of assessing the performance of treatments concerns the testing conditions. As referred above, the water regime can be very variable and can affect the persistence and the performance of treatments. Two extreme situations can be considered. On the one hand, there are tests based on depositing a sample of organisms on the treated material surface and assessing their evolution. This will be hardly related to the actual conditions on the field where the growth of biocoatings is usually related to repeating episodes of moisture. There are tests with permanent or repeatedly contact of treated surfaces with water. These tests will correspond to harsher testing conditions. Immersion in seawater can be especially harsh since, besides effects on the coating, the presence of significant amounts of salts can promote the effects of salt weathering, a process with marked effects on stone surfaces [13]. It could help to frame these issues in order to use an analogy with probability hypothesis testing concepts [14] frequently used in the assessment of medical treatments. In this regard, two opposite situations (i.e., type I and type II errors) can be taken into consideration. Type I error corresponds to wrongly rejecting the null hypothesis of no effect, meaning that it would be wrong to consider that the treatment has an effect on the problem. In the present discussion, this is favored by less demanding testing conditions such as exposure to a moist atmosphere. On the other hand, there is the type II error of wrongly not rejecting the null hypothesis, which in this study will mean that the effects of the treatment are overlooked. This is favored by harsher testing conditions (e.g., water dripping, sprinkling, runoff, or total immersion), as far as they are significantly harsher than real field conditions. The analogy with probability testing also serves as a reminder that decreasing the hazard of one error type increases the hazard of committing the other.

Another question concerns the organisms used, as they might present different responses to the treatments. The studies analyzed here concern bacteria, unicellular fungi, and algae. No references were found to NPs used on natural stone or other inorganic porous materials for protection against moss.

One might be tempted to give special value to tests in the field, especially in actual examples of stone elements. However, there are several limitations for this approach, namely, the necessary time to obtain results, variations in the conditions that affect biological colonization, and the overlapping of other pollutants that might hinder the assessment of biological growth. Currently there are very few studies concerning field tests.

11.4 Parameters for Assessing Treatments

Given the great diversity of properties involved, herein, the techniques and parameters used to assess the effects of treatments are discussed, both in terms of efficiency against biological colonization and impacts on the materials.

The assessment of biological development in the analyzed studies included some straightforward parameters such as visual evidence (covered area) or the amount of biomass. There are also some variants on this general procedure such as the Braun-Blanquet index [15], which consists of assessments using area grids defining levels of the covered area: +, below 1%; (1) between 1 and 5%; (2) between 5 and 25%; (3) between 26 and 50%; (4) between 51 and 75%; and (5) between 76 and 100%.

Another parameter concerning the development of microorganisms is the colony-forming units (CFUs), corresponding to the number of viable microorganism cells in a sample (i.e., cells that can multiply under controlled conditions). There are several methods to assess the number of microorganisms in a sample. The plate count method [16] is one of the most used for microorganisms, and it is performed by plating aliquots of prepared solutions with the microorganisms into appropriate count agar plates. These plates are then optimally incubated, and the colonies are observed on the plates, and then the numbers of CFUs are counted, assuming that every colony is separated and founded by a single viable microbial cell.

In some studies, the presence of microorganisms can also be assessed by indicators such as the presence of extracellular polymeric substances (EPS) that are secreted by microorganisms [17]. EPS are polymers synthesized by microorganisms (biopolymers) that participate in the formation of microbial aggregates and that are frequently responsible for binding cells and other particulate materials together and to the substratum (hence promoting adhesion to the substrate). Microorganisms (either prokaryotic or eukaryotic) can live and grow in aggregated forms where they are embedded in a matrix of EPS. These are usually referred to as biofilms, and mixed populations of these organisms are ubiquitously adhered to a solid surface, at solid–water, water–air, and solid–air interfaces. Here, EPS are mainly responsible for the structural and functional integrity of biofilms and determine their physicochemical and biological properties.

There are other properties that are used for the assessment of results that are not specific to organisms. Color changes are a relevant issue for stone treatments, both in terms of desired effects (cleaning) and undesired effects. The assessments of color changes in the literature considered in this chapter were done using the *CIELAB* coordinates L^*, a^*, and b^* [18], where L^* concerns lightness, and a^* and b^* correspond to dual color axis, with $-a^*$ corresponding to cyan or green and $+a^*$ to magenta or red, while $-b^*$ corresponds to blue and $+b^*$ to yellow. The assessment of change can be done by differences between two states in any of the three parameters, represented by delta notation (Δa^*, Δb^*, and ΔL^*). There could be interest in the variation of a given coordinate, for example, a^* for greenish coatings. However, since the three coordinates can show independent variations, the Euclidean norm, or length of the vector of differences in relation to a reference point, usually the non-treated surface is frequently used and is calculated by the following equation:

$$\Delta E^* = \sqrt{\left(\Delta L^*\right)^2 + \left(\Delta a^*\right)^2 + \left(\Delta b^*\right)^2} \tag{11.1}$$

Buckley and Giorgianni [18] state that a unitary ΔE^* value is just perceptible, but a great latitude in terms of the accepted limit on chromatic change represented by ΔE^* can be found. The use of these coordinates is illustrated in Fig. 11.2a, where there are two points positioned and one attempts to show ΔE^* as Euclidean length. The relation of the two points in Fig. 11.2a can illustrate both cleaning, which will correspond to going from lower-left point to upper-right one (i.e., from the darkest and greenish point to the lighter and less greenish one). A darkening effect, which can occur by applying a treatment, will correspond to going from the upper-right point to the lower-left point.

Some treatments also attempt to modify the wetting properties of the surface (which affects their water repellency). In the publications analyzed here, the parameter most frequently used is the water drop contact angle (θ), illustrated in Fig. 11.2b, which is related to the equilibrium of interfacial tensions vapor–solid, vapor–liquid, and liquid–solid at the three-phase contact point defined by a drop of liquid on a solid surface [19]. It can be related to the work necessary to separate the liquid from the solid through the Young–Dupré equation as follows:

$$W_{SL}^a = \gamma_{LV}\left(1 + \cos\theta\right) \tag{11.2}$$

where W_{SL}^a represents adhesion work of the liquid on the solid, γ_{LV} represents the liquid–vapor interfacial tension, and θ represents contact angle. The nearer θ is to

Fig. 11.2 Illustration of two parameters used to assess the effects of treatments: (**a**) CIELAB space for assessment of chromatic changes with illustration of ΔE^* as Euclidean length between two points; (**b**) drop contact angle (θ) for hydrophilic surface (drop in the left with a contact angle around 60°) and hydrophobic surface (drop in the right with a contact angle around 130°). Drawings made with LibreOffice Draw 4.2

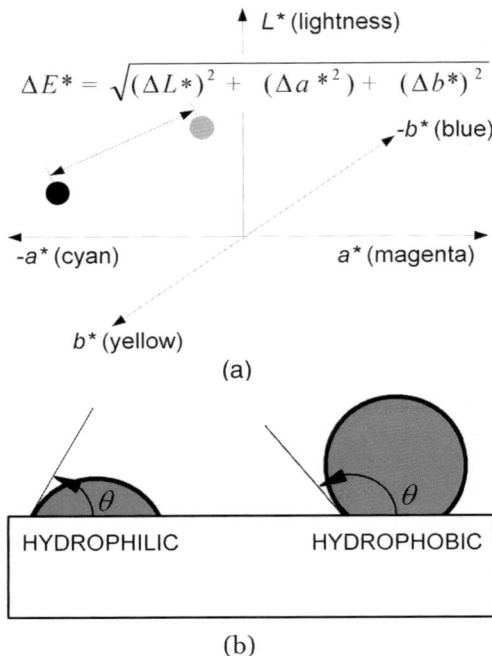

$$\Delta E^* = \sqrt{(\Delta L^*)^2 + (\Delta a^{*2}) + (\Delta b^*)^2}$$

zero, the higher the work necessary to separate the liquid from the solid will be. Below 90° (hydrophilic surfaces), the adhesion work is higher than the liquid–vapor interfacial tension. For θ above 90° (hydrophobic), the adhesion work will be lower than the liquid–vapor interfacial tension, and it will become zero for $\theta = 180°$. However, the water repellency of surfaces depends not only on a single measurement of the water drop contact angle since this parameter can show hysteresis [19], which is characterized by the difference between the advancing (when applying the drop) and receding (when removing the drop) values of the contact angle.

Another parameter that will be referred in this chapter is the sorptivity [20], with dimensions $[L][T]^{1/2}$, related to the migration of water in unsaturated porous media and representing a rate of water migration with square root of time.

11.5 Hydrophobicity

Herein, situations will be considered where the nanomaterials are used to increase hydrophobicity and are distinct from situations where NPs are mixed with hydrophobic substances (that act as binder) but do not necessarily contribute to hydrophobicity. The first case will evaluate the hydrophobic effect of the addition of nanomaterials, and in the second case, the hydrophobic effect of the binder will be assessed; therefore, different binders with the same NPs could give different results. In this context, the main goal is to obtain an impervious surface (i.e., to avoid movements from outside to the material's pore network).

Several NPs have been proposed for improving the water repellency of building materials' surfaces, but the only study concerning the hydrophobic effect of NPs on development of biocoatings is a study of SiO_2 NPs tested on limestone by Eyssautier-Chuine et al. [21]. Their work refers to a hydrophobic effect of the SiO_2 NPs, which were mixed with a binder and other additives ($AgNO_3$ and chitosan). However, tests on the development of algae films showed similar results to the untreated specimens for the SiO_2 NPs and binder mixture. It was reported by the authors that this mixture had no biocide efficacy. The biocide effects of other mixtures in this study were related to other products (not NPs) and, hence, are not included here.

11.6   Toxic Substances

Several studies were found in terms of NPs, albeit Ag and Cu NPs are clearly dominant. The collected information regarding studies in natural stone [22–32] is presented in Table 11.1, and this is the kind of treatment where the greatest diversity was found in terms of studied rock types: basalt, granite, limestone, marble, and sandstone. Hence, the main types found in the built heritage are considered, although silicate metamorphic rocks (slates, schists) are absent.

Table 11.2 presents information collected from studies on other inorganic porous materials [24, 27, 32–35], including concrete, mortars, stucco, and unglazed ceramics. The NPs are frequently applied with binders, which can cast some questions on the assessment of results since binders can also affect the materials' surface. On the other hand, there are results that suggest that the binder's effect on the substrate can be irrelevant. Veltri et al. [32] reported a drop contact angle of 61.07° (not hydrophobic) on porous ceramics after the application of the binder.

Improvements in prevention of biological development were observed for Ag NPs for bacteria and fungi in diverse substrates such as basalt, limestones, and sandstone. Although, the results obtained by Carrillo-González et al. [24] indicated a lower effect on fungi than in bacteria, but in the corrective, cleaning, and treatment, results were better for fungi than for bacteria. Data on algae growth in the harsher testing conditions reported by MacMullen et al. [34] on cement mortars showed that Ag NPs, in a polymeric binder, improve results in relation to untreated treatment and suggest a positive effect of increasing NP content. Goffredo et al. [26] showed that Ag NPs on limestone improved results in relation to untreated or other treatments considered by these authors but were not very satisfying. Other studies with algae and harsh testing conditions on other porous inorganic materials (Table 11.2) also indicate unpromising results for Ag NPs. The question of the NP carrier could also have an important role. Bellissima et al. [22] reported that the Ag NPs in a polymeric binder resisted washing, contrary to NPs alone.

Table 11.1 Summary of collected information on natural stone of NPs with antibiological action by toxicity

Reference	Product	R	Test and results
Bellissima et al. [22]	Ag NPs in binder	S	Single bacterial contamination Cell recovery clearly lower with NPs than binder alone (highest reduction achieving 81%) Results varied according to NPs content but not always in an increasing way
Essa and Khallaf [23]	Ag NPs in binders	L S	Single bacterial or fungal contamination Bacteria cell recovery reduced in relation to untreated Fungal growth lowered in relation to untreated
Carrillo-González et al. [24]	Biogenic Ag NPs in solution	B	Single bacterial or fungal contamination Bacteria and fungi growth much lower than untreated (worst result for fungi in basalt but 15% of untreated area)
Ruffolo et al. [25]	ZnO or ZnO + Ag NPs in binder	M	Field study by seawater immersion during 24 months (algae, etc.) Assessment by Braun-Blanquet index: Ag + ZnO (between levels 3 and 4), better than ZnO (level 4), binder (level 4), or untreated (level 5) No difference between two NP amount levels
Goffredo et al. [26]	Solutions of TiO_2 with Ag or Cu	L	Algae and cyanobacteria culture drip on leaning surfaces Better results with TiO_2 + Ag, but area coverage after 9 weeks near 55% TiO_2 and TiO_2 + Cu similar to untreated (around 75%)
Pinna et al. [27]	Cu NPs with consolidant or water repellent	M S	Field test during almost 2 years. Visual assessment and optical microscopy For sandstone, biological growth not detected in untreated and treated areas For marble, colonization by black fungi and green algae, better than untreated and better or similar to without NPs depending on binder
Essa and Khallaf [28]	Biogenic Cu NPs in binder	S L	Single bacterial contamination Clear reduction of bacterial cell recovery (>85%) in relation to untreated, presenting variations with bacterial strain and binder
Zarzuela et al. [29]	SiO_2/CuO NP nanocomposite	L	Single bacterial or fungi contamination Growth inhibition higher with CuO NPs but without clear relation with NP content (worst results for lowest and highest contents)
Khamova et al. [30]	Detonation nanodiamonds (DNDs) in binder	M	Field test. Fungi assessment after a year Within 1000, 700, 1000 CFU for DND 0.15%, 0.20%, and 0.25% mass, respectively, compared to 1200 CFU in binder without DND

(continued)

Table 11.1 (continued)

Reference	Product	R	Test and results
Gómez-Ortíz et al. [31]	Suspensions of Ca (OH)$_2$ particles with ZnO NPs	L	Single fungal contamination Growth inhibition both in the dark and under illumination Possible influence of mineral grain distribution and porosity in the differences in antifungal activity between types
Veltri et al. [32]	TiO$_2$, fullerenes C60, and graphene flake NPs in binder	G M	Single bacterial contamination For granite, similar to binder without NPs but fullerenes show better results For marble, better results with NPs Not affected by binder thickness (tested only with TiO$_2$ and granite) No clear difference between different types of NPs in the substrates No clear relation with NP contents Possible absorption of binder in pores prior to polymerization

NPs nanoparticles, *R* rock type, *B* basalt, *G* granite, *L* limestone, *M* marble, S sandstone, *CFU* colony-forming unit

In relation to Cu NPs, there are studies showing prevention of bioactivity in relation to bacteria and fungi in marble, limestone, and sandstone, but the description presented in Pinna et al. [27] in relation to plaster in field conditions seems to suggest similar results in treated and untreated areas. Again, results concerning algae in more water-intense environments (tested on limestone substrate) were unimpressive, being similar to untreated. In relation to other products, there were some positive results, but the scarce number of studies does not allow an extensive discussion that goes beyond the restatement of the information already available in Tables 11.1 and 11.2.

11.7   Photochemical Effects

In this case, the treatments dealt mostly with the use of TiO$_2$ NPs, with some studies considering TiO$_2$ NPs doped with other metals. These NPs are frequently mixed with binders that have a water repellency effect, which might disturb the assessment of the actual effects of the NPs on biological growth. Results on natural stone [25, 26, 31, 36–40] are presented in Table 11.3, and those obtained in other porous inorganic materials [41–48] are given in Table 11.4. The rock diversity is lower than in the previous case: limestone, marble, and sandstone (no information on igneous rocks). In terms of other porous inorganic materials, the diversity is similar, and there are results on bricks, cement, concrete, mortars, and renders. Curiously, all studies on other porous inorganic materials concern the development of algae under intense water conditions.

Table 11.2 Summary of collected information on other porous inorganic materials of NPs with antibiological action by toxicity

Reference	Product	Material	Test and results
DeMuynck et al. [33]	Ag NPs in water repellent and in water	Concrete	Algae culture sprinkled on leaning surfaces For NPs in water repellent, 100% area covered after 7 weeks for more porous concrete (74.9%) with rougher surface. Lower area and intensity for less porous concrete (6.8%) with smoother surface, much better than untreated but worse than other treatments For NPs in water, 100% area covered after 6 weeks for more porous concrete
MacMullen et al. [34]	Ag NPs in aqueous silane/siloxane emulsions	Cement mortar	Algae and cyanobacteria culture sprinkled on leaning surfaces After 4 weeks, lower area covered and lower fouling intensity for highest NP content
Carrillo-González et al. [24]	Biogenic Ag NPs as preventive and corrective treatments	Stucco	Single bacterial or fungal contamination Prevented bacterial colonization, but much lower effect on fungi growth Corrective treatment after inoculation showed in inhibiting effect on bacteria and fungi (higher for the latter)
DeNiederhäusern et al. [35]	TiO$_2$ with Ag NPs	Unglazed Ceramic	Single bacterial contamination Reduction of 100% without UV regardless of Ag amount (proportions between 1:10 and 1:30)
Pinna et al. [27]	Cu NPs mixed with traditional consolidant or water repellent	Plaster	Field test during almost 2 years. Visual assessment and optical microscopy Biological growth not detected by the naked eye or stereomicroscope, but optical microscopy showed green alga and black fungi on some samples from treated and untreated areas
Veltri et al. [32]	TiO$_2$, fullerenes C60, and graphene flake NPs in binder	Ceramic	Single bacterial contamination Better than binder alone in general, but worst for TiO$_2$ lowest content. Clearly best results with fullerenes. Other NPs generally better with increasing NP content but irregular for graphene/TiO$_2$ mixture

NPs nanoparticles

Positive results against bacteria were found even for the extreme conditions of total immersion in seawater. There are also some positive results in relation to fungi, but there are indications that porosity [38] and grain size distribution [31, 38] might have an influence on the results. Once more, the worst results were obtained for algae in more intense water conditions; in some cases, results were worse with NPs than with the binder alone, with Maury-Ramirez et al. [45] referring the possible photocatalytic degradation of the binder by TiO$_2$. Commenting on the weak effects of TiO$_2$ NPs on prevention of algae growth, Martinez et al. [47] suggested the pos-

Table 11.3 Summary of collected information on natural stone of NPs with antibiological action by photochemical effects

Reference	Product	R	Results
Poulios et al. [36]	TiO₂ NPs in aqueous solutions	M	Single bacterial and fungal contamination Concentration decreased by more than 99% after the first 10 days (higher than reduction with light without TiO₂)
Graziani et al [37]	TiO₂ NPs in aqueous solutions	L S	Algae and cyanobacteria culture sprinkling on leaning surfaces Surface covered equal or higher than 84% after 8 weeks
Goffredo et al. [26]	TiO₂ NPs in aqueous solutions	L	Algae and cyanobacteria culture drip on leaning surfaces Similar to untreated
LaRussa et al [38]	TiO₂ NPs in binder	L M	Single fungal contamination Much lower growth after 12 days than untreated but not improved by higher amount of NPs. Lower on marble but similar relation in untreated specimens (porosity of limestone 20–30 times higher than that of marble)
Ruffolo et al. [25]	TiO₂ NPs in binder	M	Field study by seawater immersion during 24 months (algae, etc.) Assessment by Braun-Blanquet index: similar to binder (level 4), better than untreated (level 5). No difference between two NP amount levels
Ruffolo et al. [39]	TiO₂ and TiO₂-Ag-doped NPs in a polymer	M	Bacterial contamination and immersion in seawater during 72 h No EPS production and no microbial colonization contrary to what happened with untreated (occasional or no patches of EPS on surface with binder alone)
Kapridaki and Maravelaki-Kalaitzaki [40]	TiO₂-SiO₂-PDMS nanocomposite	M	Cleaning of biofilm (diverse bacteria) Greater cleaning (indicated by greater chromatic change) after irradiation than untreated surfaces
Gómez-Ortíz et al. [31]	Suspensions of Ca(OH)₂ particles with TiO₂ NPs	L	Single fungal contamination Antifungal activity limited to photoperiod conditions

R rock type, *L* limestone, *M* marble, *S* sandstone, *NPs* nanoparticles, *EPS* extracellular polymeric materials

sible resistance of algae to photocatalysis effects, highlighting that most studies of photodegradation on organisms usually consider bacteria. Nonetheless, there are encouraging results related to bricks with smooth surface or low porosity (but also in one case of a high-porosity brick), the radiation used or the doping of TiO₂ with Pt and Ir. There are also indications of positive results in cleaning of diverse microorganisms, including algae and lichens.

Table 11.4 Summary of collected information on other porous inorganic materials of NPs with antibiological action by photochemical effects

Reference	Product	Material	Results
Fonseca et al. [41]	TiO_2 NPs in aqueous solutions	Mortar walls	Cleaning of biofilm (bacteria, fungi, algae, and lichen) 2 weeks after application of treatments, areas treated with TiO_2 are lighter (cleaner) than areas treated with traditional biocides
Graziani et al. [42]	TiO_2 NPs in aqueous solutions	Fired clay bricks	Algae and cyanobacteria culture sprinkling on leaning surfaces Covered area similar to untreated after 9 weeks but more effective removal
Graziani et al. [43]	TiO_2 NPs in aqueous solutions	Fired clay bricks	Algae and cyanobacteria culture sprinkling on leaning surfaces Covered area after 12 weeks lower than untreated with biggest difference for low-porosity bricks (3.55% against 24.49%) but also clearly improved for some high-porosity bricks (32.23% against 95.12)
Graziani and D'Orazio [44]	TiO_2 NPs in aqueous solutions	Fired clay bricks	Algae and cyanobacteria culture sprinkling on leaning surfaces For rough surface, similar to untreated after 12 weeks (around 95% area) For smooth surface, better (around 30% in area) than untreated (around 70% in area)
Maury-Ramirez et al. [45]	TiO_2 NP coating by vacuum saturation and TiO_2 NPs in water repellent	Autoclaved aerated concrete	Algal culture sprinkled on leaning surfaces After 7 weeks TiO_2 coating similar area than untreated TiO_2 and water repellent better but worse than water repellent alone (authors refer photocatalytic degradation of water repellent)
Radulovic et al. [46]	TiO_2 NPs in binder	Cement mortar	Algae culture sprinkled on leaning surfaces Better results with NPs than binder alone after 10 weeks with lower algal fouling for higher NPs content
Martinez et al. [47]	TiO_2 NPs in binder	Cement mortar	Algal culture runoff on leaning surfaces Worst results than water repellent alone and at the level of untreated (worse than untreated for low radiation conditions) Algae growth under capillary rising moisture Very near untreated

(continued)

Table 11.4 (continued)

Reference	Product	Material	Results
Linkous et al. [48]	TiO$_2$ with and without noble metal NPs in a binder	Portland cement	Immersion in algae contaminated water avoiding movements that promoted ablative actions TiO$_2$ better than untreated and binder alone with mixed lamp irradiation. Improved by addition of Pt and Ir with higher reduction for Pt than Ir Much worst results for fluorescent lamp only, with best inhibition results for Pt-TiO$_2$ (19%) and Ir-TiO$_2$ (28%), being slighter better than binder. TiO$_2$ alone worse than binder and even than untreated

NPs nanoparticles

11.8   Combined Effects

While in the aforementioned sections assessments of a given effect attributed to NPs were discussed, there are situations where there could be contributions from more than one effect. For example, the study of MacMullen et al. [34] assessed the biocide effect of Ag NPs on cement mortar but also reported an increase in the water drop contact (representing increasing hydrophobicity) with increasing NP content, achieving a contact angle of 140° against a value of 110° for emulsions without NPs. The results of this study also showed that while the porosity with treatments was similar to untreated, sorptivity values were lower than 10% of untreated. Radulovic et al. [46] also attributed the treatment performance on cement mortar, besides the photodegradation effect, to the contribution effect from increasing water repellency. Vidaković et al. [49], using TiO$_2$ based on layered double hydroxides, attributed a significant antifungal activity on renders to the synergy between photodegradation and biocide effects related to metal release. However, their results on bricks were not as promising. Moreover, a study conducted by Graziani et al. [50] on fired bricks showed that the addition of Cu and Ag to aqueous solutions of TiO$_2$ did not improve the antimicrobial action of the treatment coating. The study by Goffredo et al. [26] can also be considered as a combination of effects, but the results of TiO$_2$ alone (similar to untreated) suggested that the main effect was due to the biocide action of Ag NPs (Tables 11.1 and 11.3).

The work by Ruffolo et al. [25] presented a combination of all effects since it considered the application of a binder with TiO$_2$ or ZnO NPs, mixtures of both, and of each of these with Ag and Zn NPs, where the obtained data suggested a hydrophobic effect of NPs (Tables 11.1 and 11.3). Treatment results were assessed by the covered area using the Braun-Blanquet numeric index. The higher amount of Ag used was two orders of magnitude lower than either TiO$_2$ or ZnO higher amounts (due to color change hazards related to Ag NPs). Water drop angle measurements were found to be higher for NPs with a binder than for the binder alone (85°), with the higher value (105°) being obtained for the highest amount of TiO$_2$ NPs. Regarding biological growth on marble pieces immersed in seawater showed, after 2 years, the best result was obtained for TiO$_2$ mixed with Ag, with covered area at level 3 (25–50% covered area) regardless of NP amounts. The authors attributed this result to the synergy between catalytic effects of TiO$_2$ and the biocidal effect of Ag, as well as to the hydrophobicity associated with these treatments. The second best result was found for the ZnO and Ag mixture, between levels 3 and 4 (50–75% covered area). Results of TiO$_2$ or ZnO with binder were similar to those obtained with binder alone (level 4), but better than that of untreated (level 5, higher than 75% area).

11.9   Impacts on Materials

Two kinds of undesired impacts are considered: color changes and the production of soluble salts, since the latter means the introduction of what are recognized as degradation agents on building materials [13].

There are some references to the production of salts associated with these treatments. In some cases, this is attributed to the characteristics of the binder used to carry the NPs, namely, its acidity that, in the case of carbonate rocks, can lead to the production of soluble salts [51, 52]. However, in a study conducted by Aflori et al. [53], the salts' composition, calcium, and magnesium nitrates suggested that the formation of these salts was related to the nanoproduct, as silver nitrate was used for making Ag NPs; the authors furthermore reported that the percentage of nitrates was higher for Ag alone than for Ag and TiO$_2$.

The question of color changes is, unequivocally, the main concern in the studies of the application of NPs. However, this assessment for NPs might be obscured when they are mixed with other products, namely, polymers. For example, in the study by Van der Werf et al. [54], tests with ZnO NPs caused higher or lower chromatic change (in relation to binder alone) depending on the binder type and rock type. Values of ΔE^* as high as 40 [55] were reported for Cu NPs (0.28% mass content), but the same authors found that the ΔE^* value was similar to binder alone (up to 15) for 0.02% Cu NP mass content. Similar high values were found for TiO$_2$ NPs doped with Ag and Sr ($\Delta E^* = 38.3$) or Fe ($\Delta E^* = 36.5$) for the highest doping level, which became much lower for lower doping ($\Delta E^* = 5.1$ for Ag and $\Delta E^* = 8.0$ for Sr). There are several other references to the effect of the amount of NPs and the color change of the treated material, with Ag NPs deserving special concerns in terms of colorimetric changes. MacMullen et al. [34] reported increasing ΔE^* with increasing Ag NPs, with ΔE^* values between 5.35 (without Ag NPs) and 8.40 (Ag NPs < 0.5% in mass content, the highest amount studied). But there is also a study conducted by Essa and Khallaf [28], with Cu NPs, which concluded that NP amounts should be minimized because of color changes. However, this issue can help to promote certain substances, as illustrated by the reference to the minor color change associated with ZnO when compared with Cu [54, 55] or Ag [54].

The substrate chromatic characteristics should also be considered. Van der Werf et al. [54] reported that Cu NPs show a brownish color while ZnO NPs are whitish. This means a lower impact of ZnO NPs on lighter stones, but the other case might occur when applied on darker ones.

Color change related to product application can evolve with time. For example, data for Van der Werf et al. [54] for a binder with ZnO NPs showed reductions in ΔE^* after 33 days to between 35 and 17%. Similarly, Ditaranto et al. [56] reported a reduction after 2 months to around half or lower.

11.10   Impacts on the Environment

The environmental impacts of the NPs have been considered from their early recommendations and uses. For example, there are NPs that have been recommended because of being considered as environmentally friendly, with lower risks in terms of toxicity, namely, in comparison with traditional treatments, such as polydimethylsiloxane/controlled SiO_2-doped ZnO nanocomposites [57] and a binder containing organically modified boehmite nanoparticles [58, 59].

In their proposal of detonation nanodiamond (DND), Khamova et al. [30] reported that they have a "mild" biocide effect and are environmentally safe. The perception of the metals' toxicity (not specifically for NPs) can condition the amounts of NPs added, and in the study of Huang et al. [60] this allowed higher amounts for Zn than for Ag or Cu. The non-toxicity of TiO_2 NPs is frequently reported as one of its favorable characteristics [5, 25, 43, 61–65], but this issue might be more complex as it will be necessary to consider the effect of nanodimension of particles. Huang et al. [60] reported that due to their size, a small mass of NPs can have a high number of particles that can affect human health, mainly through inhalation. Zarzuela et al. [29] also stated that toxicity against microorganisms increases in NPs for a given substance. Furthermore, the toxicity of compounds can depend on the organisms. Several studies highlight that Ag has a strong microbial effect but has a low toxicity for humans [22–24, 35].

The use of NPs can also be considered to have a positive impact when they improve materials that have a lesser environmental impact and reduce the use of materials with higher impact [4, 24, 66].

Despite their increased use in building materials, there is still a lack of knowledge on the effects of NPs in the environment. Early research [67, 68] suggested the possible risks for human health and the environment as a function of both exposure and hazard. The increasing use in the construction industry implies increasing exposure, and it is clear today that outdoor use creates pollution due to degradation of nano-enabled products [69] that may even replace other previously used toxic substances in building materials [70]. However, studies are still insufficient, despite the expected negative effects on certain organisms.

The environmental impact of toxic products will depend on their release from the coatings, and the properties that make them desirable as biocide will be undesired in the environment. Data from Chen et al. [71] showed that release levels of Ag after 1 day of immersion were higher than that of the estimated world average of Ag content on streams presented in Reimann and Caritat [72], and the Ag amount in solution kept increasing for up to 30 days, which was the time interval considered in this study. While focus is being placed on coatings for porous materials, in this context, it can be interesting to refer to the study on paints with Ag NPs in outdoor exposure tests [73]. This study of Kaegi et al. [73] showed that Ag contents from the runoff of model panels decreased with time. Concentrations went from 0.145 mg/L, higher than twice the maximum value referred in Reimann and Caritat [72] for world streams, to around 0.001 mg/L, which is still around 20 times higher than the estimated world mean for streams [72]. Kaegi et al. [73] also reported that 30% of Ag NPs were released after a year of exposure. The leaching of Ag NPs was very high during the initial times of exposure, being more than the 80% of total Ag lost leached in the first eight rain events. However, microscopic studies suggested that these Ag NPs were likely transformed to less toxic forms. Additionally, leached Ag NPs were attached to the organic binder and mostly as composite colloids. In this context, the role of the carrier of NPs in fixating NPs and limiting their release to the environment should also be considered [74–78]. The study by Ditaranto et al. [55] concerning Cu NPs in a binder showed Cu content in solution of 2 ppm, higher than the maximum value reported by Reimann and Caritat [72] for streams.

11.10.1   TiO$_2$ NPs

TiO$_2$ NPs are by large the most highly produced in Europe and used in building materials [79], and their release is the most studied. Early studies, such as a model of Mueller and Nowack [80], showed that TiO$_2$ NPs could be damaging for organisms if such NPs were released in significant amounts to water systems. Kaegi et al. [81] showed that TiO$_2$ NPs can be released from façade coatings into runoff water and discharged into natural surface waters.

TiO$_2$ NPs are stable in the environment and do not undergo any dissolution reactions [82]. They include anatase and rutile (the two main crystalline phases of TiO$_2$ NPs) that have been found to be ubiquitous in water and sediments of stormwater ponds, although they usually form aggregates [70]. The reason for this is that in the environment their behavior is mainly governed by agglomeration processes that lead to sedimentation which in turn depends on the water composition, the form of TiO$_2$ NPs, and the presence of biofilms [83].

Most studies have addressed the release of TiO$_2$ NPs due to deterioration of façade coatings, but it has been observed that the most important release of TiO$_2$ NPs occurs during the application stage in the building's environment, with manual brush application causing the smallest amount of release [84]. Deterioration studies have revealed that TiO$_2$ NPs are mostly released from façade coatings as agglomerates and/or aggregates, embedded in the matrix materials, an important variable for their toxicity besides their chemical stability and morphology [79]. The degradation of TiO_2 NP coatings on façades has been demonstrated by cracking, mainly due to drying stress due to water evaporation and gradual embrittlement of the polymeric binder during its interaction with UV radiation [85]. Despite such decay, coatings resist leaching, with Ti released in water being near the background [85, 86]. However, mechanical abrasion has shown to release a considerable amount of free TiO_2 NPs in some studies, with results contrary to others that observed their release bonded to the matrix [85]. Such mechanical release has been also observed in different degrees during the period of abrasion. Shandilya et al. [85] observed four phases of release, being the initial and third ones the most important and being a function of the duration under UV, temperature, and water exposure, as observed from experimental conditions. Phase I corresponded to an increase of emitted aerosol particles' number concentration after application of the nanocoatings, while during phase II the surface abrasion takes place under a stable state until phase III arrives, when the surface material becomes exposed gradually due to the removal of the nanocoatings. Finally, in phase IV the surface material is exposed due to the release of NPs and removal of the nanocoatings. On the contrary, the release of TiO_2 NPs from photocatalytic cements occurs as free NPs, which are potentially more toxic. However, photocatalytic cements represent a minor source of release into the environment, as they are less used than coatings [79].

There are several toxicological and ecotoxicological studies that have demonstrated the toxic effects of TiO_2 NPs [85]. The toxicity of TiO_2 NPs is higher than bulk TiO_2, which is different for some other forms and increases as particle size decreases to the range of 10–40 nm [87, 88]. It has been observed that photocatalytic nano-TiO_2 is much more toxic to water organisms than photostable nano-TiO_2 [89]. The toxic effects were reported to occur at concentrations as low as 1 mg/L, but are dependent on the type of TiO_2 NPs being tested [89, 90], finding no adverse effects of photostable TiO_2 NPs and under non-illuminated conditions.

11.10.2   Ag NPs

Ag NPs have been revealed as one of the most effective biocide agents due to the broad-spectrum antimicrobial properties of Ag combined with its low toxicity [91, 92] and as such are potentially one of the most hazardous NPs to environmental systems [93]. Despite the increase of their use and published research, little is known about the mechanisms, trophic transfer, ecosystem hazards, and impact on higher organisms including toxicity for people. It has been demonstrated that Ag NPs are toxic for both aerobic and anaerobic bacteria but also for some fish, bivalves, rodents, and humans as well as being able to inactivate fungi, algae, and viruses [92, 94–98]. In vitro tests have suggested that Ag NPs may affect mitochondrial function [99, 100], cause oxidative stress and membrane damage with cellular damage [101], direct DNA damage [102], reduce photosynthesis [92], being even possible the induction of colon cancer in humans [98]. However, it is not clear if these effects are directly due to Ag NPs or free Ag^+ ions [103], and despite such toxicity, bacterial strains resistant to Ag have been found [104, 105].

Despite such effects observed from in vitro studies, the toxicity of Ag NPs in environmental systems and in vivo is limited by the solubility of Ag ions in solutions containing halides. In the presence of Cl^- ions, AgCl is quickly formed and precipitates, but also some molecules can stabilize Ag ions [106]. However, high halide concentrations result in the formation of water-soluble ionic complexes ($AgCl_2^-$ and $AgCl_3^{2-}$) that are more bioavailable, increasing the Ag toxicity to microorganisms [92, 105].

Ag NPs in paints have been reported to be among the most frequent sources of Ag NPs released to the aquatic environment [80], with likely contributions from paints applied on building façades [73].

Despite all these results, the behavior of Ag NPs in ecosystems has not received much attention. It has been observed that most Ag NPs (about 90%) are removed from water during wastewater treatment [107], fitting concentrations in wastewater model predictions [108]. Moreover, most of the released Ag NPs are found as coarse particles (fibers) larger than 450 nm, reducing their reactivity, and it has been observed that Ag NPs are rapidly converted to silver sulfide, which is much less toxic without a measurable impact [109–111]. Thus, under real-life conditions, risk assessment is more difficult as environmental ions and molecules can passivate NPs [82, 112] and more research about their in vivo effects and detection in air, water, and ecosystems is needed to assess the real impact of Ag NPs.

11.10.3 ZnO NPs

Assessing the environmental impact of ZnO NPs is difficult as it has been found in low concentrations in the environment, while Zn concentrations are comparatively higher. This is due to the high solubility of ZnO in water [82]. However, its theoretical solubility does not necessarily results in quick dissolution, as surface coatings could delay or hinder such dissolution. In this sense, Gimbert et al. [113] showed that ZnO NPs are stable in soil when the pH is above 9. In general terms, ZnO NPs have the same problems that Ag NPs because both NPs, when dissolved, can be toxic for organisms. Franklin et al. [114] showed that nano-ZnO and bulk ZnO have the same toxicity, and both are determined by dissolved Zn.

A review conducted by Gottschalk et al. [115] about the use of computer simulations for predicting environmental concentrations of ZnO NPs, among other NPs, in water in Switzerland used in several applications and possible emission processes, showed that the major NP flows come from a waste incineration plant from a landfill. The probabilistic approaches based on a same total amount of NPs released were consistent with two study cases that reported ZnO NP concentrations in aqueous systems. Measured ZnO NP concentrations [115, 116] have been found to fit modeled concentrations [108]. Nevertheless, quite a few studies have shown that agglomeration is an important parameter affecting the fate and behavior of ZnO NPs [117]. The modeled concentrations in most cases are comprised of all agglomeration states of a given NPs irrespective of the actual size of the agglomerates. This could potentially be an oversimplification of real-life scenarios that will affect the likely toxicity of ZnO NPs in aqueous systems. Key questions on the effects of ZnO NPs in the environment are also transformations. Several studies have demonstrated the sulfidation of some metallic NPs such as ZnO in wastewater. These will limit the release of ionic species [118, 119]. This is not necessarily applicable to runoff water, as this contains lower concentrations of sulfide, making sulfidation less likely, but further research will provide a more exact and precise model [70].

In a study carried out by Baalousha et al. [70], ZnO NPs were observed in the colloidal fraction of stormwater ponds. This was identified as zincite, and it was only found as suspended and sedimented particulate matter. ZnS NPs were also detected suspended in water and in the sediments. They were very small in diameter (about 5 nm) and aggregated in grains bigger than 100 nm, probably as sphalerite although other trace metals (Al, Fe, Co, Cu, and Pb) were also found associated with such aggregates. Such ZnS NPs probably result in the sulfidation of ZnO NPs via bacterial sulfate reduction, as suggested by the authors, as abiotic sulfidation occurs under anoxic conditions [120].

11.10.4   Cu NPs

Cu NPs are often referred as Cu NPs or CuO NPs. Both cases are included in this section. The behavior of Cu NPs can be compared to the bulk Cu present in water systems, but the problems that arise from such an assumption are similar to these found for other metal oxide NPs. In fact, agglomeration and speciation in such environments are predicted as important factors in assessing its effect and toxicity, and studies have taken such an approach [93, 121]. In any case, research on the toxicity and environmental behavior of Cu NPs are scarce, even compared to other metal NPs such as Ag NPs or ZnO NPs [122]. However, in the last 4 years, some studies have attempted to characterize this substance.

The toxicity of Cu NPs released in the air has been studied regarding the effects on human health but less so in water systems and for aqueous organisms such as plankton or fish [123, 124]. In plankton, it has been observed that despite a high ingestion of CuO NPs, toxicity is due in part to the dissolved fraction. Also for fungi and other organisms, it seems that the dissolved Cu is more toxic than Cu NPs [125]. However, there is still some controversy on which mechanisms cause the observed toxicity in aquatic systems.

Cu NPs have shown more inhibitory activity on bacterial than on fungal strains, being very toxic for native soil bacteria due to the interaction with cell wall components that modify cell morphology and affect the function of membrane proteins [126]. It has been observed that Cu ions are more toxic than Cu NPs for all organisms except for yeast and mammalian cells in vitro. However, the toxicity assays with mammalian cells in vitro use a serum that may disperse and coat Cu NPs and could have a different behavior than in water systems [122].

11.10.5 SiO_2 NPs

It is not easy to assess the behavior of SiO_2 NPs, as SiO_2 can correspond to several compounds with different physical and chemical behaviors. SiO_2 is usually referred to as silica, but silicon dioxide usually refers to Si–O-bounded tetrahedra that form silicate minerals that are not exactly silicon dioxide. Such minerals can be structured in different ways, from amorphous (opal-A) to crystalline molecules (quartz) with intermediate possibilities of intermediate crystallinity. Depending on the crystalline structure, its solubility and other chemical properties are very different, as well as its effects on the environment. In fact, it has been remarked that amorphous SiO_2 is less toxic than crystalline SiO_2 [88].

Few works have investigated the degradation and durability of SiO_2 NPs [127, 128], but almost none have studied their release [86]. Zuin et al. [129] observed the release in static leaching tests of paints with and without SiO_2 NPs. They observed very low concentrations of Si, mainly in the form of agglomerates with other particles. Paints were also abraded to observe the formation of particles [127], and the resulting particles were mostly micrometric. Moreover, the presence of NPs in the paint did not change the size distribution of the abraded particles compared to NP-free reference paint. In a study carried out by Al-Kattan et al. [86], it was observed that the release of SiO_2 from paints under experimental accelerated weathering conditions was very small (about 2%). The authors assessed the acceleration factors, with results corresponding to about 1–2 years of weathering under natural conditions, with dissolution being the major releasing process. No apparent trend of the concentrations of SiO_2 was observed over time, contrary to what happens with TiO_2 or Ag, but it cannot be excluded that the cause of this was that the paints used were different. Other studies on the effect of UV irradiation on SiO_2 particles have shown an increasing trend in the Si concentration on the surface with respect to the epoxy binder over time [130].

The size fractionation of SiO_2 consisted mostly of particles larger than 0.1 μm, mainly because NPs agglomerate or bound to the paint matrix. In an aged powder extraction, 10% of the released Si was present as particles smaller than 0.45 μm, embedded in a matrix, similar to previous studies with other types of NPs [73].

There are very few studies about the environmental behavior and effects in ecosystems and organisms of SiO_2 NPs. Measurements in wastewater treatment plants have shown that they are difficult to remove during water treatment [131, 132], while coated SiO_2 NPs are removed readily. Thus, their behavior in water shows that they are different from other NPs [82].

The possible toxicological effects of SiO_2 NPs have been investigated [133–135], and no differences were found between released NPs from paints with and without SiO_2 NPs. Normally bulk silica is used as a negative control in toxicity studies because of its slight influence on microorganisms. However, Van Hoecke et al. [136] reported that SiO_2 NPs are much more toxic for green algae than bulk SiO_2 and that its effects are present at concentrations of a few mg/L, a value that is well above the concentrations expected for NPs. It seems that SiO_2 NPs released during degradation of building materials behaves in a different way than pristine SiO_2 NPs [86], but this still must be discussed in future ecotoxicological studies in water. However, for their toxicity in air, only the study conducted by Saber et al. [133] found that the addition of NPs to paints does not increase the potential emission of nanodust, being probably more considerable than the toxicity of the paint matrix. This low toxicity for SiO_2 NPs is consistent with exposure models for colloidal amorphous silica [137].

11.10.6 Carbon Nanotubes (CNTs)

Few studies have focused on the potential releases of carbon nanotubes (CNTs) from NPs considering two main scenarios: due to high-energy processes (drilling, sanding, and cutting the CNT composite) and from the bound matrices due to low-energy processes (environmental degradation from UV light and weathering) The release can occur as free or agglomerated CNTs or embedded in the matrix. Weathering under intense UV light has shown scarce release [138], while abrasion increases exposure to polymer-CNTs [138], making direct release unlikely [139]. In fact, models have predicted that CNTs can be found in extremely small concentrations in the environment (0.003–0.02 ng/L) [108], coming from several products and at present not extensively used as biocides in building materials [82, 139].

Once released, their behavior in the environment is not well known due to the limited research carried out [82], but it is expected that they will not have significant effects due to their low toxicity to aquatic organisms [97, 140]. Although they are considered as persistent, some studies have demonstrated their degradation within few weeks due to enzyme-catalyzed reactions [141], and they quickly agglomerate and sediment, despite natural organic matter keeping CNT in solution [71, 142].

For modeling, it has been more difficult to build release scenarios and exposure assessments, making clear the need for both a description of standard release processes and the standardization of the reporting of release and exposure processes [139].

The knowledge on the potential toxicity of CNTs is limited by the scarce knowledge on their biopersistence [143]. CNTs are believed to be biopersistent, but recent studies have shown that they degrade [144]. It seems that they have a small effect on soils and sediments [145]. However, in aqueous systems, it seems that there are ecotoxicological effects at mg/L concentrations that can be reduced by sedimentation.

Most toxicity studies are based on their high aspect ratio, which is similar to that of asbestos [82]. However, in vitro studies have provided a lot of contradictory data, which increases the uncertainty about the risks of CNTs in nature [82]. It is known that CNTs can be vehicles for drug delivery, as they are able to cross cell membranes [146]. On the contrary, in vivo studies revealed that longer CNTs are more toxic, being more similar to asbestos, while shorter or tangled nanotubes had much less of an effect [147].

Some studies have reported an indirect toxicity due to contaminants, making toxicity negligible when contaminants are removed [148–150]. In fact, one probable source of CNTs' nanocomposite's toxicity is their potential degradation products due to weathering by photochemical reactions [151].

11.10.7 Other NPs

There are a limited number of reported studies on the environmental impact of other NPs. Some attention has been focused on the effect of DND, and CeO_2 NPs, making DND the only ones among these with applications such as a countermeasure for biological agents [30]. This is probably because it is rarely used when compared with TiO_2, SiO_2, Ag, and ZnO NPs.

Environmental issues about the preparation of DND were partially addressed by Voznyakovskii et al. [152]. It has been observed that DND particles are relatively stable in aqueous solutions but aggregate quickly in the presence of mono- or divalent salts. Even carboxyl groups formed on their surface do not alter their colloidal behavior which mainly depends on the presence of electrolytes [153, 154]. However, little is known about these NPs, and research on their environmental effects and toxicity must be improved.

11.11 Discussion

Diverse NPs have been proposed for the treatment of natural stone and other porous inorganic materials in relation to biocoatings, based on water repellency, toxicity, photochemical effects (photodegradation and hydrophilicity), or combinations of these. In general, these products consisted of the addition of NPs to an aqueous solution or a binder, which can disturb the assessment of NP effects as binders can also have desirable and undesirable effects on the substrates. In this chapter, only studies that had actual data on biological development were considered. Meta-analytic generalizations from the gathered studies must be approached with care given their small number, namely, in relation to the involved variables. For example, all data for photochemical effects on other porous inorganic materials used tests with algae under intense water conditions; however, only two of the eleven reported studies on treatments of natural stone using toxic effects have considered these conditions.

The highest number of studies was found for the application of toxic products and also with the highest diversity of substances; however, Ag NPs, followed by Cu NPs, were predominant. In relation to photochemical effects, there is a clear monopoly of TiO_2 NPs, sometimes doped with other metals (Ag, Ir, Pt). Based upon the aforementioned conditions in this chapter, only one example was found for a proposal of NPs specifically for water repellency consisting of hydrophobic SiO_2 NPs.

However, several indications of increasing water repellency were found with the addition of NPs.

In terms of the tested natural stones, limestone was the dominant rock type, but there are also diverse studies with marbles and sandstones. Only two studies with igneous rocks, one for basalt and another for granite, were found and both concerned toxic products.

Water conditions used for testing were also variable, a relevant issue considering the field conditions in which the products might be applied and that go from damp surfaces by condensation to total immersion, passing by rain wetting, dripping, sprinkling, and runoff. Linking field observations of the problem is essential in choosing suitable conditions for assessing the performance of the treatments. As mentioned above, one runs the risk of either rejecting a suitable treatment due to excessively harsh testing conditions or accepting an unsuitable treatment due to testing conditions not being harsh enough. None of the testing conditions found in the studies analyzed here can be considered harsher than possible field situations. On the other hand, testing by exposure to a moist atmosphere will be hardly related to the actual outdoor field conditions, where biocoatings' growth is usually related to repeating episodes of water contact, but it can correspond to some situations indoors. It is proposed that the burden of proof should rest on the newly proposed treatments, and hence, harsh testing conditions that involve contact with liquids should be considered the best testing conditions. Total immersion or permanent dripping, sprinkling, or runoff could be excessive in relation to many stone applications, but they can represent specific field situations, and treatments that work in these conditions should work in others, which will be coherent with the burden of proof resting on the proposed treatment. The results collected suggest that diverse nanoproducts will not be very effective in intensive water testing conditions.

Differences were also found in terms of tested microorganisms. In most of the considered studies, algae, fungi, or bacteria have been used; however, there are studies with latest two and some field studies with all aforementioned microorganisms and other organisms. Collected data further suggested that the effects on bacteria and fungi might be difficult to extrapolate to organisms such as algae. The elimination of fungi could be relevant to avoid the development of lichen biocoatings, which tend to be harder to eliminate, but algae biofilms are a very frequent problem.

Another issue concerns the NP amounts. This is a point that is affected by side effects of NPs, concerning changes to the materials, namely, chromatic changes, which impose limitations on the amounts used. In relation to the efficacy of higher amounts of NP content, the reported results have not been unanimous. This might indicate some threshold effect and can also be potentially due to the variability in experimental conditions (e.g., the unavoidable variability in properties between different specimens of the same rock).

NP amounts are linked to the main issues concerning undesired effects of these nanoproducts, both in terms of the treated substrate and the environment. In terms of impacts on the substrate, the main issue concerns the color change, which seems to be affected by NP amounts, at least up to a point, but there are evidences indicating a diminution of that impact with time. Another issue, less frequently referred, concerns the possible introduction of soluble salts, which might promote erosive decay of the substrate by crystallization. While the potential toxicity hazards of NPs are still a subject of debate and research, there is evidence suggesting that NPs could be mobilized in significant amounts by weathering agents (viz., rain) and, in this way, would affect the surrounding environment.

11.12   Conclusion

Analysis of the collected data in this work suggests that much more work is still necessary concerning the use of NPs against biocoatings on stony materials, as results related to one of the main biological agents (algae) under realistic water conditions are not very supportive. One of the main issues that needs to be addressed concerns the fixation of NPs on materials' surfaces, which needs to be balanced with the antimicrobial action. The fixation of NPs to the coating will also help to reduce potential environmental hazards, as it will limit the mobility of NPs. Another source of concern that needs further research refers to the impacts of the treatments on the substrate, especially in terms of color change.

Acknowledgments Landscapes, Heritage and Territory laboratory (Lab2PT) supported by Portuguese "Fundação para a Ciência e a Tecnologia" (FCT UID/AUR/04509/2013), Portuguese funds, and when applicable FEDER co-financing, in the aim of the new partnership agreement PT2020 and COMPETE2020—POCI 01 0145 FEDER 007528. J. Sanjurjo-Sánchez is grateful for funding from "Consolidación y estructuración de unidades de investigación competitivas—Grupo de potencial de crecimiento" (GPC2015/024), Xunta de Galicia.

References

1. Warscheid T, Braams J. Biodeterioration of stone: a review. Int Biodeter Biodegr. 2000;46(4):343–68. https://doi.org/10.1016/S0964-8305(00)00109-8.
2. Scheerer S, Ortega-Morales O, Gaylarde C. Microbial deterioration of stone monuments—an updated overview. Adv Appl Microbiol. 2009;66:97–139. https://doi.org/10.1016/S0065-2164(08)00805-8.
3. Pinna D. Coping with biological growth on stone heritage objects: methods, products, applications, and perspectives. Oakville: Apple Academic Press; 2017. ISBN: 9781771885324
4. Baglioni P, Carretti E, Chelazzi D. Nanomaterials in art conservation. Nat Nanotechnol. 2015;10(4):287–90. https://doi.org/10.1038/nnano.2015.38.
5. Banerjee S, Dionysiou DD, Pillai SC. Self-cleaning applications of TiO_2 by photo-induced hydrophilicity and photocatalysis. Appl Catal B Environ. 2015;176–177:396–428. https://doi.org/10.1016/j.apcatb.2015.03.058.
6. van Broekhuizen FA, van Broekhuizen JC. Nanotechnology in the European Construction Industry—State of the art 2009. Amsterdam: IVAM UvA BV; 2009.
7. Boostani H, Modirrousta S. Review of nanocoatings for building application. Proc Eng. 2016;145:1541–8. https://doi.org/10.1016/j.proeng.2016.04.194.

8. Sierra-Fernandez A, Gomez-Villalba LS, Rabanal ME, Fort R. New nanomaterials for applications in conservation and restoration of stony materials: a review. Mater Constr. 2017;67(325):107. https://doi.org/10.3989/mc.2017.07616.
9. ICOMOS-ISCS. Illustrated glossary on stone deterioration patterns. 2008. Available online (August 2012), at https://www.icomos.org/publications/monuments_and_sites/15/pdf/Monuments_and_Sites_15_ISCS_Glossary_Stone.pdf. Accessed August 2017.
10. Fitzner B, Heinrichs K. Damage diagnosis on stone monuments—weathering forms, damage categories and damage indices. In: Prikryl R, Viles HA, editors. Understanding and managing stone decay: The Karolinum Press; 2002. p. 11–56. http://www.stone.rwth-aachen.de/decay_diagnosis.pdf. Accessed August 2017.
11. Dorn RI. Desert rock coatings. In: Parsons AJ, Abrahams AD, editors. Geomorphology of desert environments. Dordrecht: Springer Netherlands; 2009. p. 153–86. https://doi.org/10.1007/978-1-4020-5719-9_7.
12. Bieniawski ZT. Engineering rock mass classifications: a complete manual for engineers and geologists in mining, civil and petroleum engineering. New York: Wiley; 1989. ISBN 0-471-60172-1
13. Goudie AS, Viles HA. Salt weathering hazards. New York: Wiley; 1997.
14. Marques de Sá JP. Applied statistics using SPSS, STATISTICA, MATLAB and R. Berlin. Heidelberg: Springer; 2007. ISBN 978-3-540-71971-7. https://doi.org/10.1007/978-3-540-71972-4.
15. Hurford C. Minimising Observer Error. In: Hurford C, Schneider M, editors. Monitoring nature conservation in cultural habitats. Dordrecht: Springer; 2006. p. 79–92. https://doi.org/10.1007/1-4020-3757-0_10.
16. Lee PS. Quantitation of microorganisms. In: Goldman E, Green LH, editors. Practical handbook of microbiology. Boca Ratón: CRC Press; 2009. p. 11–29. ISBN: 978-0-8493-9365-5.
17. Wingender J, Neu TR, Flemming H-C. What are bacterial extracellular polymeric substances? In: Wingender J, Neu TR, Flemming H-C, editors. Microbial extracellular polymeric substances. Berlin: Springer; 1999. p. 1–19. https://doi.org/10.1007/978-3-642-60147-7_1.
18. Buckley RR, Giorgianni EJ. CIELAB for color image encoding (CIELAB, 8-bit; domain and range, uses). In: Luo MR, editor. Encyclopedia of color science and technology. New York, NY: Springer; 2016. p. 213–21.
19. Erbil HY. Surface chemistry of solid and liquid interfaces. Oxford: Blackwell; 2006. ISBN: 978-1-4051-1968-9
20. Hall C, Hoff WD. Water transport in brick, stone and concrete. London: Spon Press; 2002. ISBN 0-419-22890-X.
21. Eyssautier-Chuine S, Vaillant-Gaveau N, Gommeaux M, Thomachot-Schneider C, Pleck J, Fronteau G. Efficacy of different chemical mixtures against green algal growth on limestone: a case study with Chlorella vulgaris. Int Biodeter Biodegr. 2015;103:59–68. https://doi.org/10.1016/j.ibiod.2015.02.021.
22. Bellissima F, Bonini M, Giorgi R, Baglioni P, Barresi G, Mastromei G, Perito B. Antibacterial activity of silver nanoparticles grafted on stone surface. Environ Sci Pollut Res. 2014;21(23):13278–86. https://doi.org/10.1007/s11356-013-2215-7.
23. Essa AMM, Khallaf MK. Biological nanosilver particles for the protection of archaeological stones against microbial colonization. Int Biodeterior Biodegrad. 2014;94:31–7. https://doi.org/10.1016/j.ibiod.2014.06.015.
24. Carrillo-González R, Martínez-Gómez MA, González-Chávez MCA, Mendoza Hernández JC. Inhibition of microorganisms involved in deterioration of an archaeological site by silver nanoparticles produced by a green synthesis method. Sci Total Environ. 2016;565:872–81. https://doi.org/10.1016/j.scitotenv.2016.02.110.
25. Ruffolo SA, Ricca M, Macchia A, La Russa MF. Antifouling coatings for underwater archaeological stone materials. Prog Org Coat. 2017;104:64–71. https://doi.org/10.1016/j.porgcoat.2016.12.004.

26. Goffredo GB, Accoroni S, Totti C, Romagnoli T, Valentini L, Munafò P. Titanium dioxide based nanotreatments to inhibit microalgal fouling on building stone surfaces. Build Environ. 2017;112:209–22. https://doi.org/10.1016/j.buildenv.2016.11.034.
27. Pinna D, Salvadori B, Galeotti M. Monitoring the performance of innovative and traditional biocides mixed with consolidants and water-repellents for the prevention of biological growth on stone. Sci Total Environ. 2012;423:132–41. https://doi.org/10.1016/j.scitotenv.2012.02.012.
28. Essa AMM, Khallaf MK. Antimicrobial potential of consolidation polymers loaded with biological copper nanoparticles. BMC Microbiol. 2016;16(1). https://doi.org/10.1186/s12866-016-0766-8.
29. Zarzuela R, Carbù M, Gil MLA, Cantoral JM, Mosquera MJ. CuO/SiO$_2$ nanocomposites: a multifunctional coating for application on building stone. Mater Des. 2017;114:364–72. https://doi.org/10.1016/j.matdes.2016.11.009.
30. Khamova TV, Shilova OA, Vlasov DY, Ryabusheva YV, Mikhal'chuk VM, Ivanov VK, Frank-Kamenetskaya OV, Marugin AM, Dolmatov VY. Bioactive coatings based on nanodiamond-modified epoxy siloxane sols for stone materials. Inorg Mater. 2012;48(7):702–8. https://doi.org/10.1134/S0020168512060052.
31. Gómez-Ortíz N, De la Rosa-García S, González-Gómez W, Soria-Castro M, Quintana P, Oskam G, Ortega-Morales B. Antifungal coatings based on Ca(OH)$_2$ mixed with ZnO/TiO$_2$ nanomaterials for protection of limestone monuments. ACS Appl Mater Interfaces. 2013;5(5):1556–65. https://doi.org/10.1021/am302783h.
32. Veltri S, Sokullu E, Barberio M, Gauthier MA, Antici P. Synthesis and characterization of thin-transparent nanostructured films for surface protection. Superlattice Microst. 2017;101:209–18. https://doi.org/10.1016/j.spmi.2016.11.023.
33. DeMuynck W, Ramirez AM, De Beli N, Verstraete W. Evaluation of strategies to prevent algal fouling on white architectural and cellular concrete. Int Biodeter Biodegr. 2009;63(6):679–89. https://doi.org/10.1016/j.ibiod.2009.04.007.
34. MacMullen J, Zhang Z, Dhakal HN, Radulovic J, Karabela A, Tozzi G, Hannant S, Alshehri MA, Buhé V, Herodotou C, Totomis M, Bennett N. Silver nanoparticulate enhanced aqueous silane/siloxane exterior facade emulsions and their efficacy against algae and cyanobacteria biofouling. Int Biodeterior Biodegrad. 2014;93:54–62. https://doi.org/10.1016/j.ibiod.2014.05.009.
35. DeNiederhäusern S, Bondi M, Bondioli F. Self-cleaning and antibacteric ceramic tile surface. Int J Appl Ceram Technol. 2013;10(6):949–56. https://doi.org/10.1111/j.1744-7402.2012.02801.x.
36. Poulios I, Spathis P, Grigoriadou A, Delidou K, Tsoumparis P. Protection of marbles against corrosion and microbial corrosion with TiO$_2$ coatings. J Environ Sci Health A. 1999;34(7):1455–71. https://doi.org/10.1080/10934529909376905.
37. Graziani L, Quagliarini E, D'Orazio M. Superfici autopulenti e biocide nel restauro archeologico di pietre e laterizi. Firenze University Press; 2016. 10.13128/RA-19508.
38. LaRussa MF, Ruffolo SA, Rovella N, Belfiore CM, Palermo AM, Guzzi MT, Crisci GM. Multifunctional TiO$_2$ coatings for cultural heritage. Prog Org Coat. 2012;74(1):186–91. https://doi.org/10.1016/j.porgcoat.2011.12.008.
39. Ruffolo AS, Macchia A, La Russa MF, Mazza L, Urzì C, De Leo F, Barberio M, Crisc GM. Marine antifouling for underwater archaeological sites: TiO$_2$ and Ag-doped TiO$_2$. Int J Photoenergy. 2013;2013:1–6. https://doi.org/10.1155/2013/251647.
40. Kapridaki C, Maravelaki-Kalaitzaki P. TiO$_2$-SiO$_2$-PDMS nano-composite hydrophobic coating with self-cleaning properties for marble protection. Prog Organ Coat. 2013;76(2–3):400–10. https://doi.org/10.1016/j.porgcoat.2012.10.006.
41. Fonseca AJ, Pina F, Macedo MF, Leal N, Romanowska-Deskins A, Laiz L, Gómez-Bolea A, Saiz-Jimenez C. Anatase as an alternative application for preventing biodeterioration of mortars: evaluation and comparison with other biocides. Int Biodeter Biodegr. 2010;64(5):388–96. https://doi.org/10.1016/j.ibiod.2010.04.006.

42. Graziani L, Quagliarini E, Osimani A, Aquilanti L, Clementi F, Yéprémian C, Lariceia V, Amoroso S, D'Orazio M. Evaluation of inhibitory effect of TiO_2 nanocoatings against micro-algal growth on clay brick façades under weak UV exposure conditions. Build Environ. 2013;64:38–45. https://doi.org/10.1016/j.buildenv.2013.03.003.
43. Graziani L, Quagliarini E, Osimani A, Aquilanti L, Clementi F, D'Orazio M. The influence of clay brick substratum on the inhibitory efficiency of TiO_2 nanocoating against biofouling. Build Environ. 2014;82:128–34. https://doi.org/10.1016/j.buildenv.2014.08.013.
44. Graziani L, D'Orazio M. Biofouling prevention of ancient brick surfaces by TiO_2-based nano-coatings. Coatings. 2015;5(3):357–65. https://doi.org/10.3390/coatings5030357.
45. Maury-Ramirez A, De Muynck W, Stevens R, Demeestere K, De Belie N. Titanium dioxide based strategies to prevent algal fouling on cementitious materials. Cem Concr Compos. 2013;36:93–100. https://doi.org/10.1016/j.cemconcomp.2012.08.030.
46. Radulovic J, MacMullen J, Zhang Z, Dhakal HN, Hannant S, Daniels L, Elford J, Herodotou C, Totomis M, Bennett N. Biofouling resistance and practical constraints of titanium dioxide nanoparticulate silane/siloxane exterior facade treatments. Build Environ. 2013;68:0150–8. https://dci.org/10.1016/j.buildenv.2013.07.001.
47. Martinez T, Bertron A, Escadeillas G, Ringot E. Algal growth inhibition on cement mortar: efficiency of water repellent and photocatalytic treatments under UV/VIS illumination. Int Biodeter Biodegr. 2014;89:115–25. https://doi.org/10.1016/j.ibiod.2014.01.018.
48. Linkous CA, Carter GJ, Locuson DB, Ouellette AJ, Slattery DK, Smitha LA. Photocatalytic inhibition of algae growth using TiO_2, WO_3, and cocatalyst modifications. Environ Sci Technol. 2000;34(22):4754–8. https://doi.org/10.1021/es001080.
49. Vidaković AM, Ranogajec JG, Markov SL, Lončar ES, Hiršenberger HM, Sever Škapin A. Synergistic effect of the consolidant and the photocatalytic coating on antifungal activity of porous mineral substrates. J Cult Herit. 2017;24:1–8. https://doi.org/10.1016/j.culher.2016.11.005.
50. Graziani L, Quagliarini E, D'Orazio M. The role of roughness and porosity on the self-cleaning and anti-biofouling efficiency of TiO_2-Cu and TiO_2-Ag nanocoatings applied on fired bricks. Constr Build Mater. 2016;129:116–24. https://doi.org/10.1016/j.conbuildmat.2016.10.111.
51. Simionescu B, Aflori M, Olaru M. Protective coatings based on silsesquioxane nanocomposite films for building limestones. Constr Build Mater. 2009;23(11):3426–30. https //doi.org/10.1016/j.conbuildmat.2009.06.032.
52. Munafò P, Goffredo GB, Quagliarini E. TiO_2-based nanocoatings for preserving architectural stone surfaces: an overview. Constr Build Mater. 2015;84:201–18. https://doi.org/10.1016/j.conbuildmat.2015.02.083.
53. Aflori M. Simionescu B, Bordianu I-E, Sacarescu L, Varganici C-D, Doroftei F, Nicolescu A, Olaru M. Silsesquioxane-based hybrid nanocomposites with methacrylate units containing titania and/or silver nanoparticles as antibacterial/antifungal coatings for monumental stones. Mater Sci Eng B. 2013;178(19):1339–46. https://doi.org/10.1016/j.mseb.2013.04.004.
54. Van der Werf ID, Ditaranto N, Picca RA, Sportelli MC, Sabbatini L. Development of a novel conservation treatment of stone monuments with bioactive nanocomposites. Heritage Sci. 2015;3(1). https://doi.org/10.1186/s40494-015-0060-3.
55. Ditaranto N, Loperfido S, van der Werf I, Mangone A, Cioffi N, Sabbatini L. Synthesis and analytical characterisation of copper-based nanocoatings for bioactive stone artworks treatment. Anal Bioanal Chem. 2011;399(1):473–81. https://doi.org/10.1007/s00216-010-4301-8.
56. Ditaranto N, van der Werf ID, Picca RA, Sportelli MC, Giannossa LC, Bonerba E, Tantillo G, Sabbatini L. Characterization and behaviour of ZnO-based nanocomposites designed for the control of biodeterioration of patrimonial stoneworks. New J Chem. 2015;39(9):6836–43. https://doi.org/10.1039/C5NJ00527B.
57. Selim MS, Shenashen MA, Elmarakbi A, Fatthallah NA, Hasegawa S, El-Safty SA. Synthesis of ultrahydrophobic and thermally stable inorganic–organic nanocomposites for self-cleaning foul release coatings. Chem Eng J. 2017;320:653–66. https://doi.org/10.1016/j.cej.2017.03.067.

58. Esposito Corcione C, De Simone N, Santarelli ML, Frigione M. Protective properties and durability characteristics of experimental and commercial organic coatings for the preservation of porous stone. Prog Org Coat. 2017;103:193–203. https://doi.org/10.1016/j.porgcoat.2016.10.037.
59. Esposito Corcione C, Manno R, Frigione M. Sunlight curable boehmite/siloxane-modified methacrylic nano-composites: an innovative solution for the protection of carbonate stones. Prog Org Coat. 2016;97:222–32. https://doi.org/10.1016/j.porgcoat.2016.04.037.
60. Huang H-L, Lin C-C, Hsu K. Comparison of resistance improvement to fungal growth on green and conventional building materials by nano-metal impregnation. Build Environ. 2015;93:119–27. https://doi.org/10.1016/j.buildenv.2015.06.016.
61. Quagliarini E, Bondioli F, Goffredo GB, Cordoni C, Munafò P. Self-cleaning and de-polluting stone surfaces: TiO$_2$ nanoparticles for limestone. Constr Build Mater. 2012;37:51–7. https://doi.org/10.1016/j.conbuildmat.2012.07.006.
62. Goffredo GB, Quagliarini E, Bondioli F, Munafò P. TiO$_2$ nanocoatings for architectural heritage: self-cleaning treatments on historical stone surfaces. Proc Inst Mech Engrs Pt N J Nanoeng Nanosyst. 2014;228(1):2–10. https://doi.org/10.1177/1740349913506421.
63. LaRussa MF, Macchia A, Ruffolo SA, De Leo F, Barberio M, Barone P, Crisci GM, Urzì C. Testing the antibacterial activity of doped TiO$_2$ for preventing biodeterioration of cultural heritage building materials. Int Biodeter Biodegr. 2014;96:87–96. https://doi.org/10.1016/j.ibiod.2014.10.002.
64. Fiorentino S, Grillini GC, Vandini M. The National Monument to Francesco Baracca in Lugo di Romagna (Ravenna, Italy): materials, techniques and conservation aspects. Case Studies Constr Mater. 2015;3:19–32. https://doi.org/10.1016/j.cscm.2015.05.003.
65. Tobaldi DM, Graziani L, Seabra MP, Hennetier L, Ferreira P, Quagliarini E, Labrincha JA. Functionalised exposed building materials: self-cleaning, photocatalytic and biofouling abilities. Ceram Int. 2017;43(13):10316–25. https://doi.org/10.1016/j.ceramint.2017.05.061.
66. Pal S, Contaldi V, Licciulli A, Marzo F. Self-cleaning mineral paint for application in architectural heritage. Coatings. 2016;6(4):48. https://doi.org/10.3390/coatings6040048.
67. Oberdörster G, Ferin J, Finkelstein G, Wade P, Corson N. Increased pulmonary toxicity of ultrafine particles? II. Lung lavage studies. J Aerosol Sci. 1990;21(3):384–7.
68. Oberdörster G, Ferin J, Gerlein R, Soderholm SC, Finkelstein G. Role of the alveolar macrophage in lung injury: studies with ultrafine particles. Environ Health Perspect. 1992;97:193–9.
69. Wiek A, Guston D, van der Leeuw S, Selin C, Shapira P. Nanotechnology in the city: sustainability challenges and anticipatory governance. J Urban Technol. 2013;20:45–62. https://doi.org/10.1080/10630732.2012.735415.
70. Baalousha M, Yang Y, Vance ME, Colman BP, McNeal SXJ, Blaszczak J, Steele M, Bernhardt E, Hochella MF. Outdoor urban nanomaterials: the emergence of a new, integrated, and critical field of study. Sci Total Environ. 2016;557–558:740–53. https://doi.org/10.1016/j.scitotenv.2016.03.132.
71. Chen KL, Smith BA, Ball WP, Fairbrother DH. Assessing the colloidal properties of engineered nanoparticles in water: case studies from fullerene C-60 nanoparticles and carbon nanotubes. Environ Chem. 2010;7:10–27. https://doi.org/10.1071/EN09112.
72. Reimann C, de Caritat P. Chemical elements in the environment: factsheets for the geochemist and environmental scientist. Berlin: Springer; 1998.
73. Kaegi R, Sinnet B, Zuleeg S, Hagendorfer H, Mueller E, Vonbank R, Boller M, Burkhardt M. Release of silver nanoparticles from outdoor facades. Environ Pollut. 2010;158:2900–5. https://doi.org/10.1016/j.envpol.2010.06.009.
74. Quagliarini E, Bondioli F, Goffredo GB, Licciulli A, Munafò P. Self-cleaning materials on architectural heritage: compatibility of photo-induced hydrophilicity of TiO$_2$ coatings on stone surfaces. J Cult Herit. 2013;14(1):1–7. https://doi.org/10.1016/j.culher.2012.02.006.
75. Mendoza C, Valle A, Castellote M, Bahamonde A, Faraldos M. TiO$_2$ and TiO$_2$-SiO$_2$ coated cement: comparison of mechanic and photocatalytic properties. Appl Catal B Environ. 2015;178:155–64. https://doi.org/10.1016/j.apcatb.2014.09.079.

76. Colangiuli D, Calia A, Bianco N. Novel multifunctional coatings with photocatalytic and hydrophobic properties for the preservation of the stone building heritage. Constr Build Mater. 2015;93:189–96. https://doi.org/10.1016/j.conbuildmat.2015.05.100.
77. Gherardi F, Colombo A, D'Arienzo M, Di Credico B, Goidanich S, Morazzoni F, Simonutti R, Toniolo L. Efficient self-cleaning treatments for built heritage based on highly photo-active and well–dispersible TiO$_2$ nanocrystals. Microchem J. 2016;126:54–62. https //doi. org/10.1016/j.microc.2015.11.043.
78. Lettieri M, Calia A, Licciulli A, Marquardt AE, Phaneuf RJ. Nanostructured TiO$_2$ for stone coating: assessing compatibility with basic stone's properties and photocatalytic effectiveness. Bull Eng Geol Environ. 2017;76(1):101–14. https://doi.org/10.1007/s10064–015–082C–z.
79. Bossa N, Chaurand P, Levard C, Borschneck D, Miche H, Vicente J, Geantet C, Aguerre-Chariol O, Michel FM, Rose J. Environmental exposure to TiO$_2$ nanomaterials incorporated in building material. Environ Pollut. 2017;220:1160–70. https://doi.org/10.1016/j. envpol.2016.11.019.
80. Mueller NC, Nowack B. Exposure modeling of engineered nanoparticles in the environment. Environ Sci Technol. 2008;42:4447–53. https://doi.org/10.1021/es7029637.
81. Kaegi R, Ulrich A, Sinnet B, Vonbank R, Wichser A, Zuleeg S, Simmler H, Brunner S, Vonmont H, Burkhardt M, Boller M. Synthetic TiO$_2$ nanoparticle emission from exterior facades into the aquatic environment. Environ Pollut. 2008;156:233–9. https://doi. org/10.1016/j.envpol.2008.08.004.
82. Som C, Wick P, Krug H, Nowack B. Environmental and health effects of nanomaterials in nanotextiles and façade coatings. Environ Int. 2011;37:1131–42. https://doi.org/10.1016/j. envint.2011.02.013.
83. Battin TJ, Kammer FVD, Weilhartner A, Ottofuelling S, Hofmann T. Nanostructured TiO$_2$: transport behavior and effects on aquatic microbial communities under environmental conditions. Environ Sci Technol. 2009;43:8098–104. https://doi.org/10.1021/es9017046.
84. Ferrari AM, Pini M, Neri P, Bondioli F. Nano-TiO$_2$ coatings for limestone: which sustainability for cultural heritage? Coatings. 2015;5:232–45. https://doi.org/10.3390/coatings5030232.
85. Shandilya N, Le Bihan O, Bressot C, Morgeneer M. Emission of titanium dioxide nanoparticles from building materials to the environment by wear and weather. Environ Sci Technol. 2015;49 2163–70. https://doi.org/10.1021/es504710p.
86. Al-Kattan A, Wichser A, Vonbank R, Brunner S, Ulrich A, Zuin S, Arroyo Y, Golanski L, Nowack B. Characterization of materials released into water form paint containing nano-SiO$_2$. Chemosphere. 2015;119:1314–21. https://doi.org/10.1016/j.chemosphere.2014.02.005.
87. Chang X, Zhang Y, Tang M, Wang B. Health effects of exposure to nano-TiO$_2$: a meta-analysis of experimental studies. Nanoscale Res Lett. 2013;8(1):1–10. https://doi. org/10.1186/1556-276X-8-51.
88. Jones W, Gibb A, Bust P. Managing the unknown—addressing the potential health risks on nanomaterials in the built environment. Constr Manag Econ. 2017;35(3):122–36. https://doi. org/10.1080/01446193.2016.1241413.
89. Hund-Rinke K, Simon M. Ecotoxic effect of photocatalytic active nanoparticles TiO$_2$ on algae and daphnids. Environ Sci Pollut Res. 2006;13:225–32. https://doi.org/10.1065/ espr2006.06.311.
90. Aruoja V, Dubourguier HC, Kasemets K, Kahru A. Toxicity of nanoparticles of CuO, ZnO and TiO$_2$ to microalgae *Pseudokirchneriella subcapitata*. Sci Total Environ. 2009;407:1461–8. https://doi.org/10.1016/j.scitotenv.2008.10.053.
91. Monteiro DR, Gorup LF, Takamiya AS, Ruvollo AC, de Camargo ER, Barbosa DB. The growing importance of materials that prevent microbial adhesion: antimicrobial effect of medical devices containing silver. Int J Antimicrob Agents. 2009;34:103–10. https://doi. org/10.1016/j.ijantimicag.2009.01.017.
92. Marambio-Jones C, EMV H. A review of the antibacterial effects of silver nanomaterials and potential implications for human health and the environment. J Nanopart Res. 2010;12 1531–51. https://doi.org/10.1007/s11051-010-9900-y.

93. Navarro E, Piccapietra F, Wagner B, Marconi F, Kaegi R, Odzak N, Sigg L, Behra R. Toxicity of silver nanoparticles to *Chlamydomonas reinhardtii*. Environ Sci Technol. 2008;42:8959–64. https://doi.org/10.1021/es801785m.
94. Choi O, Hu Z. Size dependent and reactive oxygen species related nanosilver toxicity to nitrifying bacteria. Environ Sci Technol. 2008;42:4583–8. https://doi.org/10.1021/es703238h.
95. Panyala NR, Peña-Méndez EM, Havel J. Silver or silver nanoparticles: a hazardous threat to the environment and human health? J Appl Biomed. 2008;6:117–29.
96. Lu L, Sun R, Chen R, Hui C, Ho C, Luk J, Lau G, Che C. Silver nanoparticles inhibit hepatitis B virus replication. Antivir Ther. 2008;13:253–62.
97. Asharani P, Wu Y, Gong Z, Valiyaveettil S. Toxicity of silver nanoparticles in zebrafish models. Nanotechnology. 2008;19:1–8. https://doi.org/10.1088/0957-4484/19/25/255102.
98. Elechiguerra J, Burt J, Morones J, Camacho-Bragado A, Gao X, Lara H, Yacaman M. Interaction of silver nanoparticles with HIV-1. J Nanobiotechnol. 2005;3:6. https://doi.org/10.1186/1477-3155-3-6.
99. Hsin Y, Chena C, Huang S, Shih T, Lai P, Chueh P. The apoptotic effect of nanosilver is mediated by a ROS- and JNK-dependent mechanism involving the mitochondrial pathway in NIH3T3 cells. Toxicol Lett. 2008;179:130–9. https://doi.org/10.1016/j.toxlet.2008.04.015.
100. Hussain S, Hess K, Gearhart J, Geiss K, Schlager J. In vitro toxicity of nanoparticles in BRL 3A rat liver cells. Toxicol In Vitro. 2005;19:975–83. https://doi.org/10.1016/j.tiv.2005.06.034.
101. Arora S, Jain J, Rajwade J, Paknikar K. Cellular responses induced by silver nanoparticles: in vitro studies. Toxicol Lett. 2008;179:93–100. https://doi.org/10.1016/j.toxlet.2008.04.009.
102. Asharani PV, Mun GLK, Hande MP, Valiyaveettil S. Cytotoxicity and genotoxicity of silver nanoparticles in human cells. ACS Nano. 2009;3:279–90. https://doi.org/10.1021/nn800596w.
103. Limbach LK, Wick P, Manser P, Grass RN, Bruinink A, Stark WJ. Exposure of engineered nanoparticles to human lung epithelial cells: influence of chemical composition and catalytic activity on oxidative stress. Environ Sci Technol. 2007;41:4158–63. https://doi.org/10.1021/es062629t.
104. Gupta A, Silver S. Molecular genetics—silver as a biocide: will resistance become a problem? Nat Biotechnol. 1998;16:888. https://doi.org/10.1038/nbt1098-888.
105. Silver S, Phung LT, Silver G. Silver as biocides in burn and wound dressings and bacterial resistance to silver compounds. J Ind Microbiol Biotechnol. 2006;33:627–34. https://doi.org/10.1007/s10295-006-0139-7.
106. Gupta A, Maynes M, Silver S. Effects of halides on plasmid-mediated silver resistance in *Escherichia coli*. Appl Environ Microbiol. 1998;64:5042–5.
107. Tiede K, Boxall ABA, Wang XM, Gore D, Tiede D, Baxter M, et al. Application of hydrodynamic chromatography-ICP-MS to investigate the fate of silver nanoparticles in activated sludge. J Anal At Spectrom. 2010;25:1149–54. https://doi.org/10.1039/B926029C.
108. Gottschalk F, Sonderer T, Scholz RW, Nowack B. Modeled environmental concentrations of engineered nanomaterials (TiO$_2$, ZnO, Ag, CNT, fullerenes) for different regions. Environ Sci Technol. 2009;43:9216–22. https://doi.org/10.1021/es9015553.
109. Choi O, Cleuenger TE, Deng BL, Surampalli RY, Ross L, Hu ZQ. Role of sulfide and ligand strength in controlling nanosilver toxicity. Water Res. 2009;43:1879–86. https://doi.org/10.1016/j.watres.2009.01.029.
110. Kim B, Park C-S, Murayama M, Hochella MF. Discovery and characterization of silver sulfide nanoparticles in final sewage sludge products. Environ Sci Technol. 2010;44:7509–14. https://doi.org/10.1021/es101565j.
111. Luther GW, Rickard DT. Metal sulfide cluster complexes and their biogeochemical importance in the environment. J Nanopart Res. 2005;7:389–407. https://doi.org/10.1007/s11051-005-4272-4.
112. Nowack B. Nanosilver revisited downstream. Science. 2010;330:1054–5. https://doi.org/10.1126/science.1198074.

113. Gimbert LJ, Hamon RE, Casey PS, Worsfold PJ. Partitioning and stability of engineered ZnO nanoparticles in soil suspensions using flow field-flow fractionation. Environ Chem. 2007;4:8–10. https://doi.org/10.1071/EN06072.
114. Franklin NM, Rogers NJ, Apte SC, Batley GE, Gadd GE, Casey PS. Comparative toxicity of nanoparticulate ZnO, bulk ZnO, and $ZnCl_2$ to a freshwater microalga (*Pseudokirchneriella subcapitata*): the importance of particle solubility. Environ Sci Technol. 2007;41:8484–90. https://doi.org/10.1021/es071445r.
115. Gottschalk F, Sun T, Nowack B. Environmental concentrations of engineered nanomaterials: review of modeling and analytical studies. Environ Pollut. 2013;181:287–300. https://doi.org/10.1016/j.envpol.2013.06.003.
116. Majedi SM, Lee HK, Kelly BC. Chemometric analytical approach for the cloud point extraction and inductively coupled plasma mass spectrometric determination of zinc oxide nanoparticles in water samples. Anal Chem. 2012;84(15):6546–52. https://doi.org/10.1021/ac300833t.
117. Scheckel KG, Luxton TP, El Badawy AM, Impellitteri CA, Tolaymat TM. Synchrotron speciation of silver and zinc oxide nanoparticles aged in a kaolin suspension. Environ Sci Technol. 2010;44:1307–12. https://doi.org/10.1021/es9032265.
118. Kim B, Levard C, Murayama M, Brown GE, Hochella MF. Integrated approaches of X-ray absorption spectroscopic and electron microscopic techniques on zinc speciation and characterization in a final sewage sludge product. J Environ Qual. 2014;43:908–16. https://doi.org/10.2134/jeq2013.10.0418.
119. Levard C, Hotze EM, Lowry GV, Brown GE. Environmental transformations of silver nanoparticles: impact on stability and toxicity. Environ Sci Technol. 2012;46:6900–14. https://doi.org/10.1021/es2037405.
120. Ma R, Levard C, Michel FM, Brown GE, Lowry GV. Sulfidation mechanism for zinc oxide nanoparticles and the effect of sulfidation on their solubility. Environ Sci Technol. 2013;47 2527–34. https://doi.org/10.1021/es3035347.
121. Heidmann I. Metal oxide nanoparticle transport in porous media—an analysis about (un) certainties in environmental research. J Phys Conf Ser. 2013;429:012042. https://doi.org/10.1088/1742–6596/429/1/012042.
122. Bondarenko O, Juganson K, Ivask A, Kasemets K, Mortimer M, Kahru A. Toxicity of Ag, CuO and ZnO nanoparticles to selected environmentally relevant test organisms and mammalian cells in vitro: a critical review. Arch Toxicol. 2013;87:1181–200. https://doi.org/10.1007/s00204-013-1079-4.
123. Villarreal FD, Das GK, Abid A, Kennedy IM, Kültz D. Sublethal effects of CuO nanoparticles on Mozambique tilapia (*Oreochromis mossambicus*) are modulated by environmental salinity. PLOS One. 2014;9(2):e88723. https://doi.org/10.1371/journal.pone.0088723.
124. Adam N, Leroux F, Knapen D, Bals S, Blust R. The uptake and elimination of ZnO and CuO nanoparticles in Daphnia magna under chronic exposure scenarios. Water Res. 2015;68:249. https://doi.org/10.1016/j.watres.2014.10.001.
125. Bao S, Lu Q, Fang T, Dai H, Zhang C. Assessment of the toxicity of CuO nanoparticles by using *Saccharomyces cerevisiae* mutants with multiple genes deleted. Appl Environ Microbiol. 2015;81:8098–107. https://doi.org/10.1128/AEM.02035-15.
126. Concha-Guerrero S, EM SB, Piñón-Castillo HA, Tarango-Rivero SH, Caretta CA, Luna-Velasco A, Duran R, Orrantia-Borunda E. Effect of CuO nanoparticles over isolated bacterial Strains from agricultural soil. J Nanomater. 2014;2014:1. https://doi.org/10.1155/2014/148743.
127. Koponen IK, Jensen KA, Schneider T. Sanding dust from nanoparticle containing paints: physical characterisation. J Phys Conf Ser. 2009;151:012048.
128. Scrinzi E, Rossi S, Kamarchik P, Deflorian F. Evaluation of durability of nano-silica containing clear coats for automotive applications. Prog Org Coat. 2011;71:384–90. https://doi.org/10.1016/j.porgcoat.2011.04.009.

129. Zuin S, Gaiani M, Ferrari A, Golanski L. Leaching of nanoparticles from experimental water-borne paints under laboratory test conditions. J Nanopart Res. 2014;16:2185. https://doi.org/10.1007/s11051-013-2185-1.
130. Nguyen T, Pellegrin B, Bernard C, Gu X, Gorham JM, Stutzman P, Shapiro A, Byrd E, Chin J. Direct evidence of nanoparticle release from epoxy nanocomposites exposed to UV radiation; nanotechnology 2010: advanced materials, CNTs, particles, films and composites—Technical proceedings of the 2010 NSTI nanotechnology conference and expo, vol. 1. NSTI-Nanotech; 2010. p. 724–7.
131. Chang MR, Lee DJ, Lai JY. Nanoparticles in wastewater from a science-based industrial park—coagulation using polyaluminum chloride. J Environ Manag. 2007;85:1009–14. https://doi.org/10.1016/j.jenvman.2006.11.013.
132. Jarvie HP, Al-Obaidi H, King SM, Bowes MJ, Lawrence MJ, Drake AF, et al. Fate of silica nanoparticles in simulated primary wastewater treatment. Environ Sci Technol. 2009;43:8622–8.
133. Saber AT, Koponen IK, Jensen KA, Jacobsen NR, Mikkelsen L, Möller P, Loft S, Vogel U, Wallin H. Inflammatory and genotoxic effects of sanding dust generated from nanoparticle-containing paints and lacquers. Nanotoxicology. 2011;6:1–13. https://doi.org/10.3109/17435390.2011.620745.
134. Irfan A, Sachse S, Njuguna J, Pielichowski K, Silva F, Zhu H. Assessment of nanoparticle release from polyamide 6- and polypropylene–silicon composites and cytotoxicity in human lung A549 cells. J Inorg Organomet Polym. 2013;23:861–70. https://doi.org/10.1007/s10904-013-9856-3.
135. Kaiser J-P, Roesslein M, Diener L, Wic P. Human health risk of ingested nanoparticles that are added as multifunctional agents to paints: an in vitro study. PLoS One. 2013;8:e83215. https://doi.org/10.1371/journal.pone.0083215.
136. Van Hoecke K, De Schamphelaere KAC, Van der Meeren P, Lucas S, Janssen CR. Ecotoxicity of silica nanoparticles to the green alga *Pseudokirchneriella subcapitata*: importance of surface area. Environ Toxicol Chem. 2008;27:1948–57. https://doi.org/10.1897/07-634.1.
137. Michel K, Scheel J, Karsten S, Stelter N, Wind T. Risk assessment of amorphous silicon dioxide nanoparticles in a glass cleaner formulation. Nanotoxicology 2013; 7:974–988. https://dx.doi.org/10.3109%2F17435390.2012.689881
138. Wohlleben W, Brill S, Meier MW, Mertler M, Cox G, Hirth S, von Vacano B, Strauss V, Treumann S, Wiench K, Ma-Hock L, Landsiedel R. On the lifecycle of nanocomposites: comparing released fragments and their in-vivo hazards from three release mechanisms and four nanocomposites. Small. 2011;7:2384–95. https://doi.org/10.1002/smll.201002054.
139. Nowack B, David RM, Fissan H, Morris H, Shatking JA, Stintz M, Zepp R, Brouwer D. Potential release scenarios for carbon nanotubes used in composites. Environ Int. 2013;59:1–11. https://doi.org/10.1016/j.envint.2013.04.003.
140. Handy RD, van der Kammer F, Lead JR, Hassellöv M, Owen R, Crane M. The ecotoxicology and chemistry of manufactured nanoparticles. Ecotoxicology. 2008;17:287–314. https://doi.org/10.1007/s10646-008-0199-8.
141. Allen BL, Kichambare PD, Gou P, Vlasova II, Kapralov AA, Konduru N, Kagan VE, Star A. Biodegradation of single-walled carbon nanotubes through enzymatic catalysis. Nano Lett. 2008;8:3899–903. https://doi.org/10.1021/nl802315h.
142. Hyung H, Fortner JD, Hughes JB, Kim JH. Natural organic matter stabilizes carbon nanotubes in the aqueous phase. Environ Sci Technol. 2007;41:179–84. https://doi.org/10.1021/es061817g.
143. Wick P, Clift MJD, Rösslein M, Rothen-Ruthishauser B. A brief summary of carbon nanotubes science and technology: a health and safety perspective. ChemSusChem. 2011;4:905–11. https://doi.org/10.1002/cssc.201100161.
144. Kagan VE, Konduru NV, Feng WH, Allen BL, Conroy J, Volkov Y, Vlasova I, Belikova NA, Yanamala N, Kapralov A, Tyurina YY, Shi J, Kisin ER, Murray AR, Franks J, Stolz D, Gou P, Klein-Seetharaman J, Fadeel B, Star A, Shedova AA. Carbon nanotubes degraded

by neutrophil myeloperoxidase induce less pulmonary inflammation. Nat Nanotechnol. 2010;5:354–9. https://doi.org/10.1038/nnano.2010.44.

145. Petersen EJ, Zhang LW, Mattison NT, O'Carroll DM, Whelton AJ, Uddin N, Nguyen T, Huang Q, Henry TB, Holbrook D, Chen KL. Potential release pathways, environmental fate, and ecological risks of carbon nanotubes. Environ Sci Technol. 2011;45:9837–56. https://doi.org/10.1021/es201579y.

146. Kostarelos K, Lacerda L, Pastorin G, Wu W, Wieckowski S, Luangsivilay J, Godefroy S, Pantarotto D, Briand J–P, Muller S, Prato M, Bianco A. Cellular uptake of functionalized carbon nanotubes is independent of functional group and cell type. Nat Nanotechnol. 2007;2(2):108–13. https://doi.org/10.1038/nnano.2006.209.

147. Poland CA, Duffin R, Kinloch I, Maynard A, Wallace WAH, Seaton A, Stone V, Brown S, MacNee W, Donaldson K. Carbon nanotubes introduced into the abdominal cavity of mice show asbestos-like pathogenicity in a pilot study. Nat Nanotechnol. 2008;3:423–8. https://doi.org/10.1038/nnano.2008.111.

148. Kagan VE, Tyurina YY, Tyurin VA, Konduru NV, Potapovich AI, Osipov AN, Kisin ER, Schwegler-Berry D, Mercer R, Castranova V, Shvedova AA. Direct and indirect effects of single walled carbon nanotubes on RAW 264.7 macrophages: role of iron. Toxicol Lett. 2006;165:88–100. https://doi.org/10.1016/j.toxlet.2006.02.001.

149. Pulskamp K, Diabate S, Krug HF. Carbon nanotubes show no sign of acute toxicity but induce intra-cellular reactive oxygen species in dependence on contaminants. Toxicol Lett. 2007;168:58–74. https://doi.org/10.1016/j.toxlet.2006.11.001.

150. Wörle-Knirsch JM, Pulskamp K, Krug HF. Oops they did it again! Carbon nanotubes hoax scientists in viability assays. Nano Lett. 2006;6:1261–8. https://doi.org/10.1021/nl060177c.

151. Nguyen T, Pellegrin B, Bernard C, Gu X, Gorham JM, Stutzman P, Stanley D, Shapiro A, Byrd E, Hettenhouser R, Chin J. Fate of nanoparticles during life cycle of polymer nanocomposites. J Phys Conf Ser. 2011;304. https://doi.org/10.1088/1742-6596/304/1/012060.

152. Voznyakovskii AP, Shumilov FA, Ibatullin AK, Shugalei IV. Environmental issues related to preparation of detonation nanodiamonds. Surface and functionalization. Russ J Gen Chem. 2012;82(13):2253–5. https://doi.org/10.1134/S1070363212130117.

153. Desai C. Stability and precipitation of diverse nanoparticles. PhD thesis, New Jersey Institute of Technology; 2013.

154. Desai C, Chena K, Mitra S. Aggregation behavior of nanodiamonds and their functionalized analogs in an aqueous environment. Environ Sci Process Impacts. 2014;16(3):518–23. https://doi.org/10.1039/C3EM00378G.

Chapter 12
Preserving Cultural Heritage Stone: Innovative Consolidant, Superhydrophobic, Self-Cleaning, and Biocidal Products

Rafael Zarzuela, Manuel Luna, Luis A.M. Carrascosa, and Maria J. Mosquera

12.1 Introduction

Stone structures, such as buildings, monuments, sculptures, and archeological remains, are vulnerable to deterioration as they are constantly exposed to outdoor environments. Stone decay is caused by the combined, and often synergistic, action of biological, physical, and chemical agents. The promoted structural damage entails high maintenance and restoration costs. Most popular decay mechanisms are associated with thermal and freeze/thaw cycles [1, 2], abrasion by sand [3], acid rain [4], salt crystallization [5], and microbial colonization (by both chemical [6, 7] and physical [8] mechanisms). Apart from structural damage, the decay agents can lead to evident aesthetical alterations, specifically, the deposition of atmospheric contaminants (soot, dirt particles, etc.) and staining from pigment-rich organisms [3, 9], and even acts of vandalism (graffiti) are a major concerns in cultural heritage preservation.

Moreover, a synergistic effect exists between the different decay mechanisms. For example, damage caused by physical–chemical agents accentuates biological decay and vice versa [10–12]. Therefore, the development of multifunctional conservation treatments that combine either consolidant, hydrophobic, self-cleaning, or antifouling effects has become promising alternatives to the application of consecutive treatments.

R. Zarzuela • M. Luna • L.A.M. Carrascosa • M.J. Mosquera (✉)
TEP-243 Nanomaterials Group, Dto. Química-Física, Facultad de Ciencias, Campus Universitario Río San Pedro, Universidad de Cádiz, Puerto Real, Cádiz, Spain
e-mail: mariajesus.mosquera@uca.es

© Springer International Publishing AG 2018   259
M. Hosseini, I. Karapanagiotis (eds.), *Advanced Materials for the Conservation of Stone*, https://doi.org/10.1007/978-3-319-72260-3_12

Nowadays, there is a variety of consolidant, hydrophobic, and biocide products, all of which are commercially available and reported in scientific literature [13]. However, most of these products have not been specifically developed for preserving such elements of cultural heritage. Consequently, they are plagued by limited performance and structural drawbacks, specifically resin-based coating products (e.g., urethanes, acrylic resins) which have been widely used as consolidant and hydrophobic treatments on various substrates [14]. However, they present some disadvantages when applied on stone, namely, poor penetration, weak interaction with the substrate, and a sharp decrease in water vapor permeability. Another popular strategy is the use of products based on nanoparticles (NPs) dispersed in a solvent [15]. Some examples include $Ca(OH)_2$ [16] and SiO_2 NPs for consolidation, hydrophobic SiO_2 NPs, TiO_2 NPs for self-cleaning and biocidal effect [17], as well as biocide metallic NPs [18]. The main disadvantage of these treatments lies in their low durability compared with other products due to poor penetration and interaction with the stone. A different strategy consists of the application of products obtained by a sol–gel route [19]. Commonly, SiO_2 oligomer/monomer-based products (e.g., tetraethoxysilane (TEOS)) are widely used for the consolidation of stone materials. Their advantages are well-known. Their low viscosity allows them to penetrate deeply into porous stone, and, after polymerization which occurs upon contact with environmental moisture, a stable gel with a silicon–oxygen backbone (with a similar composition to most stone materials) is formed. However, they suffer from disadvantages such as their poor interaction with carbonate minerals [20] and form cracks caused by the action of the high capillary pressure supported during their drying process [21]. Finally, another disadvantage is the requirement for most of the available products to be dissolved in volatile organic compounds (VOCs), which pose an environmental and human health risk.

Herein, the development of an innovative sol–gel route for preserving cultural heritage stonework is discussed. This is a surfactant-assisted sol–gel synthesis to produce, in situ, mesoporous nanomaterials on stone substrates, with long-term consolidant performance [22–24]. The drawbacks associated with the commercial products such as cracking, short-lasting effectiveness, poor penetration depth, low adherence, low chemical affinity to the stones, and high content in VOCs have been eliminated or minimized in the synthesis routes developed. Moreover, by simple chemical modifications of this synthesis route, hydrophobic, water-repellent [25, 26], self-cleaning [27, 28], and biocidal [29] properties can be incorporated into the product.

This chapter reviews the research carried out in the laboratory with the objective of obtaining innovative nanomaterials with applications in stone conservation. First, the mechanism by which the proposed strategy prevents cracking is described, and then the effectiveness of these materials as stone consolidants is examined. Finally, the different chemical modifications adopted in order to achieve multifunctional treatments and the effectiveness of these new materials are discussed.

Fig. 12.1 (**a**) Photograph of the commercial consolidant Tegovakon and (**b**) photograph of a xerogel prepared by the developed route

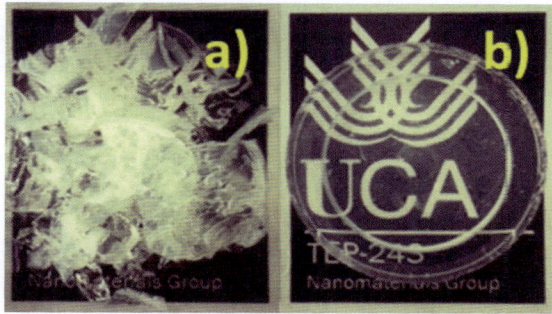

12.2   Inverse Micelle Mechanism Producing Crack-Free Xerogels

The starting point of this research was focused on the drying process of two popular commercial consolidant products, Wacker OH100 and Tegovakon, produced by Wacker Chemie and Evonik, respectively [21]. These products consist of a silica monomer/oligomer and a neutral catalyst. The results showed that both consolidants generate compact microporous xerogels with pores less than 1 nm, which are prone to cracking (Fig. 12.1a). The gel cracking is a consequence of the capillary pressure produced by the solvent evaporation in the pores of the material [21]. The capillary pressure is described by the Young–Laplace equation (Eq. (12.1)) which establishes that the capillary pressure increases as the pore size is reduced. Therefore, the microporosity of these commercial products promotes a capillary pressure higher than the xerogel structural resistance.

$$p_{\mathrm{c}} = \frac{2\gamma \cos\theta}{r_{\mathrm{p}}} \tag{12.1}$$

where p_{c} is capillary pressure, γ is surface tension, θ is meniscus curvature angle, and r_{p} is pore radius.

The synthesis strategy for preventing xerogel cracking is based on the use of n-octylamine [30], which plays two important roles: (1) it reduces the surface tension, and (2) it promotes the formation of a particulated xerogel with a mesoporous structure. This combination of surface tension reduction and increasing of pore size results in a lowering of capillary pressure that allows monolithic xerogels to be obtained (Fig. 12.1b). Additionally, n-octylamine also acts as the basic catalyst for the sol–gel reaction.

Recently, the authors have proposed an inverse micelle mechanism to explain the formation of crack-free, SiO$_2$ xerogels produced by this route [31]. More specifically, an aqueous n-octylamine dispersion with a concentration above its critical

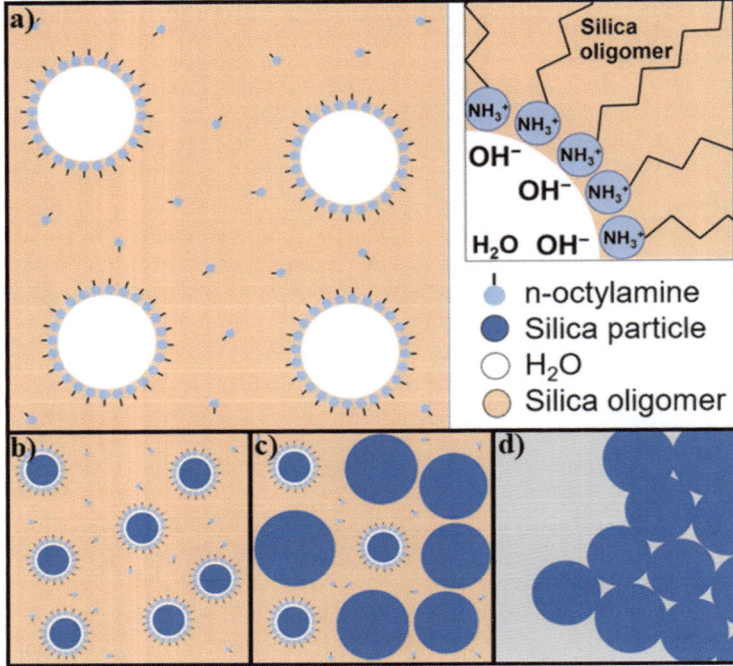

Fig. 12.2 Schematic representation of the sequences involved in the inverse micelle mechanism (**a**) formation of the inverse micelles, (**b**) SiO₂ seeds formation, (**c**) growth of the SiO₂ seeds, (**d**) packing of the SiO₂ particles

micelle concentration is mixed with a silica oligomer under ultrasonic agitation. The proposed mechanism takes place according to the following sequences detailed below.

12.2.1 Formation of a Microemulsion Containing n-Octylamine Inverse Micelles

Water is immiscible with the SiO₂ oligomer, due to the nonpolar nature of the latter, but the SiO₂ oligomer and the *n*-octylamine aqueous dispersions produce stable microemulsions under ultrasound homogenization. This homogeneous system is constituted by *n*-octylamine inverse micelles containing encapsulated water which are dispersed in the SiO₂ oligomer (Fig. 12.2a). The interaction between the water and the SiO₂ oligomer in the micelle interphase produces the hydrolysis of ethoxy groups from the SiO₂ oligomer. Next, the hydroxyl groups increase the polarity of the oligomer molecules, facilitating their diffusion within the inverse micelles. In addition, *n*-octylamine produces a basic pH within the micelles promoting the condensation of the hydrolyzed oligomer. Thus, the inverse micelles act as nanoreactors for the SiO_2 formation.

12.2.2 Formation of SiO_2 Seeds Within the Inverse Micelles

The reactions of hydrolysis and condensation continue in the micelles until stable SiO_2 seeds are formed (Fig. 12.2b). This step takes place until the initial water included in the system is totally consumed or is insufficient to maintain the micelles. Therefore, the number of seeds formed is determined by the original number of micelles in the system.

12.2.3 Growth of SiO_2 Seeds in the SiO_2 Oligomer Media

The SiO_2 seed growth continues due to the condensation of additional SiO_2 oligomers around them (Fig. 12.2c). During this stage, the formation of new SiO_2 seeds is prevented because this process is thermodynamically disfavored. The seeds have a high surface/bulk ratio, and their formation increases the system energy, whereas the growth of existing particles reduces their surface/bulk ratio. The absence of new seed means that the growth and, thereby, the size of the particles depend on the number of SiO_2 seeds formed in the previous stage.

12.2.4 Packing of the SiO_2 Nanoparticles and Formation of the Mesoporous Xerogel

The aggregation of SiO_2 particles gives rise to a xerogel composed of close-packed amorphous SiO_2 particles (Fig. 12.2d). The xerogel pore structure is due to the interstitial holes produced by the packing of the SiO_2 particles. Thus, the pore size distribution and the textural properties of the xerogel are directly related to the size of the SiO_2 NFs. Therefore, it can be concluded that the xerogel nanostructure, responsible for crack prevention, is directly related to the amount of water and n-octylamine employed in the sol–gel synthesis.

12.3 Consolidant Products

A consolidant product must fulfill a series of criteria in order to be considered acceptable in the context of cultural heritage preservation [32]. The treatment should (1) increase the cohesion of mineral grains and the mechanical properties of stone, (2) penetrate deep into the structure, (3) have a good adhesion to the stone substrate (chemical compatibility), and (4) avoid significant alterations of color and water vapor permeability of the stone.

In a preliminary study carried out by the authors, the consolidant performance of the products, prepared according to the previously described synthesis route, was evaluated [30], on a biocalcareous stone, commonly used in monumental buildings in southern Spain, and was employed in the construction of the Seville Cathedral. For comparison, the commercial consolidant Tegovakon V100 from Evonik was also evaluated. The compressive strength measurements revealed a 20–45% increase in the strength of the stones treated with the product prepared in the laboratory, whereas the commercial product had no discernible effect. Additionally, the decrease in water vapor diffusivity after the treatment was lower than 25%, and the color change was not significant (<3).

In order to accelerate the sol–gel transition, the synthesis was modified by increasing the amount of n-octylamine in the starting sol [24]. The consolidant effectiveness was evaluated on the same biocalcarenite and compared with a commercial consolidant (TV 100, Evonik) by a drilling resistance test. The commercial product had almost no consolidant effect on the stone, whereas the stone samples treated with the product prepared in the laboratory significantly increased the drilling resistance of the stone (up to 500%). Moreover, this increase was evident across the full depth of tested stone (30 mm), demonstrating that the University of Cadiz (UCA) product efficiently penetrated into the pore structure of the stone.

In a recent work, the consolidant product was applied during the restoration of a granitic Romanesque church in Galicia, Spain (Church of Santa María del Campo) [33] (Fig. 12.3). The performance and possible negative effects of the experimental product were compared with two commercial products, Paraloid B-82 (acrylic-based) and Estel 1000 (TEOS-based product). In a previous laboratory validation, the UCA product promoted a significantly higher resistance against salt crystallization than the commercial products. The UCA product was applied on different areas of the monument by brush. The consolidant effectiveness was evaluated, in situ, by a peeling test, showing an average 83% decrease of weight loss compared to the

Fig. 12.3 Application and in situ evaluation of the UCA consolidants in the Church of Santa María del Campo (Galicia, Spain)

untreated stone. The color variation was 3 (being below 5, the human perception threshold limit), and the decrease in water permeability was 31% (50% being the maximum value allowed), thus proving the suitability of the products for the conservation of the monumental building.

12.4   Consolidant/Hydrophobic and Superhydrophobic Products

Water is the main vehicle for degradation agents in both cultural heritage and modern building materials [34–36]. Thus, the development of products which prevent water penetration inside the porous structure of stones is an important issue. A hydrophobic surface should have a static water contact angle (CA) value higher than 90°. According to Young's equation, it is necessary to decrease surface energy to achieve this hydrophobic behavior [37]. In the case of a superhydrophobic surface, a static water CA above 150° and a high repellence characterized by a CA hysteresis (difference between advancing and receding CA values) lower than 10° [38, 39] should be achieved. Considering that reducing surface energy does not generate CA values higher than 130° [40], it is necessary to create a roughness which is capable of trapping air beneath water droplets (Cassie–Baxter state), increasing the CA and repellence [41, 42].

In order to protect stone building materials against water damage, several routes for synthesizing hydrophobic and superhydrophobic products have been developed by the authors. In a first approach [22], a hydrophobic/consolidant product was prepared by adding hydroxyl-terminated polydimethylsiloxane (PDMS) to the previously described synthesis route. The sol–gel reaction was catalyzed by the nonionic surfactant n-octylamine, which also plays a valuable role preventing cracking. Another role played by n-octylamine is the promotion of co-condensation between the SiO_2 oligomer and PDMS, which permits a homogeneous ORMOSIL (organically modified SiO_2) to be obtained. Regarding the effectiveness on stone, the synthesized sol was applied onto the previously described biocalcareous stone, giving rise to a CA value of around 130° due to the role played by PDMS in reducing surface energy. Regarding the evaluation of negative effects, the reduction of water vapor permeability was around 35%, whereas a color change of around 2 was measured.

The previous synthesis was modified by adding a higher quantity of PDMS in order to increase stain resistance [23]. The obtained sol was applied onto a pure limestone from the Estepa quarry. An increase of between 100 and 300% in drilling resistance was achieved for the experimental product compared with the untreated stone, whereas the commercial product (Silres BS OH100, Wacker Chemie) barely showed any consolidant effect. Moreover, PDMS played a valuable role in reducing surface energy and, therefore, gave rise to hydrophobic properties. Again, an improvement of hydrophobicity was observed when PDMS concentration was increased, with static CA values of around 140° for the highest proportion (56%). This effect is due to the combination of (1) PDMS reducing surface energy and (2) an increase in the stone roughness enhancing hydrophobicity, with a low water repellence (demonstrating a Wenzel state creation) [43]. These results were compared with a hydrophobic commercial product (Silres BS290, Wacker Chemie), which produced CA values of around 90°, again demonstrating the greater effectiveness of the new products. In addition, a suitable stain resistance was achieved, no negative effect was observed, and the reduction of vapor permeability was lower than 50%. Regarding color change, the treatment gave values lower than 5.

In addition, the consolidant product previously described in the consolidant products section [24] was also modified by adding PMDS (from 2 to 10% v/v) in the sol. The obtained sols were applied onto the biocalcareous stone previously described. In this case, the drilling resistance test showed that the addition of PDMS reduced the consolidant ability of these products. Nevertheless, they still demonstrated an increase in drilling resistance of around 100%, being higher than the commercial product (TV 100, Evonik) which practically did not produce consolidation, as previously explained. Regarding the hydrophobic properties, the addition of PDMS to the experimental products developed by the authors increased CA values to around 110°.

The previous synthesis was modified in order to achieve superhydrophobicity [25] by adding SiO_2 NPs in a similar approach than that previously developed by Manoudis et al. [44]. Fumed SiO_2 NPs of around 40 nm were added to the previously developed sol, and it was applied onto sandstone. Atomic force microscopy (AFM) and scanning electron microscopy (SEM) measurements demonstrated the formation of a coating composed of closely packed particles (Fig. 12.4a, b). The evaluation of wetting properties showed static CA values higher than 150° and CA hysteresis values below 7° (Fig. 12.4c). This effect is due to the combination of reducing surface energy (i.e., the presence of PDMS) and the roughness induced by SiO_2 NPs which creates air pockets between the water and the solid surface, promoting a Cassie–Baxter state [45].

The previous product was also applied on clay roof tiles and its effectiveness, and durability was evaluated [26]. The product conferred superhydrophobic properties (static CA of around 150° and hysteresis of around 7°) to the roof tiles, due to the combination of low surface energy and roughness, as previously explained. After being subjected to accelerated aging in a climatic chamber, the treated samples conserved high static CA values. However, a decrease in repellent properties was observed due to the loss of the Cassie–Baxter topography. In addition, all the treated samples showed color changes lower than 5.

Recently, an amphiphobic (superhydrophobic and oleophobic) coating was produced on sandstone, by means of a double-step process [46]. First, a sol containing a SiO_2 oligomer, 40-nm SiO_2 particles, and an aqueous solution of n-octylamine was applied on the selected stone. After the sol–gel transition took place, a SiO_2 nanocomposite of densely packed particles was produced, giving rise to a Cassie–Baxter state. Then, a hydrolyzed fluorinated alkoxysilane was applied onto the previously treated surface in order to reduce the surface energy. The treated surfaces

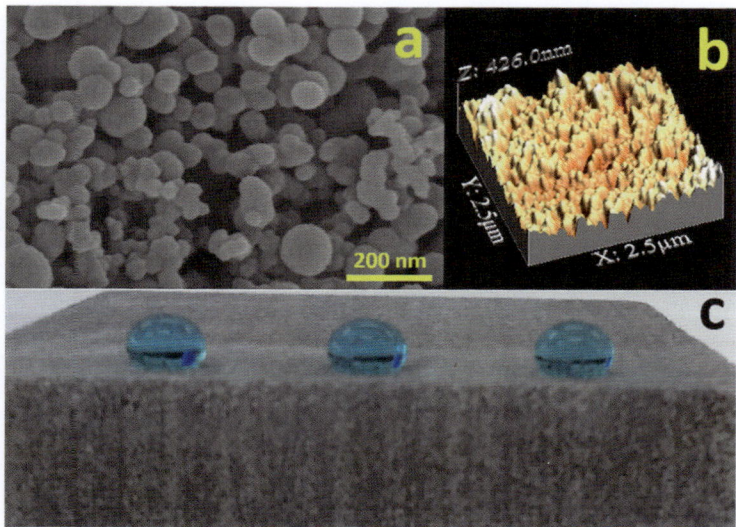

Fig. 12.4 (a) SEM, (b) 3D AFM images of superhydrophobic product on the sandstone under study, (c) water droplets on the superhydrophobic stone. "Reprinted with permission from Construction and Building Materials, Vol 76, I. De Rosario, F. Elhaddad, A. Pan, R. Benavides, T. Rivas, M.J. Mosquera "Effectiveness of a novel consolidant on granite: Laboratory and in situ results", pages 140–149, Copyright (2014), Elsevier"

showed superhydrophobic properties (static CA of around 160° and hysteresis of around 6°) with self-cleaning performance. Moreover, the application of a fluorinated compound allowed oleophobic surfaces (CA values with oil droplets of over 90°) to be obtained. Regarding the durability of this treatment, a peeling test was performed to test the adhesion of the coating to the stone. The surfaces demonstrated long-lasting properties due to the effective grafting between the SiO_2 nanocomposites with the stone surface.

12.5 Photocatalytic Consolidants with Self-Cleaning Properties

Since the photocatalytic effect of TiO_2 was discovered by Honda and Fujishima [47], it has been widely employed to produce building materials with self-cleaning and decontaminating properties [48]. A common strategy to achieve these photocatalytic materials is the application of TiO_2 NP dispersion on building substrates [49–54], but these types of coatings have an associated important drawback related to their durability. Commonly, the coatings constituted exclusively by NPs have low adherence and form cracked films, limiting their durability [55–58].

To resolve this problem, TiO$_2$ NPs were included in the sol–gel synthesis to produce TiO$_2$–SiO$_2$ xerogels that combine the TiO$_2$ photocatalytic properties with the great affinity and adhesion between the building substrate and the synthesized xerogels. The initial work conducted by the authors in this field consisted of the incorporation of commercial TiO$_2$ NPs in the sol–gel route (containing *n*-octylamine) to synthesize products containing different TiO$_2$ proportions [27]. These products were applied on a limestone for their evaluation and were compared with the treatment using an aqueous dispersion of TiO$_2$ NPs and other TiO$_2$/SiO$_2$ products prepared without *n*-octylamine. Both TiO$_2$ NPs and the product without *n*-octylamine produced cracked films on the stone, and these coatings were almost totally removed during an adhesion test, whereas the product containing *n*-octylamine produced a homogenous, crack-free coating with resistance to the adhesion test. The self-cleaning activity of the samples was evaluated by a methylene blue (MB) degradation test, with all samples containing TiO$_2$ showing a considerable MB decoloration (Fig. 12.5). Comparing the different treatments, it was found that for 800 h of UV illumination, the treatments with only TiO$_2$ NPs and the treatments containing *n*-octylamine showed similar MB degradations, whereas the treatments without *n*-octylamine showed the slowest degradations. Thus, *n*-octylamine also improves the self-cleaning activity due to the structure that it produces having a large pore volume that allows the diffusion of the reactant species into the photoactive centers. Later, this TiO$_2$/SiO$_2$ structure produced in the presence of *n*-octylamine was widely studied by transmission electron microscopy (TEM) [28]. This complete characterization study demonstrated that SiO$_2$ and TiO$_2$ are maintained as independent domains. The nanomaterial is composed of SiO$_2$ NPs of nearly uniform size, created by an inverse micelle mechanism due to the effect of *n*-octylamine. The preformed TiO$_2$ NPs are integrated into the SiO$_2$ aggregates. It was also demonstrated that some of the TiO$_2$ remains outside the SiO$_2$ matrix as individual NPs, which permits a direct photodegradation. These features favor the fast removal of MB due to the synergistic effect of its adsorption into the mesoporous SiO$_2$ structure and its photodegradation by TiO$_2$ NPs.

The TiO$_2$/SiO$_2$ product with the best performance (containing 2% w/v of commercial TiO$_2$NPs) was also compared with two commercial products, a consolidant (TV 100, Evonik) and a photocatalytic product (E503, Nanocer) [59]. The results of this study showed that the experimental product developed by the authors and the commercial photocatalytic had similar self-cleaning activities, with the experimental product demonstrating higher adhesion. The experimental product also produced a greater increase in the stone's resistance compared with the commercial consolidant, confirming its use as a multifunctional treatment for stone preservation, providing consolidant and self-cleaning properties.

The following study evaluated the influence of the type of TiO$_2$ NPs and their loading on the properties of the resultant TiO$_2$/SiO$_2$ materials [60]. Regarding the type of TiO$_2$ NPs, it was found that the best self-cleaning activity corresponded with the particles that had the larger size and the sharper shape. These particles with an initial average size of 20 μm were significantly reduced (at nanometric scale) after ultrasonic stirring during the synthesis. Their higher activity was attributed to the

Fig. 12.5 Evolution of MB stains during the self-cleaning test. Photographs for (**a**) untreated stone, (**b**) treated with TiO$_2$/SiO$_2$ product, and (**c**) evolution of MB degradation

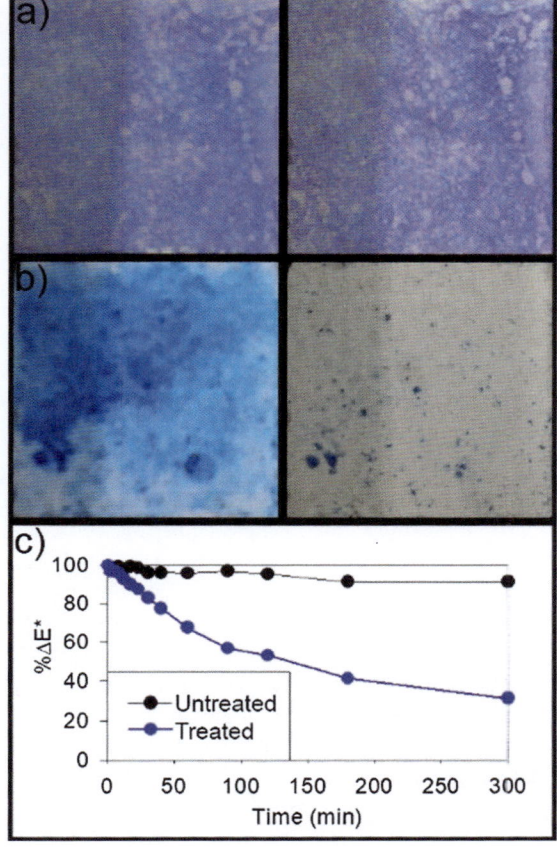

surface of the TiO$_2$ NPs being more available in the structure of the materials constituted by this type of NP. The sequence of TiO$_2$ loading showed an unexpected behavior, the TiO$_2$ increasing from 1 to 4% w/v improved the self-cleaning effectiveness considerably, but the treatments with 10% w/v exhibited the lowest photoactivity. It was correlated that this activity fell with a drastic pore volume reduction for the xerogels containing 10% w/v TiO$_2$. This behavior reconfirmed that the matrix where the particles are embedded needs a pore structure with good connectivity to achieve a correct photoactivity. Additionally, the use of 10% w/v TiO$_2$ presented other drawbacks such as low adherence and inhibition of the catalytic effect of *n*-octylamine during sol–gel transition, thus increasing the gel times.

The last goal of the study was to improve the efficiency of TiO$_2$ photoactivity for real applications or rather in the presence of visible light as the amount of UV light from sunlight is limited. Silver doping was chosen as a strategy for improving the TiO$_2$ photoactivity due to two facts: (1) the noble metal NPs reduce the electron–hole recombination rate, and (2) their localized surface plasmon resonance extends the photocatalyst activation to visible light [61]. A silver precursor solution was added to the TiO$_2$/SiO$_2$ sol–gel synthesis, which was reduced in situ, giving rise to silver NPs dispersed in the xerogel structure [62]. The obtained materials had an important visible light absorption, and the treated stones showed greater self-cleaning activities compared with their counterparts without silver.

Currently, the use of gold NPs is being studied to improve the effectiveness of this product, and evaluation of the self-cleaning performance against real staining agents, such as soot, and the decontaminant activity is also being investigated by NO photodegradation.

12.6   Consolidant/Biocide Products

The biological growth on stones can be inhibited by the incorporation into the SiO$_2$ matrix of bioactive components such as organic biocides [63], photocatalytic materials based on TiO$_2$ [54, 64] and ZnO [65, 66], and products incorporating biocidal metals (Ag, Cu, etc.) in the form of their salts [67, 68] or as NPs [18, 69]. Multifunctional treatments (i.e., consolidant, hydrophobic, and biocide properties) have been developed by adding CuO or Ag NPs to the previously described synthesis route. Copper oxide and silver NPs show some promising advantages as an inexpensive alternative to other biocidal agents. Compared to organic biocides (e.g., benzalkonium chloride, quaternary ammonium salts, etc.), they are more stable against sunlight and oxidation. In addition, they are compatible with hydrophobic treatments, unlike the commonly used organic quaternary ammonium salts [70]. In contrast to TiO$_2$ and ZnO particles, they are equally effective in darkness and light conditions. Another advantage, related to their stability, is the ease of immobilizing them in a matrix which allows a controlled release of the biocide species [71].

In a recent study [29], it was reported the synthesis and characterization of multifunctional (consolidant and biocidal) CuO/SiO$_2$ nanocomposites for the conservation of stone building materials. The nanocomposites integrate commercial CuO NPs (0.00–0.35% w/v) in a SiO$_2$-based matrix prepared according to previous studies (TEM image in Fig. 12.6a) and were applied on a limestone. The biocidal activity of the nanocomposites was measured against two reference laboratory organisms: Gram-negative bacteria (*E. coli*) and a yeast (*S. cerevisiae*). The cell growth in the stones was inhibited by 87% for *E. coli* and 80% for *S. cerevisiae*, with respect to the control samples, with the products containing 0.15% w/v CuO (Fig. 12.6b). However, the stones treated with products containing a higher quantity of CuO (0.35% w/v) showed lower inhibition (37% and 20%, for *E. coli* and *S. cerevisiae*, respectively) due to the agglomeration and precipitation of CuO. The consolidation effectiveness was measured by the Vickers hardness test and drilling resistance. The surface of the stones treated with CuO/SiO$_2$ nanocomposites presented up to a 25%

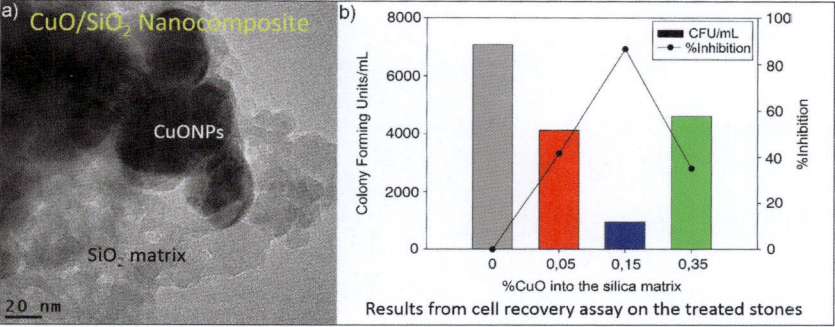

Fig. 12.6 (**a**) TEM image of the CuO/SiO$_2$ consolidant/biocide product (**b**): results from the biocide test against *E. coli*. "Reprinted with permission from Materials & Design, Vol 114, R. Zarzuela, M. Carbú, M.L.A. Gil, J.M. Cantoral, M.J. Mosquera, "CuO/SiO$_2$ nanocomposites: A multifunctional coating for application on building stone", pages 364–372, Copyright (2016), Elsevier"

increase in the Vickers hardness test. The drilling resistance values increased up to 45% for the treated stones. The hydrophobic properties were evaluated by measuring the static CA and water absorption by capillarity. For the treated stone samples, all static CA values were above 90°, and the total water uptake decreased over 95%. The hydrophobic nature was due to the presence of ethoxy groups, as confirmed by Fourier transform infrared spectroscopy (FTIR) measurements.

In a recent work conducted by the authors, the use of surface-modified SiO$_2$/Ag NPs for the production of treatments with combined superhydrophobic and biocide effects has been studied. The modification of the surface of the SiO$_2$NPs with an aminoalkyl-alkoxysilane increases the stability of the Ag NPs during the synthesis and promotes a more homogeneous distribution, leading to improved biocide effectiveness.

12.7 Conclusion

TEOS-based products are commonly employed for preserving cultural heritage stone structures. A well-known drawback of these conservation products is their tendency to crack during their drying into the stone structure. It is obvious that a cracking material cannot suitably preserve the treated substrate. A simple and low-cost route to obtain crack-free materials has been developed. Specifically, a silica precursor and a surfactant are mixed under ultrasonic agitation. The micelles created by the surfactant act as nanoreactors, producing a structure composed of nearly uniform silica nanoparticles. This mesoporous structure prevents material cracking because it reduces capillary pressure during gel drying. Due to their crack-free structure, these mesoporous materials show an improved consolidant performance and better adherence to the stone substrates. Additionally, this surfactant-assisted sol–gel strategy provides the advantage of a high versatility. By simple chemical modifications (addition of PDMS, SiO_2 NPs, fluorinated compounds, TiO_2 NPs, CuO NPs, etc.) of the process, it is possible to obtain multifunctional products with a combination of consolidant and hydrophobic/superhydrophobic, oleophobic, photocatalytic, and biocide performance.

Acknowledgments We wish to express our gratitude for the financial support from the Spanish Government MINECO/FEDER-EU (MAT2013-42934-R) and the Regional Government of Andalusia (Group TEP-243). R. Zarzuela and M. Luna would also like to thank the Spanish Government for their predoctoral grants (FPU14/02054 and BES-2014-068031).

References

1. Chang D, Liu J. Review of the influence of freeze-thaw cycles on the physical and mechanical properties of soil. Sci Cold Arid Reg. 2013;5(4):457–60.
2. Hall K, Thorn CE. Thermal fatigue and thermal shock in bedrock: an attempt to unravel the geomorphic processes and products. Geomorphology. 2014;206:1–13.
3. Shi XJ, Shi XF. Numerical prediction on erosion damage caused by wind-blown sand movement. Eur J Environ Civ Eng. 2014;18:550–66.
4. El-Gohary MA. A holistic approach to the assessment of the groundwater destructive effects on stone decay in Edfu temple using AAS, SEM-EDX and XRD. Environ Earth Sci. 2016;75:13.
5. Dragovich D, Egan M. Salt weathering and experimental desalination treatment of building sandstone, Sydney (Australia). Environ Earth Sci. 2011;62:277–88.
6. Friedrich EWE. Solubilization, transport and deposition of mineral cations by microorganisms-efficient rock weathering agents. In: Drever JI, editor. The Chemistry of Weathering. 1st ed. Dordrecht: Springer Netherlands; 1985;161–173.
7. Tiano P. Biodegradation of cultural heritage: decay mechanisms and control methods. CNR-Centro di Stud Sulle Cause Deperimento e Metod Conserv Opere d'Arte. 2001;9:1–37.
8. Diakumaku E, Gorbushina AA, Krumbein WE, et al. Black fungi in marble and limestones—an aesthetical, chemical and physical problem for the conservation of monuments. Sci Total Environ. 1995;167:295–304.
9. Saiz-Jimenez C. Microbial melanins in stone monuments. Sci Total Environ. 1995;167:273–86.
10. Randazzo L, Montana G, Alduina R, et al. Flos Tectorii degradation of mortars: an example of synergistic action between soluble salts and biodeteriogens. J Cult Herit. 2015;16:838–47.
11. Zanardini E, Abbruscato P, Ghedini N, et al. Influence of atmospheric pollutants on the biodeterioration of stone. Int Biodeter Biodegr. 2000;46:8305.
12. Miller AZ, Sanmartín P, Pereira-Pardo L, et al. Bioreceptivity of building stones: a review. Sci Total Environ. 2012;426:1–12.
13. Doehne E, Price CA. Stone conservation: An Overwiew of Current Research. 2nd ed. Los Angeles: Getty Publications; 2010.
14. Sang YK, Man CS, Un YK, Hyung JK. Conservation study of stones by using acrylic monomer. Polymer. 2008;32:213–8.
15. Barberio M, Veltri S, Imbrogno A, Stranges F, Bonano A, Antici P. TiO_2 and SiO_2 nanoparticles film for cultural heritage: conservation and consolidation of ceramic artifacts. Surf Coat Technol. 2015;271:174–80.
16. Costa D, Rodrigues JD. Consolidation of a porous limestone with nanolime. 12th International Congress on the Deterioration and Conservation of Stone.

17. Ivask A, George S, Bondarenko O, Kahru A. Nano-antimicrobials. 1st ed. Berlin: Springer; 2012.
18. Carrillo-González R, Martínez-Gómez MA, González-Chávez MDCA, Mendoza Hernández JC. Inhibition of microorganisms involved in deterioration of an archaeological site by silver nanoparticles produced by a green synthesis method. Sci Total Environ. 2016;565:872–81.
19. Wheeler G. Alkoxysilanes and the consolidation of stone. 1st ed. Los Angeles: Getty Publications; 2005.
20. Pinto APF. Rodrigues JD. Stone consolidation: the role of treatment procedures. J Cult Herit. 2008;9:38–53.
21. Scherer GW. Recent progress in drying of gels. J Non Cryst Solids. 1992;147–148:363–74.
22. Mosquera MJ, de los Santos DM, Rivas T. Surfactant-synthesized ormosils with application to stone restoration. Langmuir. 2010;26:6737–45.
23. Illescas JF. Mosquera MJ. Surfactant-synthesized PDMS/silica nanomaterials improve robustness and stain resistance of carbonate stone. J Phys Chem C. 2011;115:14624–34.
24. Illescas JF. Mosquera MJ. Producing surfactant synthesized nanomaterials in situ on a building susbstrate, without volatile organic compounds. Appl Mater Interfaces. 2012;4:4259–69
25. Facio DS, Mosquera MJ. Simple strategy for producing superhydrophobic nanocomposite coatings in situ on a building substrate. ACS Appl Mater Interfaces. 2013;5:7517–26.
26. Carrascosa LAM, Facio DS, Mosquera MJ. Producing superhydrophobic roof tiles. Nanotechnology. 2016;27:95604.
27. Pinho L, Mosquera MJ. Titania-silica nanocomposite photocatalysts with application in stone self-cleaning. J Phys Chem C. 2011;115:22851–62.
28. Pinho L, Hernández-Garrido JC, Calvino JJ, Mosquera MJ. 2D and 3D characterization of a surfactant-synthesized TiO_2–SiO_2 mesoporous photocatalyst obtained at ambient temperature. Phys Chem Chem Phys. 2013;15:2800–8.
29. Zarzuela R, Carbú M, Gil MLA, et al. CuO/SiO_2 nanocomposites: a multifunctional coating for application on building stone. Mater Des. 2017;114:364–72.
30. Mosquera MJ, de los Santos DM, Montes A, Valdez-Castro L. New nanomaterials for consolidating stone. Langmuir. 2008;24:2772–8.
31. Facio DS, Luna M, Mosquera MJ. Facile preparation of mesoporous silica monoliths by an inverse micelle mechanism. Microporous Mesoporous Mater. 2017;247:166–76.
32. Rodrigues JD, Grossi A. Indicators and ratings for the compatibility assessment of conservation actions. J Cult Herit. 2007;8:32–43.
33. De Rosario I, Elhaddad F, Pan A, et al. Effectiveness of a novel consolidant on granite: laboratory and in situ results. Construct Build Mater. 2015;76:140–9.
34. Figueiredo MO, Silva TP, Veiga JP. Analysis of degradation phenomena in ancient, traditional and improved building materials of historical monuments. Appl Phys A Mater Sci Process. 2008;92:151–4.
35. Charola AE. Acid rain effects on stone monuments. J Chem Educ. 1987;64:436.
36. Bravo AH, Soto AR, Sosa ER, et al. Effect of acid rain on building material of the El Tajín archaeological zone in Veracruz, Mexico. Environ Pollut. 2006;144:655–60.
37. Young T. An essay on the cohesion of fluids. Philos Trans R Soc Lond A. 1805;95:65–87.
38. Gao L, McCarthy TJ. A perfectly hydrophobic surface ($\theta(A)/\theta(R) = 180°/180°$). J Am Chem Soc. 2006.128:9052–3.
39. Gao L, McCarthy TJ. Wetting 101°. Langmuir. 2009;25:14105–15.
40. Lafuma A. Quéré D. Superhydrophobic states. Nat Mater. 2003;2:457–60.
41. Cassie ABD, Baxter S. Wettability of porous surfaces. Trans Faraday Soc. 1944;40:546–51.
42. Sun T, Feng L, Gao X, Jiang L. Bioinspired surfaces with special wettability. Acc Chem Res. 2005;38:644–52.
43. Wenzel RN. Resistance of solid surfaces to wetting by water. Ind Eng Chem. 1936;28:988–94.
44. Manoudis PN, Papadopoulou S, Karapanagiotis I, et al. Polymer-silica nanoparticles composite films as protective coatings for stone-based monuments. J Phys Conf Ser. 2007;61:1361–5.
45. Bhushan B, Her EK. Fabrication of superhydrophobic surfaces with high and low adhesion inspired from rose petal. Langmuir. 2010;26:8207–17.

46. Facio DS, Carrascosa LAM, Mosquera MJ. Producing lasting amphiphobic building surfaces with self-cleaning properties. Nanotechnology. 2017;28:265601.
47. Fujishima A, Honda K. Electrochemical photolisis of water at a semiconductor electrode. Nature. 1972;238:37–8.
48. Chen J, Poon CS. Photocatalytic construction and building materials: from fundamentals to applications. Build Environ. 2009;44:1899–906.
49. Gherardi F, Colombo A, D'Arienzo M, et al. Efficient self-cleaning treatments for built heritage based on highly photo-active and well-dispersible TiO_2 nanocrystals. Microchem J. 2016;126:54–62.
50. Liu Q, Liu Q, Zhu Z, et al. Application of TiO_2 photocatalyst to the stone conservation. Mater Res Innov. 2015;19:S8-51–4.
51. Bergamonti L, Alfieri I, Lorenzi A, et al. Nanocrystalline TiO_2 coatings by sol–gel: photocatalytic activity on Pietra di Noto biocalcarenite. J Sol-Gel Sci Technol. 2015;75:141–51.
52. Licciulli A, Calia A, Lettieri M, et al. Photocatalytic TiO_2 coatings on limestone. J Sol-Gel Sci Technol. 2011;60:437–44.
53. Quagliarini E, Bondioli F, Goffredo GB, et al. Smart surfaces for architectural heritage: preliminary results about the application of TiO_2-based coatings on travertine. J Cult Herit. 2012;13:204–9.
54. Quagliarini E, Bondioli F, Goffredo GB, et al. Self-cleaning and de-polluting stone surfaces: TiO_2 nanoparticles for limestone. Construct Build Mater. 2012;37:51–7.
55. Mendoza C, Valle A, Castellote M, et al. TiO_2 and TiO_2–SiO_2 coated cement: comparison of mechanic and photocatalytic properties. Appl Catal Environ. 2015;178:155–64.
56. Rao KVS, Subrahmanyam M, Boule P. Immobilized TiO_2 photocatalyst during long-term use: decrease of its activity. Appl Catal Environ. 2004;49:239–49.
57. Calia A, Lettieri M, Masieri M. Durability assessment of nanostructured TiO_2 coatings applied on limestones to enhance building surface with self-cleaning ability. Build Environ. 2016;110:1–10.
58. Borsoi G, Veiga R, Silva AS. Effect of nanostructured lime-based and silica-based products on the consolidation of historical renders. In: University of West Scotland, editors. Proceedings of the 3rd historic mortars conference. Glasgow; 2013.
59. Pinho L, Elhaddad F, Facio DS, Mosquera MJ. A novel TiO_2–SiO_2 nanocomposite converts a very friable stone into a self-cleaning building material. Appl Surf Sci. 2013;275:389–96.
60. Pinho L, Mosquera MJ. Photocatalytic activity of TiO_2-SiO_2 nanocomposites applied to buildings: influence of particle size and loading. Appl Catal Environ. 2013;134–135:205–21.
61. Zhou X, Liu G, Yu J, Fan W. Surface plasmon resonance-mediated photocatalysis by noble metal-based composites under visible light. J Mater Chem. 2012;22:21337–54.
62. Pinho L, Rojas M, Mosquera MJ. Ag-SiO_2-TiO_2 nanocomposite coatings with enhanced photoactivity for self-cleaning application on building materials. Appl Catal Environ. 2014;178:144–54.
63. Eyssautier-Chuine S, Gommeaux M, Moreau C, et al. Assessment of new protective treatments for porous limestone combining water-repellency and anti-colonization properties. Q J Eng Geol Hydrogeol. 2014;47:177–87.
64. Perkas N, Lipovsky A, Amirian G, et al. Biocidal properties of TiO_2 powder modified with Ag nanoparticles. J Mater Chem B. 2013;1:5309.
65. Ruffolo SA, La Russa MF, Malagodi M, et al. ZnO and $ZnTiO_3$ nanopowders for antimicrobial stone coating. Appl Phys A Mater Sci Process. 2010;100:829–34.
66. Ditaranto N, van der Werf ID, Picca RA, et al. Characterization and behaviour of ZnO-based nanocomposites designed for the control of biodeterioration of patrimonial stoneworks. New J Chem. 2015;39:6836–43.
67. Arreche R, Bellotti N, Blanco M, Vázquez P. Improved antimicrobial activity of silica–Cu using a heteropolyacid and different precursors by sol–gel: synthesis and characterization. J Sol-Gel Sci Technol. 2015;75:374–82.

68. Eyssautier-Chuine S, Vaillant-Gaveau N, Gommeaux M, et al. Efficacy of different chemical mixtures against green algal growth on limestone: a case study with Chlorella vulgaris. Int Biodeter Biodegr. 2015;103:59–68.
69. MacMullen J, Zhang Z, Dhakal HN, et al. Silver nanoparticulate enhanced aqueous silane/ siloxane exterior facade emulsions and their efficacy against algae and cyanobacteria biofouling. Int Biodeter Biodegr. 2014;93:54–62.
70. Moreau C, Vergès-Belmin V, Leroux L, et al. Water-repellent and biocide treatments: assessment of the potential combinations. J Cult Herit. 2008;9:394–400.
71. Young M, Santra S. Copper (cu)–silica nanocomposite containing valence-engineered Cu: a new strategy for improving the antimicrobial efficacy of Cu biocides. J Agric Food Chem. 2014;62:6043–52.

Chapter 13
Antimicrobial Properties of Nanomaterials Used to Control Microbial Colonization of Stone Substrata

B.O. Ortega-Morales, M.M. Reyes-Estebanez, C.C. Gaylarde, J.C. Camacho-Chab, P. Sanmartín, M.J. Chan-Bacab, C.A. Granados-Echegoyen, and J.E. Pereañez-Sacarias

13.1   Introduction

Subaerial biofilms are microbial communities that colonize solid surfaces exposed to the atmosphere [1]. Subaerial biofilms associated with rock substratum colonize surfaces as epilithic growths and as endolithic communities grow within fissures, cracks, and pores in the rock matrices. The extent to which surfaces are colonized by microbes depends on climatic parameters (e.g., temperature, light, and humidity), the composition of the mineral substrate, and its intrinsic properties.

The concept of bioreceptivity arose from a joint consideration of these factors. Guillitte [2] first introduced this concept, describing it as "the ability of a material to be colonized by living organisms," which does not necessarily imply biodegradation of the material. The term bioreceptivity encompasses the properties of a material that contributes to its colonization by the development of microorganisms.

B.O. Ortega-Morales (✉) • M.M. Reyes-Estebanez • J.C. Camacho-Chab
M.J. Chan-Bacab • J.E. Pereañez-Sacarias
Departamento de Microbiología Ambiental y Biotecnología (DEMAB),
Universidad Autónoma de Campeche, San Francisco de Campeche, Mexico
e-mail: beortega@uacam.mx

C.C. Gaylarde
Department of Microbiology and Plant Biology, University of Oklahoma, Norman, OK, USA

P. Sanmartín
Departamento de Edafoloxía e Química Agrícola, Facultade de Farmacia,
Universidade de Santiago de Compostela, Santiago de Compostela, Spain

C.A. Granados-Echegoyen
Centro de Desarrollo Sustentable y Aprovechamiento de la Vida Silvestre (CEDESU),
Universidad Autónoma de Campeche, San Francisco de Campeche, Mexico

© Springer International Publishing AG 2018   277
M. Hosseini, I. Karapanagiotis (eds.), *Advanced Materials for the Conservation of Stone*, https://doi.org/10.1007/978-3-319-72260-3_13

Various studies have been carried out with the aim of establishing the properties that generally affect bioreceptivity, as not all materials are susceptible to colonization by the same agents. Although not all authors agree about which factor has the greatest influence [3], it is generally considered that the most important properties affecting bioreceptivity are the chemical and mineral composition of the rock, the porosity, and the roughness of the rock surface. Several types of bioreceptivity can be defined according to the state of the material under study [2, 3]. Primary bioreceptivity indicates the initial potential of the material to be colonized. It refers to the intrinsic potential of the unaltered recently extracted or newly produced material. Secondary bioreceptivity is observed over time when a material has already been altered by the action of external factors (e.g., salts, thermal alteration, biological colonization, etc.). Tertiary bioreceptivity is defined as the type of bioreceptivity created when the material is modified by human activity such as conservation practices.

13.2   Control of Microbial Biodeterioration and Nanoscience Applied to Built Cultural Heritage

It is now accepted that microorganisms may contribute to the aesthetic and physical deterioration of stone monuments and statues. Nevertheless, recent research shows that the association between biofilms and stone weathering is not necessarily detrimental. Some studies point toward a negligible effect of biofilms or even a protective role on the underlying material [4]. Once an appropriate assessment has indicated that microbial biodeterioration is taking place, it is a common practice to implement a set of strategies that include mechanical removal, biophysical eradication, and chemical control of biofilms, either alone or in combination. The application of biocides to control microbial colonization on the stone is by far the most widespread approach and dominates in the literature [5].

Nanoparticles (NPs) have recently received attention from the conservation and restoration communities as alternatives to biocides. Nanomaterials have both photocatalytic and catalytic properties, which were first used for the consolidation of degraded stone or to enhance the properties of painted walls. According to Baglioni and co-workers, the first time that nanoscience was applied in the field of cultural heritage conservation dates back to the end of 1990 when calcium hydroxide nanoparticles were used to replace old polymer resins covering mural painting [6]. Both photocatalytic and catalytic properties help to remove dirt and polluting agents from heritage surfaces and are thus attractive self-cleaning and de-polluting agents; these properties may also include biocidal activity, as in the case of titanium dioxide (TiO_2) [7].

This contribution briefly reviews the antimicrobial properties of NPs, discusses the limitations and advantages of NP-based treatments, and highlights the importance of choosing appropriate model organisms for testing and community-based approaches.

13.3   Antimicrobial Properties of NPs Used on Stone Substrate

13.3.1   Fundamentals of Antimicrobial Activity of NPs

NPs provide an alternative to antibiotics to control gram-positive and gram-negative bacteria. The mode of action of these nanomaterials includes induction of oxidative stress, the release of metal ions, and nonoxidative mechanisms; processes can occur simultaneously. The major processes responsible for the antibacterial effects displayed by NPs include interactions with DNA and proteins, as well as penetration of cell membrane [8].

These processes are sequentially triggered, usually starting when NPs bind electrostatically to the cell wall and membranes, altering the function of membrane, leading to depolarization and loss of integrity, which causes problems in transport, energy transduction and, finally, cell death. Disruption of the respiratory chain induces a burst of ROS causing damage to macromolecules involving lipid peroxidation, enzyme inhibition, and alterations to RNA and DNA (Fig. 13.1). In some cases, the toxicity of the nanoparticles is induced by visible or UV light (photocatalytic effect) [9], which also generates ROS. Other effects of NPs include induction of nitrogen reactive species (NRS) and programmed cell death [10, 11].

Ag NPs have shown antimicrobial effects against bacteria and yeasts species. Relatively, Ag NPs have a low toxicity in our cells when compared to other metals [10]. The size and shape of the NPs influence their ability to interact with the bacte-

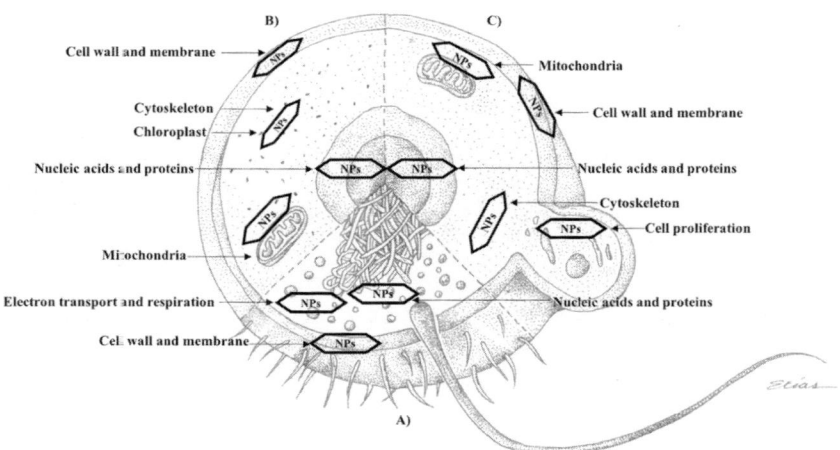

Fig. 13.1 Hypothetical model microbial cell. (**a**) Bacterium, (**b**) algae, and (**c**) fungi-yeast, depicting cellular components interacting with NPs

rial surface and release of Ag+ into solution. Spherical NPs (1–10 nm) can effectively attach to the cell membrane's surface and disrupt permeability and respiration [12].

Ag NPs with positive charge attach to the bacterial membrane by electrostatic interactions, while negatively charged attach to P- and S-containing molecules. Ag ions are then released into the cell and inhibit respiratory enzymes, which promoted the generation of ROS and consequently damage of the cell membrane, disruption of membrane morphology increases, and permeability, leading to leaking of the internal components and resulting in cell death [13]. In order to avoid this, Ag may be complexed inside the cells, or the permeability of Ag may be reduced by combining with a mechanism to pump Ag out of the cell [13]. Nano-silver ions are also the centers of catalytic activity to activate oxygen in air or water, leading to the production of ROS, which causes lipid peroxidation [9] and also prevents the bacterial growth or kills those [8].

In the case of TiO$_2$ NPs, UV- or visible-light-activated TiO$_2$ catalyzes the cleavage of water into hydrogen and oxygen and produces ROS in solution. ROS generated through photoactivation were suggested to be responsible for the TiO$_2$ antimicrobial efficiency. ROS damage the cell membrane and disrupt essential membrane-bound proteins, in addition to creating single-stranded or double-stranded breaks in DNA, rendering it incapable of replication [12].

On bacterial surfaces, Cu NPs bind, as they have a high affinity for constitutive amine and carboxyl groups, causing the release of internal ions. DNA macromolecules interact with these ions that are intercalated in their chains. Since ROS can be generated both by CuO and Cu+/Cu^{2+} ions, their effects are greater than with Ag NPs alone. In addition, Cu and CUO NPs can also induce the formation of ROS in extracellular environments. For this reason, it is considered that the released Cu ions are responsible for the microbial effect of these NPs, but the non-selectivity and toxicity to eukaryotic cells is a latent problem [13].

UV photocatalysis of ZnO NPs produces a strong bactericidal effect. Indeed, the innovation in the Zn NPs goes through the formulation of ZnO particles and the inclusion of Zn in polymers and other materials to improve their efficiency and range of action [13]. Hydroxyl radicals are generated at the surface of excited ZnO, when electrons are released from the water molecule and/or the hydroxide ions. These electrons can reduce to O$_2$ producing the superoxide anion. These ROS cause damage to membranes, DNA by strand breakage or oxidized nucleotides, as well as oxidation of protein catalytic centers. They can also cause external damage, as these ROS, given their negative charge, cannot pass through the membrane. They can be combined, however, with H+, creating hydrogen peroxide, which can penetrate the membrane, causing internal damage and cell death. In addition, ZNO NPs also can generate ROS and bactericidal properties in the complete absence of light [13].

13.3.2   Antimicrobial Activity of TiO₂ NPs

TiO_2 NPs are widely used to inhibit microbial growth because of their broad-spectrum biocidal activity, and they have been the subject of considerable research on stone cultural heritage science (Table 13.1). The biocidal efficiency of TiO_2 NPs has been recently studied and partially reviewed by Batista and Munafó [14].

The nanometric form of TiO_2 is probably the most commonly used nanomaterial. Antimicrobial activity of TiO_2-based coatings has been demonstrated using a range of microorganisms, testing methods, experimental conditions, and time of trial [7] (Table 13.1). As stated previously, the biocidal effect of TiO_2 is seen under UV irradiation, causing the formation of ROS, and thus oxidative stress that attacks membrane lipids, producing damage to the membrane or DNA level [15]. The microbial groups most studied as monospecies cultures are fungi. For example, TiO_2 has been evaluated against *Aspergillus niger* (*A. niger*) or, as a mixed ZnO/TiO_2 formulation, on calcareous stone (limestone and marble coupons) [16, 17]. Both studies revealed good antifungal properties against *A. niger*, also delaying the onset of recolonization [17] or quantitatively reducing fungal coverage [16]. It is difficult to compare the efficiency of treatments since the conditions of exposure and methods of assessment varied, but it could be speculated that the synergistic effect of ZnO/TiO_2 would make this formulation more efficient and more versatile. Goméz-Ortiz [16] showed that TiO_2 was only effective under light conditions, while the mixed formulations were also effective in the dark, suggesting their potential application for indoor non-illuminated settings. Antimicrobial activity has also been ascribed to ZnO; coated nanosystems based on $Ca(OH)_2$–50% ZnO and pure zincite nanoparticulate films were effective against microorganisms, indicating that Zn NPs alone could be promising. An interesting finding of both studies is that the porosity of the substratum played a role in efficiency; more compact material performed better, presumably by allowing greater contact of NPs with fungal cells which are able to penetrate porous rock more readily [16, 17].

Phototrophs are also affected by TiO_2-based coatings. In a reported study on the biocidal effect of mixed formulation platinum-loaded titanium oxide (TiO_2/Pt) on microbial communities by Matsunaga et al. [18], the bacteria *Lactobacillus acidophilus* (*L. acidophilus*) and *Escherichia coli* (*E. coli*), the yeast *Saccharomyces cerevisiae* (*S. cerevisiae*), and the alga *Chlorella vulgaris* (*C. vulgaris*) were evaluated as individual cultures. Synergism was sought by mixing TiO_2 with platinum (Pt) loaded to increase the photoelectrochemical reaction. The approach proved valuable as in a short time (60–120 min), cell survival was reduced for all organisms, except for the alga *C. vulgaris*. This was in agreement with the previous work that found TiO_2 NPs particularly effective against gram-positive and gram-negative bacteria [19]. Further confirmation comes from another laboratory-based experiment that showed the ability of these nanomaterials to prevent fouling by the gram-negative bacterium *Stenotrophomonas maltophilia* (*S. maltophilia*) and the gram-positive *Micrococcus luteus* (*M. luteus*), each deposited individually on marble samples under conditions representing an underwater environment. There was equal activity against both bacteria, especially in the first stages of biofilm formation [20].

Table 13.1 Biocidal treatments using TiO₂ NPs to inhibit biological growth on stone materials

Substrate	Organisms involved	Type of NPs	Testing conditions (time and exposure)	Testing methods	Reference
Marble and limestone	*Aspergillus niger*	TiO₂ anatase	Outdoors, 8 days	SEM, CAM, C	[18]
Clay brick	*Chlorella mirabilis, Chroococcidiopsis fissurarum*	TiO₂	Water runoff test in laboratory conditions 10 weeks	C, DIA, CSLM	[39]
Limestone	*Aspergillus niger, Penicillium oxalicum* tested individually	ZnO/TiO₂	Laboratory conditions; 21 days	MCT, SEM, XRD	[17]
Mortar	*Lactobacillus acidophilus, Escherichia coli, Saccharomyces cerevisiae, and Chlorella vulgaris*, tested individually	TiO₂/Pt	Laboratory conditions, 120 min	MCT, CoA-C	[19]
Cover slips	*Escherichia coli, Candida albicans,* tested individually	Methacrylate + TiO₂ or Ag NPs	Laboratory conditions 24 h	ADT	[21]
Mortar slabs (sand/cement)	*Volvox, Chlorella, Aphanothece,* and *Pleurococcus,* bacteria and protozoa community	Silane, polyhydroxymethylsiloxane, emulsifier POE, and TiO anatase or ZnO emulsions	In laboratory conditions, antifouling rig setup for 8 weeks	ACI, WCA, C, Ph-S	[43]
Stone	*Stenotrophomonas maltophilia* and *Micrococcus luteus*	TiO₂, Ag, Fe, Sr alone, and mixed TiO₂ + Ag, Fe, Sr; TiO₂ + Ag + Sr, TiO₂ + Fe + Sr, TiO₂ + Ag + Fe	In vitro diffusion test on agar	ADT, CTM	[22]
Brick substrata	*Chlorella mirabilis, Chroococcidiopsis fissurarum*	TiO₂ + Cu, TiO₂ + Ag	Accelerated water runoff test, 11 weeks	C, MM	[1]
Travertine	*Chlorella* sp., *Klebsormidium* sp., *Phormidium* sp., *Chlorogloeopsis* sp. mixture	TiO₂, TiO₂ + Ag NPs, TiO₂ + Cu NPs	Accelerated water runoff test, 9 weeks	C, S, SEM-EDX, DIA	[15]

Plaster	Bacteria (*Bacillus paralicheniformis*, *Rhodococcus trifolii*, *Microbacterium chocolatum*, *M. kitamiense*, *Nesterenkonia sandarakina*, and *Parapedobacter koreensis*) Fungi (*Citiophora*, *Devriesia* sp., *Cladosporium* spp., *Aureobasidium pullulans*, *Cephalosporium* sp.) Phototrophs (*Chlorella* sp., *Leptolyngbya* sp., *Chlorella*-like)	Pretreatment biotin + ethanol 5% then TiO$_2$, TiO$_2$ + Ag NPs + nanosilica	In situ on wall of stone monument, 8 months	EM, SEM-EDX, MCT, MI	[23]

ACI agar culture identification, *ADT* agar diffusion test, *C* colorimetry, *CAM* contact angle measurement, *CoA-C* acetyl coenzyme A concentration, *CSLM* confocal scanning laser microscopy, *CTM* contact time microscopy, *DIA* digital image analysis, *EM* epifluorescence microscopy, *GC-MS* gas chromatography-mass spectrometry, *MCT* microbial culture techniques, *MI* molecular identification, *MM* mathematical model, *PhS* photospectroscopy, *S* spectrophotometry, *SEM* scanning electron microscopy, *SEM-EDX* scanning electron microscopy, energy-dispersive X-ray spectroscopy, *WCA* water contact angle, *XRD* X-ray diffraction

Aflori et al. (2013) [21] tested the ability of nanomaterials containing Ag NPs with or without TiO$_2$ to prevent colonization of substrates by the bacterium *Escherichia coli* (*E. coli*) and the fungus *Candida albicans* (*C. albicans*) in laboratory conditions using cover slips. The presence of TiO$_2$ improved the biocidal effect of Ag NPs, even without UV illumination. The Ag NPs alone were equally effective against the bacteria and the fungus, whereas TiO$_2$ plus Ag was more effective in preventing fungal growth than in preventing bacterial growth.

In a more complex study in a terrestrial outdoor environment, La Russa [22] tested the biocidal action of NPs (Ag, Fe, Sr, and TiO$_2$) and doped metal nanocomposites (TiO$_2$, Ag-TiO$_2$, Sr-TiO$_2$, Fe-TiO$_2$, Ag-Sr-TiO$_2$, and Ag-Fe-TiO$_2$). In all cases, the presence of the metal increased the capacity of TiO$_2$ to absorb visible light. The most effective materials were Ag-TiO$_2$ and Ag-Sr-TiO$_2$, and the least effective was Ag-Fe-TiO$_2$. The authors attributed the differences in efficacy to the ionic radius of the metal targets. Strontium (Sr) and Ag have the largest radii, larger than that of the titanium to which the doping metal is adsorbed. In contrast, the radius of iron is smaller than that of titanium, and the metal is inserted deeply into the TiO$_2$ structure, rendering it inactive.

Few studies have been carried out under field trials and mixed species/experimental conditions. One of these was carried out by Ruffolo [23], who evaluated the effect of a nanocomposite of TiO$_2$ on a multispecies microbial community of bacteria, algae, and fungi over 8 months. They found that in these in situ conditions, variable results were obtained. For example, *Bacillus paralicheniformis* (*B. paralicheniformis*) was isolated in all treatments tested, while *Paracoccus caeni* (*P. caeni*) and *Microbacterium chocolatum* (*M. chocolatum*) were isolated from one sample only, and *Pseudonocardia* sp., *Rhizobium trifolii* (*R. trifolii*), *Microbacterium* sp., and *Bacillus* sp. from one other sample, even though all the samples were exposed to the same environmental conditions. The time selected for the experiment was appropriate to show that the nanocomposite was effective, retarding growth in treated areas. However, the efficiency of the nanomaterial in these high-humidity conditions is questioned, since recolonization was found to begin in the treated areas after 4 months. More long-term in situ tests on microbial communities are necessary.

Fonseca and co-workers [24] pioneered the approach of carrying out both laboratory and on-site studies. They tested the ability of pure TiO$_2$ anatase and iron-doped anatase (Fe-TiO$_2$) to prevent colonization of mortar slabs by a mixed culture of two green microalgae *Stichococcus bacillaris* (*S. bacillaris*) and *Chlorella ellipsoidea* (*C. ellipsoidea*) and a cyanobacterium *Gloeocapsa dermochroa* (*G. dermochroa*) in the laboratory and to prevent fouling by lichens and phototrophic microorganisms on walls of the Palácio Nacional da Pena (Sintra, Portugal). Both the pure and the Fe-doped TiO$_2$ anatases were effective, and better results were achieved than with conventional biocides.

13.3.3 Antimicrobial Activity of Ag NPs and Mixed Formulations

A variety of products exist for the conservation of stone, paper, and mortars, the active principles being based on NPs and nanoemulsions of hydroxides of calcium, magnesium, strontium, ferrite, silicon oxide, magnesium, zinc, and Ag.

Certain nanoproducts can produce a biocidal effect themselves, and these are useful for the control of the biodeterioration of stone materials. The sensitivity to Ag NPs was evaluated with the major microbial contaminants from various objects, using 32 bacterial and fungal strains [25]. The size of the silver nanoparticles Ag NPs produced under chemical synthesis was 10–100 nm, defining its effective concentration for the removal of microorganisms present on the surface of the objects at 45 ppm. The results of the experiment allowed the elimination of 94% of the microorganisms present, with the exception of *Bacillus subtilis* (*B. subtilis*) and *Staphylococcus xylosus* (*S. xylosus*) [25] In the same year, a coating containing approximately 0.0001–0.01% dry weight of nano-sized Ag demonstrated its efficacy in laboratory (agar plate) studies on painted fiberglass panels inoculated with either *Aureobasidium pullulans* (*A. pullulans*) or a mixture of *Alternaria alternata* (*A. alternata*) and *Penicillium pinophilum* (*P. pinophilum*) [26]. A year later in a study conducted in Brazil [26], the addition of an aqueous solution of 10% nano-silver product (SOLTICIDE G-30) into preformed gypsum panels resulted in the reduction in discoloration caused by *Cladosporium* sp. [26]. Table 13.2 summarizes the biocidal activity of Ag NPs.

In another study, Shirakawa and co-workers [26] evaluated the effectivity of Ag NPs against *Cladosporium* sp. They found that the effectivity of the treatments was reached at a concentration of 5–15 ppm. The use of Ag NPs at a concentration of 2% or 3% in the silicone paint was particularly noteworthy. The greatest efficiency of the protective coating was noted in a mixed system, Ag NPs (1%) and ACTICIDE® EPW (0.1%), an aqueous phase biocide composed of a combination of benzimidazole carbamate, 2-octyl-isothiazolin-3-one (OIT), and a urea derivative; this achieved 100% inhibition of tested fungal growth [27].

Roy et al. [28] published a study on the synthesis of Ag NPs by microorganisms. The extracellular enzyme nitrate reductase, produced by the fungus *Aspergillus foetidus* (*A. foetidus*) MTCC8876, was utilized for NP production [25]. In an alternative microorganism-linked synthesis, Ag NPs were produced using the biogenic volatiles of the bacterium *Nesterenkonia halobia* (*N. halobia*); their antimicrobial activities were evaluated against the gram-positive bacterium *Streptomyces parvullus* (*S. parvullus*) and the fungus *A. niger*. The Ag NPs were mixed with two types of consolidation polymers and used to coat the external surfaces of sandstone and limestone blocks. The stones treated with silicon polymer loaded with Ag NPs showed an elevated antimicrobial potential against *A. niger* and *S. parvullus* [29]. In a similar study with fungi, the effectiveness of Ag NPs was proven with an amount of 5–15 ppm [27]. Silver NPs produced by microorganisms and another bio-method were screened against 19 fungal isolates from Meymand rocks and historic village. They were able to impact on 68% of fungal isolates [30].

Table 13.2 Biocidal treatments using Ag NPs to inhibit biological growth on stone materials

Substrate	Organisms involved	Type of NPs	Testing conditions (time and exposure)	Testing methods	Reference
Stucco, basalt, and calcite	Bacteria (23), fungi (14), *Pectobacterium carotovorum* *Alternaria alternata*	Ag NPs by green synthesis	Laboratory conditions 72 h, 120 days	CCP, DIA	[33]
Gypsum	*Cladosporium* sp.	Ag NPs	Laboratory conditions 5 weeks	C, SEM-EDX	[26]
Limestone	*Chlorella vulgaris*	Tetraetoxysilane +AgNO$_3$ or chitosan or hydrophobic silica	Accelerated test 4 weeks	C, F (chlorophyll)	[38]
Cover slips	*Escherichia coli*, *Candida albicans*	Methacrylate + TiO$_2$ or Ag NPs	In vitro, 24 h	ADT	[21]
Travertine	*Chlorella* sp., *Klebsormidium* sp., *Phormidium* sp., *Chlorogloeopsis* sp. mixture	TiO$_2$, TiO$_2$ + Ag NPs, TiO$_2$ + Cu NPs	Accelerated water runoff test 9 weeks	C, S, SEM-EDX, DIA	[15]
Stone	*Stenotrophomonas maltophilia* and *Micrococcus luteus*	TiO$_2$, Ag, Fe, Sr alone, and TiO$_2$ + Ag, Fe, Sr; TiO$_2$ + Ag + Sr; TiO$_2$ + Fe + Sr; TiO$_2$ + Ag + Fe	In vitro diffusion test on agar	ADT, CTM	[22]
Concrete	*Chlorella vulgaris*	Water repellents (stearates, silanes, biocides, 3-trimethoxy silyl propyl dimethyl octadecyl ammonium chloride, zeolite, 2,3,5,6, tetrachloro-4-methylsulfonyl-pyridine, and Ag NPs	Modular setup in laboratory conditions 2 weeks	VA, C	[37]

(continued)

Table 13.2 (continued)

Substrate	Organisms involved	Type of NPs	Testing conditions (time and exposure)	Testing methods	Reference
Plaster	Bacteria (*Bacillus paralicheniformis*, *Rhodococcus trifolii*, *Microbacterium chocolatum*, *M. kitamiense*, *Nesterenkonia sandarakina*, and *Parapedobacter koreensis*), fungi (*Ciliophora*, *Devriesia* sp., *Cladosporium* spp., *Aureobasidium pullulans*, *Cephalosporium* sp.), phototrophs (*Chlorella* sp., *Leptolyngbya* sp., *Chlorella*-like)	Pretreatment biotin + ethanol 5% then TiO_2, TiO_2 + Ag NPs + nanosilica	In situ on wall stone monument, 8 months	EM, SEM-EDX MCT, MI	[23]

*CCP*ᶦchromatic changes photograph, *F* fluorimetry (chlorophyll A), *VA* visual assessment

Numerous studies have demonstrated that the use of plants in the synthetic process offers greater advantages over other biological processes. Plant extracts function as synthesis-inducing agents, giving the NPs increased stability and durability. Kalishwaralal et al. [31] studied the anti-biofilm activity of Ag NPs against *Pseudomonas aeruginosa* (*P. aeruginosa*) and *Staphylococcus epidermidis* (*S. epidermidis*) biofilms. They demonstrated that the NPs could effectively block the synthesis of extracellular polymeric substances (EPS). Green synthesis using the aqueous fruit extract of *Aegle marmelos* (*A. marmelos*) allowed the generation of Ag NPs that have the ability to block EPS or biofilm production by the bacterial isolates [32].

Leaf aqueous extracts from two species, *Foeniculum vulgare* (*F. vulgare*) and *Tecoma stans* (*T. stans*), were used in Ag NP synthesis [32]. These NPs were tested in situ to protect stucco, basalt, and calcite materials. Ag NPs from *F. vulgare* were more effective for in vitro microbial growth inhibition than those from *T. stans*. The use of Ag NPs as a preventive or corrective treatment decreased microbial colonization in three kinds of stone, and so it was recommended to use this class of NPs as antimicrobial agents to prevent future mechanisms of biodeterioration [33].

In a study comparing the biological resistance of green and conventional building materials before and after nano-metal treatment to improve fungal growth resistance, *Aspergillus brasiliensis* (*A. brasiliensis*) or *Penicillium funiculosum* (*P. funiculosum*) was inoculated on samples, and their growth was visually evaluated according to ASTM G21–09 [33]. Without nano-metals, green materials were not more prone to fungal growth than conventional ones. After nano-metal treatment at the highest selected dose, the observed order of fungal growth resistance was nano-zinc = nano-copper > nano-silver for wooden flooring and green wooden flooring; nano-zinc > nano-silver = nano-copper for gypsum board; nano-zinc > nano-silver > nano-copper for gypsum board, calcium silicate board, and green calcium silicate board; nano-silver > nano-copper = nano-zinc for mineral fiber ceiling; and nano-silver > nano-copper > nano-zinc for green mineral fiber ceiling [34].

A silver-silica nanocomposite-based geopolymer antibacterial mortar has been developed by simple adsorption of silver in a suitable amount of a colloidal silica suspension [35]. The silver NPs (3–7 nm) were attached to the surface of 20–50 nm-sized silica NPs. Mechanical strength, durability, and mechanistic antibacterial activity of the silver-silica nanocomposite-modified geopolymer mortar (GMAg-Si) were investigated and compared to nanosilica-modified geopolymer mortar and control cement mortar. Inhibition and mortality of gram-positive and gram-negative bacteria were assessed in liquid cultures. At 6% (w/w), the GMAg-Si (cured at ambient temperature) showed substantial improvement in mechanical strength, durability, and antibacterial properties. ROS generation and cell wall rupture, as observed by fluorescence microscopy and field emission scanning electron microscopy (FESEM), have been suggested as reasons for the antibacterial efficacy of the GMAg-Si [35].

Three nanomaterials, (Ag, TiO_2, and CuO), at three concentrations (5, 10, 15 μg/mL), were evaluated for inhibition of bacterial and fungal growth [35]. Three fungi, *A. niger*, *A. flavus*, and *A. fumigatus*, and three gram-positive bacillary bacteria were isolated from three ancient Egyptian funeral masks. Ag NPs exhibited activity against *A. niger*, *A. flavus*, and *A. fumigatus* at 15 μg/mL. Ag NP reactivity with bacteria was higher than for fungi [36].

De Muynck et al. [37] performed one of the first studies comparing the antimicrobial properties of nanomaterials with those of conventional biocides. They applied Ag NPs to concrete materials to prevent colonization by phototrophs. However, the treatment was not as effective as expected such as proved by Eyssautier-Chuine et al. [38]. Moreover, the treated samples showed very noticeable color changes, with pronounced darkening of the surface. Graziani and D'Orazio [39] found that, in addition to UV light, the efficacy of the nanocoating was influenced by the properties of the material (roughness and porosity). They also applied a suspension of the same organisms as used in the previous study [40] to ancient bricks and found that the nanocoating was effective when combined with UV-A light, except on rough areas of the substrate. In an attempt to overcome this problem, the research group tested nanostructured solutions of TiO_2-Ag and TiO_2-Cu. MacMullen [41] tested the ability of a combination of Ag, TiO_2, and ZnO NPs in silane/siloxane emulsions applied to mortars to prevent fouling by algae (mainly *Chlorella vulgaris*), cyanobacteria (*Synechococcus*), bacteria, and protozoa, all isolated from Canoe Lake in Portsmouth (UK). The bioreceptivity of the material seemed to be reduced by the presence of Ag NPs, which also enhanced water repellent façade treatments.

Within the framework of COMAS (Consevazione in situ dei Manufatti Archeologici Sommersi) project, aimed at preventing the deterioration of underwater archaeological artifacts, Ruffolo et al. [23] tested the ability of TiO_2, ZnO, and

Ag NPs dispersed in siloxane wax to prevent fouling by algae, barnacles, bryozoans, and marine worms. They carried out laboratory and in situ studies over a period of 2 years in the underwater archaeological park of Baia (Naples, Italy). After 2 years, all treatments had reduced the colonization to 25–50% of that in untreated samples. The best result was achieved with TiO$_2$ mixed with Ag, while the poorest results were obtained with ZnO. The efficacy of the products was not increased by applying larger quantities. All treatments seemed to prevent colonization by endolithic species, which are more powerful degradation agents than epilithic species.

13.3.4 Antimicrobial Activity of ZnO NPs, CuO NPs, and Mixed Formulations

Cu NPs have been poorly explored for the protection and conservation of stone substrates. However, they have good antimicrobial properties either alone or together with other elements, and have the ability to generate multiple toxic effects, such as the formation of ROS, lipid peroxidation, protein oxidation, and DNA degradation, which was detected in *E. coli* as a model microorganism [42].

Cu NPs have been studied for their activity against phototrophs, especially the chlorophyte *Chlorella*, a common microalga on various stone surfaces. This microalga has been shown to be susceptible to Cu NPs, either as an individual [1] or in a phototrophic microbial community [15]. In both cases accelerated runoff tests were carried out over a reasonable period of exposure (9–11 weeks).

Other microbial groups less studied are bacteria and yeasts. Zazuela et al. [43] evaluated the activity of a CuO/SiO$_2$ nanocomposite to protect stone surfaces and showed a decrease in growth of *E. coli* (CECT 01) and the yeast *S. cerevisiae*. Essa and Khallaf [44] showed the effect of nanocomposites consisting of consolidant polymers (acrylates and silicon) and Cu NPs on target microorganisms such as *Aspergillus* spp., *Candida albicans*, and *Penicillium chrysogenum*, considered as cosmopolitan fungi, or the pathogens *Fusarium solani* (plants) and *C. albicans* (human). Ditaranto et al. [45] used as consolidant the water repellent commercial product (ESTEL 1100) as dispersing medium for Cu NPs. The authors inoculated *Arthrobacter histinolovorans* (ATCC 11442) and evaluated this product applied to calcareous stone and characterized the surface physical properties. The nanocoating strongly inhibited this soil bacterium. However, most microorganisms tested are not common inhabitants of stone surfaces and could provide an unrealistic response in comparison with the behavior of autochthonous microorganisms growing in stone communities. All studies on antimicrobial properties of Cu NPs on the stone substrate are based on microscopy (optical, SEM), microbial culture techniques, and colorimetry, which are enough to demonstrate a positive effect (Table 13.3). However, other interesting techniques to evaluate the efficiency of nanocomposites, especially those with water-repellent properties, include water absorption by contact sponge method (WACSN) [46], water absorption by capillarity (WAC), and

Table 13.3 Biocidal treatments using Cu NPs to inhibit biological growth on stone materials

Type of rock	Organisms involved	Type of NPs	Testing conditions (time and exposure)	Testing methods	Reference
Calcareous	*Arthrobacter histidinolovorans*	ESTEL 1100 + Cu NPs	In vitro laboratory conditions	MCT, SP, C, SEM-EDX	[45]
Sandstone, marble, plaster	*Aspicilia calcarea, Aspicilia contorta* ssp. *hoffmanniana, Caloplaca aurantia, Caloplaca crenularia, Diploicia canescens, Diploschistes actinostomus, Diplotomma ambiguum, Lecanora campestris, Lecanora pruinosa, Lecanora muralis, Parmelia loxodes, Parmelina tiliacea, Physcia adscendens, Tephromela atra, Verrucaria nigrescens, Xanthoria elegans, Xanthoria parietina*	Tetraethylorthosilicate, methylethoxy polysiloxane, Paraloid B72, tributyltin oxide, dibutyltin dilaurate, Cu NPs	In situ, 33 months	OM, C, FTIR, WACSM	[46]
Clay brick	*Chlorella mirabilis, Chroococcidiopsis fissurarum*	TiO_2 + Cu, TiO_2 + Ag	In laboratory conditions by water runoff test 11 weeks	C, MM	[1]
Limestone, sandstone	*Escherichia coli* Z1, *Pseudomonas aeruginosa, Micrococcus luteus, Streptomyces parvulus* and *Bacillus subtilis, Aspergillus niger, A. flavus, Penicillium chrysogenum, Fusarium solani, Alternaria solani*	Methyl and ethyl acrylate, silicon, silane/siloxane + $CuSO_4$ and Cu NPs	In laboratory conditions on microbial culture techniques 24 h	MCT, SEM, EDX	[44]

(continued)

Table 13.3 (continued)

Type of rock	Organisms involved	Type of NPs	Testing conditions (time and exposure)	Testing methods	Reference
Limestone	*Escherichia coli*, *Saccharomyces cerevisiae*	CuO/SiO$_2$	Laboratory conditions 15 days	MCT, C, SEM, WAC WCA	[43]
Traventine	*Chlorella* sp., *Klebsormidium* sp., *Fhormidium* sp., *Chlorogloeopsis* sp.	TiO$_2$, TiO$_2$ + Ag NPs, TiO$_2$ + Cu NPs	Accelerated runoff test 9 weeks	C, S, MCT, SEM, EDX, DIA	[15]

FTIR Fourier transform-infrared spectroscopy, *MM* mathematical model, *WAC^5* water absorption by capillarity, *WACSM^4* water absorption by contact sponge method

water contact angle (WCA) [8]. In situ experimentation studies are relatively rare. Pinna et al. [46] compared the performance of traditional treatments (tetraethylorthosilicate, methylethoxy polysiloxane, Paraloid B72, tributyltin oxide, dibutyltin dilaurate) and Cu NPs in the prevention of recolonization of sandstone, marble, and plaster by crustose and foliose lichens in the archaeological area of Fiesole, Italy. The changes in the bioreceptivity of treated materials were monitored over a period of almost 3 years. The Cu NPs plus water-repellent material yielded good results in terms of preventing biological colonization. In addition, a strengthener and water-repellent material to the Cu NPs also helped to prevent recolonization of the surfaces.

The antimicrobial activities of Zn NPs used individually or as nanocomposite with other agents have also received scant attention (Table 13.4). The principal biological target to test these NPs is fungi, especially those belonging to the genus *Aspergillus* and *Penicillium oxalicum* [8, 17, 20, 47–50], which are known stone inhabitants and deteriogens. The principal natural stones evaluated for protection using ZnO-NPs are calcareous [8, 23]. Using a microbial community model, Zang and co-workers [20] studied the biocidal effect of Zn and Ti nano-oxide silane/siloxane emulsions in laboratory conditions. They evaluated an alga-dominated microbial community and observed that these nanomaterials only slightly inhibited the growth of phototrophs.

13.4 Discussion

Stones are colonized by complex microbial communities [51]. Their composition depends on climate and microclimate, as well as the intrinsic properties of the substrate. Tertiary bioreceptivity is determined by the influence of any human activity that interferes with the material such as consolidation, cleaning, or antimicrobial treatment based on NPs and biocides.

Table 13.4 Biocidal treatments using Zn NPs to inhibit biological growth on stone materials

Substrate	Organisms involved	Type of NPs	Testing conditions (time and exposure)	Testing methods	Reference
Limestone	*Aspergillus niger* or *Penicillium oxalicum*	Ca(OH)$_2$.ZnO, Ca(OH)$_2$.TiO	Laboratory conditions, 28 days	MCT, SEM	[48]
Mortar slabs (sand/cement)	*Volvox, Chlorella, Aphanothece,* and *Pleurococcus,* bacteria and protozoa community	Silane, polyhydroxymethylsiloxane, emulsifier POE, and TiO anatase or ZnO emulsions	Laboratory conditions antifouling rig setup 8 weeks	MCT, WCA, C, Pn-S	[20]
Glass slides and Limestone	*Aspergillus niger* or *Penicillium oxalicum*	Ca[(OH)$_3$]$_2$.2H$_2$O	In vitro, 21 days, Monospecies assay	OM, SEM	[17]
Calcareous	*Aspergillus niger*	ESTEL 1000, 1100, SILO 111, ESTEL 1000 + ZnO, ESTEL 1100 + ZnO, SILO111 + ZnO, ESTEL 1100 + Cu, SILO III + Cu	In situ, 6 months	C. SEM-EDX, MI, MCT	[49]
Agar plate	*Alternaria alternata, Aspergillus niger, Penicillium chrysogenum,* and *P. pinophilum* mixed	ZnO	In laboratory conditions 10 days	BP, MI	[47]
Limestone, dolostone, and glass slides	*Aspergillus niger, Penicillium oxalicum, Paraconiothyrium* sp., *Pestalotiopsis maculans*	MgO, Zn, ZnO/MgO	In laboratory conditions	OM, ESEM-BSE, FESEM, DIA, C	[50]

BP biomass protein of exopolymeric substances (EPS), *ESEM-BSE* environmental backscattered electron scanning microscopy, *FESEM* field emission scanning electron microscopy, *OM* optical microscopy, *VCS* virtual crosshatching system

The use of nanomaterials to control microbial colonization has increased considerably in recent years. The nanometric form of TiO_2 is probably the most commonly used nanomaterial. TiO_2 NPs are characterized by their broad-spectrum biocidal activity and by being nontoxic, highly photoreactive, chemically stable, and inexpensive. However, the reactivity of nanoscale TiO_2 is light dependent. The biocidal activity of the material is greater under UV-A light than under visible light (natural or artificial), and the material is ineffective indoors and in closed environments with no light. In some studies, the TiO_2 matrix has been added to other metal NPs, including Ag [1, 14, 21–23], Cu [1, 8, 14, 46], and Pt [19].

Ag, Zn, and Cu NPs have also been shown to be effective in microbial control, either used as sole biocide agents or in mixed formulations. It is difficult to assess the effectiveness of treatments and perform direct comparisons between the published studies due to the heterogeneity of methods employed to test antimicrobial activity and the range of microorganisms tested. However, it appears that synergism is an important feature of the reviewed studies. The use of standard methods and model organisms would improve our understanding of their biocidal activity, specificity, and their interaction with the substrate.

The application of a nanomaterial that displays antimicrobial properties should reduce the bioreceptivity of a substrate. On the other hand, the tertiary bioreceptivity of a substrate should be lower than the primary bioreceptivity. In this respect, numerous laboratory studies have been conducted to evaluate the inhibitory efficacy of novel products over several weeks in controlled colonization tests with microbial communities. In particular, such studies have used bacteria, which tend to grow faster than algae and cyanobacteria, even though the latter is considered to be the most abundant colonizers of façades that usually appear before other species such as fungi, lichens, mosses, and other bryophytes. In situ (field) studies in this area are much less frequent. Indeed, the first report of a case study of the direct use of a TiO_2 coating biocide on cultural heritage buildings has been published very recently. Ruffolo et al. [23] conducted an 8-month-long study to monitor phototrophic and chemoorganotrophic colonization in the archaeological site of "Villa dei Papiri," Ercolano (Naples, Italy). These researchers found that TiO_2 NPs (alone or combined with Ag) enhanced the effect of the previously applied organic conventional biocide (biotin R). Four months after application of nanoproducts, no recolonization of the surfaces was observed. By the end of the study, the recolonization rate was higher in damper areas of the surface, suggesting the presence of water reduces the effectivity of the nanomaterial.

The characteristics of the target microorganisms influence the response to NPs. The most important of these characteristics are structural properties such as the complexity and thickness of the cell envelope. Thus, for example, fungi are usually less affected by TiO_2 than the structurally more complex bacteria and viruses [7], although Shirakawa et al. [26] demonstrated the efficacy of TiO_2 against fungal colonization on the modern glass in a 5-month field study in Sao Paulo, Brazil. Since the activity of antimicrobials is so dependent on the structure and activity of the target organisms, it is of maximum importance that they are tested in the laboratory against a wide range of organisms before being transferred to field trials carried out under the actual conditions and on the same substrates as their intended use. Model organisms for testing need to be representative inhabitants of the stone subaerial environment, which generally have thick capsules, are highly pigmented, overcome desiccation by producing osmolytes, exhibit a mucoid phenotype, and possess low surface-to-volume ratios (reduced contact with the environment) [52, 53]. Low surface-to-volume ratio reduces biocide uptake. Similarly, thick capsules and sheaths would make it difficult for NPs to enter the cells. Displaying a mucoid phenotype, indicative of EPS production, would be valuable to microorganisms as a barrier for NP contact or even to reduce their activity. Few organisms of the reviewed studies possess these attributes. The reported studies should, therefore, be considered cautiously to extrapolate efficiency for outdoor building surfaces, as NPs may have a low efficiency.

Microbes generally prefer to colonize damp substrates. Nanomaterials formulated with the aim of preventing biofouling therefore often include hydrophobic properties in addition to biocidal and sometimes strengthening properties. The hydrophobic properties also reinforce the biocidal properties by preventing direct contact between organisms and the surface to be colonized [7]. Nanomaterials with these properties will also help protect mineral surfaces, as water is one of the main factors involved in the decay of such substrates. In planning a field trial to test such new protective treatments, the most stringent conditions should be used; hot and humid environments are preferred. Hence companies producing and testing antimicrobials generally use test sites in the humid tropics and suitable sites containing culturally important buildings abound in, for example, India, Latin America, and the Far East.

The potential of NPs in the field of conservation of cultural heritage has been established in a number of ways, consolidating decayed materials, enhancing and de-polluting surfaces, self-cleaning, or as a biocide in biodeterioration [7, 8, 53, 54]. There is limited understanding of the environmental fate of NPs after release from treated surfaces and the impact on surrounding nontarget organisms and ecological processes. Gladis et al. [55] previously stressed the importance of ecotoxicological assessments of active agents under development. The release of NPs may result when the coatings are not fixed adequately to stone or when the durability of materials is not sufficiently effective to remain adhered to the rock over a long period of time [56–59]. Using runoff experimental setups were able to demonstrate unequivocally direct release of Ag NPs and TiO_2 NPs from façade paints (aged and new coatings) and their transport into surface waters and soils. Research on risk assessment of environmental impact and human health issues associated with the release of NPs into the surrounding environment in field studies is essential [7].

13.5   Conclusion

In this chapter, the antimicrobial properties of major types of NPs (TiO_2, Ag, Cu, and Zn) and their mixed formulations that have been reported for stone conservation were reviewed. The heterogeneity of testing methods and microorganisms tested render comparisons difficult. However, some patterns arise. TiO_2 NPs are the most studied type of NPs displaying biocidal properties. Comparatively, little is known about the potential of Cu and Zn NPs, although they appear to be particularly useful for indoor surfaces lacking light. There are no clear specific activities against microbial groups, as several NPs appear to exert toxic activity using similar modes of action. There is an overrepresentation among studies of medical or soilborne microorganisms, which are not representative of the stone subaerial environment. Also, as most studies are carried out under lab conditions using mono- or dual species (two organisms), there is not currently a solid body of knowledge on the effectiveness of NPs in field studies for use on building surfaces. Further studies need to be carried out using more complete and realistic approaches. These approaches include the use of standardized testing methods, preferably community-based techniques such as phospholipid fatty acids combined with molecular biology, a combination that provides information about biomass, diversity, and function of the communities. The use of model organisms from subaerial habitats and long-term field trials are also necessary.

Acknowledgments The authors are grateful to Elías García-López for the fine illustration. This work was supported by CONACYT Ciencia Básica 2016 grant 257449 "Influencia de tratamientos con nano y biomateriales en la colonización microbiana de roca monumental" to Benjamín Otto Ortega Morales. Patricia Sanmartín is financially supported by a postdoctoral contact within the framework of the 2011–2015 Galician Plan for Research, Innovation and Growth, Plan 12C, Modality B (2016 Call).

References

1. Graziani L, Quagliarini E, D'Orazio M. The role of roughness and porosity on the self-cleaning and anti-biofouling efficiency of TiO-Cu and TiO-Ag nanocoatings applied on fired bricks. Construct Build Mater. 2016;129:116–24.
2. Guillitte O. Bioreceptivity: a new concept for building ecology studies. Sci Total Environ. 1995;167:215–20.
3. Miller A, Sanmartín P, Pereira-Pardo L, Dionísio A, Saiz-Jimenez C, Macedo M, Prieto B. Bioreceptivity of building stones: a review. Sci Total Environ. 2012;426:1–12.
4. Pinna D. Biofilm and lichens on stone monuments: do they damage or protect? Front Microbiol. 2014;5:133.
5. Pinna D. Coping with biological growth on stone heritage objects: methods, products, applications, and perspectives. 1st ed. New York: CRC Press/Taylor & Francis Group; 2017.
6. Baglioni P, Chelazzi D, Giorgi R. Consolidation of wall paintings and stone. Nanotech Conserva Cultural Heritage. 2014. https://doi.org/10.1007/978-94-017-9303-2_2.
7. Munafò P, Goffredo G, Quagliarini E. TiO$_2$-based nanocoatings for preserving architectural stone surfaces: an overview. Construct Build Mater. 2015;84:201–18.
8. Whang L, Hu C, Shao L. The antimicrobial activity of nanoparticles: present situation and prospects for the future. Int J Nanomedicine. 2017;12:1227–49.
9. Amabye TG, et al. Antibacterial activities of nanoparticles of titanium dioxide, intrinsic and doped with indium and iron. J Med Chem Toxicol. 2016;1(1):1–7.
10. Ebrahiminezhad A, Javad Raee M, Manafi Z, Sotoodeh Jahromi A, Ghasemi Y. Ancient and novel forms of silver in medicine and biomedicine. J Adv Med Sci Appl Technol. 2016;2(1):122–8.
11. Beyth N, Houri-Haddad Y, Domb A, Khan W, Hazan R. Alternative antimicrobial approach: Nano-antimicrobial material. Evid Based Complement Alternat Med. 2015;2015:246012. https://doi.org/10.1155/2015/246012.

12. Miller KP, Wang L, Benicewicz BC, Decho AW. Inorganic nanoparticles engineered to attack bacteria. Chem Soc Rev. 2015;44:7787–07.
13. Kurtjak M, Aničić N, Vukomanovicć M. Inorganic nanoparticles: innovative tools for antimicrobial agents. In: Kumavath RN, editor. Antibacterial agents. London: InTech; 2017.; Chapter 3. https://doi.org/10.5772/67904.
14. Batista GG, Munafó P. Preservation of historical stone by TiO$_2$ nanocoatings. Coatings. 2015;5:222–31.
15. Batista GG, Accoroni S, Totti C, Romagnoli T, Valentini L, Manufó P. Titanium dioxide based nanotreatment to inhibit microalgal fouling on building stone surfaces. Build Environ. 2017;112:209–22.
16. Chen P, Taniguchi A. Detection of DNA damage response caused by different forms of titanium dioxide nanoparticles using sensor cells. J Biosen Bioelectron. 2012;3:129.
17. Gómez-Ortíz NM, De la Rosa García SC, González-Gómez WS, Soria-Castro M, Quintana P, Oskam G, Ortega-Morales BO. Antifungal coating base on Ca(OH)$_2$ mixed with ZnO/TiO$_2$ nanomaterials for protection of limestone monuments. Appl Mater Interfaces. 2013;5:1556–65.
18. La Russa MF, Ruffolo SA, Rovella N, Belfiore CM, Palermo AM, Guzzi MT, Crisci GM. Multifunctional TiO coatings for cultural heritage. Prog Org Coat. 2012;74(1):186–91.
19. Matsunaga T, Tomoda R, Nakajima T, Wak H. Photoelectrochemical sterilization of microbial cells by semiconductor powders. FEMS Microbiol Lett. 1985;29:211–4.
20. Zhang Z, MacMullen J, Dhakal HN, Redulovic J, Herodotou C, Totomis M, Bennett N. Biofouling resistance of titanium dioxide and zinc oxide nanoparticulate silane/siloxane exterior façade treatments. Build Environ. 2013;59:47–55.
21. Aflori M, Simionescu B, Bordianu IE, Sacarescu L, Varganici CD, Doroftei F, Nicolescu A, Olaru M. Silsesquioxane-based hybrid nanocomposites with methacrylate units containing titania and/or silver nanoparticles as antibacterial/antifungal coatings for monumental stones. Mater Sci Eng B. 2013;178:1339–46.
22. La Russa MF, Macchia A, Ruffolo SA, De Leo F, Barberio M, Barone P, Crisci GM, Urzi C. Testing the antibacterial activity of doped TiO$_2$ for preventing biodeterioration of cultural heritage building materials. Int Biodeter Biodegr. 2014;96:87–96.
23. Ruffolo SA, De Leo F, Ricca M, Arcudi A, Silvestri C, Bruno L, Urzi C, La Russa MF. Medium-term in situ experiment by using organic biocides and titanium dioxide for the mitigation of microbial colonization on stone surfaces. Int Biodeter Biodegr. 2017;123:17–26.
24. Fonseca AJ, Pina F, Macedo MF, Leal N, Romanowska-Deskins A, Laiz L, Gómez-Bolea A, Saiz-Jimenez C. Anatase as an alternative application for preventing biodeterioration of mortars: evaluation and comparison with other biocides. Int Biodeter Biodegr. 2010;64:388–96.
25. Martínez-Gómez MA, González-Chavez MC, Mendoza-Hernández JC, Carrillo-González R. Nanopartículas para el control del biodeterioro en monumentos históricos. Mundo Nano. 2013;6:23–34.
26. Shirakawa MA, Gaylarde CC, Sahão HD, Lima JRB. Inhibition of Cladosporium growth on gypsum panels treated with nanosilver particles. Int Biodeter Biodegr. 2013;85:57–61.
27. Banach M, Szczygłowska R, Pulit J, Bryk M. Building materials with antifungal efficacy enriched with silver nanoparticles. Chem Sci J. 2014;5:085.
28. Roy N, Gaur A, Jain A, Bhattacharya S, Rani V. Green synthesis of silver nanoparticles: an approach to overcome toxicity. Environ Toxicol Pharmacol. 2013;36(3):807–12.
29. Dasan K. History of antifouling coating and future prospects for nanometal/polymer coatings in antifouling technology. In: Dasan K, editor. Eco-friendly nano-hybrid materials for advanced engineering applications. Waretown: Apple Academic Press; 2016. p. 381–400.
30. Parsia P, Khaleghi M, Madani M. Assessment of the antifungal effect of silver Nano- particles produced by pseudomonas sp1 on screened fungus in Meymand Historic Village. Int. J Nanosci Nanotechnol. 2016;10:97–102.
31. Kalishwaralal K, BarathMani Kanth S, Pandian SR, Deepak V, Gurunathan S. Silver nanoparticles impede the biofilm formation by Pseudomonas aeruginosa and staphylococcus epidermidis. Colloids Surf B Biointerfaces. 2010;79(2):340–4.

32. Nithya Deva Krupa A, Raghavan V. Biosynthesis of silver nanoparticles using Aegle marmelos (Bael) fruit extract and its application to prevent adhesion of bacteria: a strategy to control microfouling. Bioinorg Chem Appl. 2014;2014:949538. https://doi.org/10.1155/2014/949538.

33. Carrillo-González R, Martínez-Gómez MA, González-Chávez MC, Mendoza HJC. Inhibition of microorganismos involved in deterioration of an archeological site by silver produced by green synthesis method. Sci Total Environ. 2016;565:872–81.

34. Huang HL, Li CC, Hsu K. Comparison of resistance improvement to fungal growth on green and conventional building materials by nano-metal impregnation. Build Environ. 2015;93:119–27.

35. Adak D, Sarkar M, Maiti M, Tamang A, Mandaj S, Chattopadyay B. Anti-microbial efficiency of nano silver–silica modified geopolymer mortar for eco-friendly green construction technology. RSC Adv. 2015;5:64037–45.

36. Helmi FM, Ali NM, Ismael SM. Nanomaterials for the inhibition of microbial growth on ancient Egyptian funeral masks. Mediterranean Archeol Archaeometry. 2015;15:87–95.

37. De Muynck W, Maury-Ramirez A, De Belie N, Verstraete W. Evaluation of strategies to prevent algal fouling on white architectural and cellular concrete. Int Biodeter Biodegr. 2009;63:679–89.

38. Eyssautier-Chuine S, Vaillant-Gaveau N, Gommeaux M, Thomachot-Schneider C, Pleck J, Fronteau G. Efficacy of different chemical mixtures against green algal growth on limestone: a case study with *Chlorella vulgaris*. Int Biodergr Biodeterioration. 2015;103:59–68.

39. Graziani L, D'Orazio M. Biofouling prevention of ancient brick surfaces by TiO_2-based nano-coatings. Coatings. 2015;5:357–65.

40. Graziani L, Quagliarini E, Osimani A, Aquilanti L, Clementi C, Yéprémian C, Larricia V, Amoroso S, D'Orazio M. Evaluation of inhibitory effect of TiO_2 nanocoatings against micro-algal growth on clay brick façades under weak UV exposure conditions. Build Environ. 2013;64:38–45.

41. MacMullen J, Zhang Z, Dhakal HN, Radulovic J, Karabela A, Tozzi G, Hannant S, Alshehri MA, Buhé V, Herodotou C, Totomis M, Bennett N. Silver nanoparticulate enhanced aqueous silane/siloxane exterior façade emulsions and their efficacy against algae and cyanobacteria biofouling. Int Biodeter Biodegr. 2014;93:54–62.

42. Chatterjee AK, Chakraborty R, Basu T. Mechanism of antibacterial activity of copper nanoparticles. Nanotechnology. 2014;25(13):1–12.

43. Zarzuela R, Carbú M, Gil MLA, Cantoral JM, Mosquera MJ. CuO/SiO nanocomposites: a multifunctional coating for application on building stone. Mater Des. 2017;114:364–72.

44. Essa AMM, Khallaf MK. Antimicrobial potential of consolidation polymers loaded with biological copper nanoparticles. BMC Microbiol. 2016;16:144.

45. Ditaranto N, Loperfido S, Van der Werf ID, Mangone A, Cioffi N, Sabbatini L. Synthesis and analytical characterization of copper-based nanocoatings for bioactive stone artworks treatment. Annal Bioanal Chem. 2011;399:473–81.

46. Pinna D, Salvadori B, Galeotti M. Monitoring the performance of innovative and traditional biocides mixed with consolidants and water-repellents for the prevention of biological growth on stone. Sci Total Environ. 2012;423:132–41.

47. Gambino M, Ahmed MAA, Villa F. Zinc oxide nanoparticles hinder fungal biofilm development in an ancient Egyptian tomb. Int Biodeter Biodegr. 2017;122:92–9.

48. Gómez-Ortíz NM, González-Gómez WS, De la Rosa García SC, Oskam G, Quintana P, Soria-Castro M, Gómez-Cornelio S, Ortega-Morales BO. Antifungal activity of $Ca[Zn(OH)_3]_2$-$2H_2O$ coating for the preservation of limestone monuments: an in vitro study. Int Biodeter Biodegr. 2014;91:1–8.

49. Van der Werf ID, Ditaranto N, Picca RA, Sportelli MC, Sabbatini L. Development of a novel conservation treatment of stone monuments with bioactive nanocomposites. Herit Sci. 2015;3:29.

50. Sierra-Fernández A, De la Rosa-García S, Gómez-Villalba LS, Gómez-Cornelio S, Rabanal ME, Fort R, Quintana P. Synthesis, photocatalytic, and antifungal properties of MgO, ZnO

and Zn/Mg oxide nanoparticles for the protection of calcareous stone heritage. Appl Mater Interfaces. 2017;9(29):24873–86.

51. Gutarowska B, Celikkol-Aydin S, Bonifay V, Otlewsk A, Aydin E, Oldham AL, Brauew JL, Duncan KE, Adamiak J, Sunner JA, Beech IB. Metabolomic and high-throughput sequencing analysis—modern approach for the assessment of biodeterioration of materials from historic buildings. Front Microbiol. 2015. https://doi.org/10.3389/fmicb.2015.00979.

52. Gorbushina AA. Life on the rocks. Environ Microbiol. 2007;9(7):1613–31.

53. Baglioni M, Benavides YJ, Berti D, Giorgi R, Keiderling U, Baglioni P. An amine-oxide surfactant-based microemulsion for the cleaning of work of art. J Colloid Interface Sci. 2015;440:204–10.

54. La Russa MF, Ruffolo SA, de Buerfo MA, Ricca M, Belfiore CM, Pezzino A, Crisci GM. The behaviour of consolidanted Neopolitan yellow Tuff against salt weathering. Bull Eng Geol Environ. 2017;76(1):115–24.

55. Gladis F, Eggert A, Karsten U, Schumman R. Prevention of biofilm growth on man-made surface: evaluation of antialgal activity of two biocides and photocatalytic nanoparticles. Biofouling. 2010;26(1):89–101.

56. Carmona-Quiroga PM, Jacobs RMJ, Martínez-Ramírez S, Viles HA. Durability of anti-graffiti coatings on stone: natural vs accelerated weathering. PLoS One. 2017;12(2):1–18.

57. Kaegi R, Sinnet B, Zuleeg S, Hagendorfer H, Mueller E, Vonbank R, Boller M, Burkhardt M. Release of silver nanoparticles from outdoor facades. Environ Pollut. 2010;158(9):2900–5.

58. Kaegi R, Ulrich A, Sinnet B, Vonbank R, Wichser A, Zuleeg S, Simmler H, Brunner S, Vonmont H, Burkhardt M, Boller M. Synthetic TiO$_2$ nanoparticle emission from exterior facades into the aquatic environment. Environ Pollut. 2008;156(2):233–9.

59. Shandilya N, Le Bihan O, Bressot C, Morgeneyer M. Emission of titanium dioxide nanoparticles from building materials to the environment by wear and weather. Environ Sci Technol. 2015;49(4):2163–70.

Chapter 14
Advanced and Novel Methodology for Scientific Support on Decision-Making for Stone Cleaning

E.T. Delegou, I. Ntoutsi, C.T. Kiranoudis, J. Sayas, and A. Moropoulou

14.1 Introduction

Stone cleaning is important for both aesthetical and physicochemical reasons in the course of a monument's sustainable protection. It is a totally nonreversible conservation intervention, and therefore, careful planning is required regarding what is to be removed from the surface that is to be cleaned. Besides surface composition, stone texture and cohesion, as well as stone aesthetics and color, are surface parameters that after cleaning are modified and need also to be considered. In order to study these surface modifications that stone undergoes after cleaning, prerequisites include the characterization of the stone substrate and the decay pattern in which the cleaning will be applied. In order to plan cleaning interventions on the scale of monuments, as well as to study the surface modifications that are taking place with stone cleaning, pilot cleaning interventions are applied at small areas of the monument that present different decay patterns. The development of different decay patterns on stone façades is mainly affected by stone location, orientation, and exposure to sunlight, rainwash, and wind for the given environment that the monument is located in. Thorough decay diagnosis allows for the selection of cleaning methods and techniques that are pilot applied on the monument, while assessment of cleaning tests performance follows. According to the performance of the pilot cleanings

E.T. Delegou • I. Ntoutsi • A. Moropoulou (✉)
Department of Materials Science and Engineering, School of Chemical Engineering,
National Technical University of Athens (NTUA), Athens, Greece
e-mail: amoropul@central.ntua.gr

C.T. Kiranoudis
Department of Process Analysis and Systems Design, School of Chemical Engineering,
National Technical University of Athens (NTUA), Athens, Greece

J. Sayas
Department of Geography and Regional Planning, School of Rural and Surveying
Engineering, National Technical University of Athens, Athens, Greece

© Springer International Publishing AG 2018
M. Hosseini, I. Karapanagiotis (eds.), *Advanced Materials for the Conservation
of Stone*, https://doi.org/10.1007/978-3-319-72260-3_14

per decay pattern and substrate (in cases where more than one type of stone are used on the monument façade), decision can be made for the methods that are to be applied on the monument during conservation works. Therefore, assessment of the pilot cleaning interventions is fundamental for the stone's durability and consequently for monument sustainability.

In this context, the scientific community over the last several decades has tried to establish a common framework regarding cleaning assessment methodology and criteria, including physicochemical characteristics of stone, surface morphology, and aesthetic characteristics like color [1–5]. However, objective difficulties concerning different stone types, decay patterns, and environmental conditions [6, 7], application of different types of cleaning methods [8, 9], different instrumentation used for cleaning assessment [10–12], as well as consideration differences [13–16] restrain these efforts.

Considering all the above, and in an attempt to establish a uniform and integrated methodology for the assessment of stone cleaning in terms of compatibility, the precise characteristics that a surface (for particular building material and particular decay pattern) should hold after cleaning have to be further clarified. Moreover, the specific parameters that should be measured, in order to give values to these characteristics, as well as the acceptance threshold levels of these parameters should also be determined. In parallel, the most appropriate experimental techniques for these parameters to be measured should be examined as well. Furthermore, environmental parameters like ambient conditions, location, and orientation of the surface under cleaning on a monument façade have to be considered. Therefore, in this work, a methodological approach that encompasses the above factors is suggested. Cleaning assessment criteria are set along with their critical parameters and the experimental techniques that can measure these parameters. The pursuing goals for each cleaning assessment criterion are clarified, while the cleaning acceptance threshold levels are determined by the means of a fuzzy logic model (Mamdani type). This fuzzy logic model is incorporated into a GIS platform which holds spatial data relating to the monument, environment, and the stone surface that is under investigation. Finally, the ability of the developed integrated decision-making system to constantly respond in relation to the cleaning performance and durability in a pointed spatial entity and environmental conditions is presented. The system was successfully used on a marble surface of the National Archaeological Museum (NAM), a historic building in Athens, Greece, which presented the characteristic decay of black crust.

14.2 Analytical Techniques, Materials, and Methods

The experimental techniques used for the assessment of stone cleaning can be summarized as follows:

SEM-EDX (JEOL JSM-5600, OXFORD LINKTM ISISTM 300 with energy dispersive X-ray microanalysis system; accelerating voltage, 20 KV; beam current, 0.5 nA; lifetime, 50 s; beam diameter < 2 μm) was applied on collected monument samples, cut in cross sections, before and after cleaning, to determine the chemical and mineralogical composition of the investigated marble surfaces.

Surface texture, cohesion, and microstructure of the stone's surface before and after cleaning were assessed by digital image processing (DIP) of SEM images as well as by laser profilometry (LP). In particular, DIP of SEM images was performed using the EDGE program which was developed by the US Geological Survey [17]. The marble microstructural index that was evaluated is the near-surface fracture density (FD) of the stone which is a measure of the fraction of the stone's volume filled by fractures, crevices, and pore space. The FD results are reported as the percentage of pixels identified as components of the fractures calculated up to 100 μm under the surface area [18]. FD values of the black crust were measured before cleaning in order to compare them to that of FD surface microstructure values obtained after cleaning. The FD values were used as an evaluation index of the marble fracturing after cleaning. These values can be used to determine any side effects caused by the cleaning process as well as to ascertain the degree of susceptibility of the stone's surface to further decay. In addition, 3D micro-topography plots of core samples collected from the monument were attained using the LP (Proscan 2000) with a laser triangulation sensor whose resolution in the perpendicular direction is 1 μm. The roughness parameter (Rq) was estimated at an evaluation length of 1.25 mm, with a step size of 1 μm and a cutoff filter of 0.25 mm in accordance with the standard of BS EN ISO 4288:1988, as follows:

$$Rq = \sqrt{\frac{1}{l}\int_{1}^{0} z^2(x)\,dx} \qquad (14.1)$$

Furthermore, the surface area (i.e., ratio of actual to projected area) was measured at each LP micro-topography [19]. This ratio is a geometrical descriptor of a surface and has a very close relation with some functional properties of surfaces like wear [20]. The values of these characteristics (Rq and surface area) function as assessment indexes of the surface morphology in regard to the susceptibility of the surface to further decay which includes the slow reactivation of the sulfation process and the slow adsorption of black particles.

Furthermore, colorimetry was applied before and after the pilot cleaning interventions in the field using the Dr. Lange color pen LMG 159/160 spectrophotometer to measure the L*, a*, and b* values according to CIELab color space 1976. Total color difference (ΔE) was estimated according to EN 15886:2010 and ASTM D2244:1993 by the following formula:

$$\Delta E = \sqrt{\Delta L^{*2} + \Delta a^{*2} + \Delta b^{*2}} \qquad (14.2)$$

Colorimetry was used to assess the aesthetical modifications that the architectural surfaces underwent due to cleaning interventions, reflecting the degree of black deposition removal [18].

Finally, GIS thematic maps were created in ArcMap/ArcInfo 9.2 using the CAD architectural drawings as the blueprint for the GIS base-map development.

14.3 Results and Discussion

14.3.1 Cleaning Assessment Criteria and Critical Parameters

One of the most widespread decay patterns that cleaning is appointed to remove is black crusts. They are found in urban environments at rain-sheltered areas of monument façades. Their formation, in the cases of calcareous substrates, includes the attack of the atmospheric pollutant of sulfur dioxide (SO_2) on the stone calcite ($CaCO_3$) producing gypsum ($CaSO_4 \cdot 2H_2O$) [21]. Furthermore, black depositions like soot, dust fall (particles of aluminum silicate composition), saturated and unsaturated hydrocarbons, as well as metal oxides and/or metals like Fe, Pb, Zn, Ni, and Vn develop the characteristic dark-colored areas which not only greatly affect the visual appearance of the monument but also act as catalysts for the sulfation processes, accelerating further decay [21]. Additionally, when the crust (gypsum layer plus black layer) reaches a particular thickness of several μm, which varies and depends on different factors like stone type and surface treatment (i.e., smooth, hammered, relief), pollution, temperature, and relative humidity, it loses its cohesion to the substrate leading to the detachment of parts of the black crust [22]. Thus, irreversible loss of authentic material can take place, a fact that confirms the necessity of cleaning as a conservation intervention.

The first arising question is: which part(s) of the black crust have to be removed for the resulting cleaned surface to comply with compatibility? Different approaches on this issue can be found in literature, since some researchers [11, 14, 23] present total black crust removal, both gypsum and black depositions, as accepted cleaning. This is in contrast to the predominate theory of black crust formation that proposes calcite, the original material of the stone substrate, is transformed into gypsum. In particular, sulfation starts on the stone's external surface, and a compact, firm, and clear gypsum layer is developed inward, substituting calcite [24–26]. This clear gypsum layer often consists of crypto- and/or microcrystalline gypsum and preserves the relief details that have been lost from the gypsum-calcite interface [24–26]. In parallel, gypsum formation is favored at the stone grain boundaries, meaning that its removal will facilitate the detachment of calcite grains. Finally, on the top of this clear gypsum layer, a dark gypsum layer, including the environmental black depositions, is formed [24–26]. Therefore, the clear gypsum layer is derived from the stone itself, and it should be considered authentic material that has to be preserved for a successful cleaning, while the external layer consisting of black depositions and gypsum can be removed for both aesthetical and physicochemical reasons.

Gypsum in the external black layer is believed to develop outward at the environmental interface through diffusion processes of the stone calcite grains' calcium anions [26]; but since it is mixed with black depositions in a common layer, it cannot be retained. Another important issue regarding what must not be removed from the stone surface during cleaning is the patina and polychromes which often are found along with black crusts.

Patina is a layer, usually firm and coherent, found on well-preserved stone surfaces and has been shown to protect the substrate, slowing down deterioration processes [27–29]. Therefore whenever found, and despite the debate of its origin (biological or chemical), patina should be preserved during cleaning. Oxalates and/ or phosphates are usually found in patinas, and so whenever found in a black crust–substrate system, they are considered protective layers and thus have to be preserved as well during cleaning. Polychromes and their residues are also considered patinas, since they are historic evidence that the stone surface was colored in the past and as such they should be preserved and not removed by cleaning [27]. Furthermore, residues and by-products of the applied cleaning method (especially in the case of chemical cleaning) should not be left on the surface after it has been cleaned, meaning that a thorough rinse of the surface should take place.

Optical microscopy (OM) and/or scanning electron microscopy with energy dispersion by X-ray analysis (SEM-EDX) are analytical techniques used systematically by the research community for the identification of the mineralogical and chemical composition of the stone surface before and after cleaning. When core samples are taken from the monument and are examined in cross sections by these two techniques, stratification and thickness of the black crust, as well as the detection of patina, can be accomplished before cleaning. After cleaning, study of the surface morphology regarding patina preservation, any remaining layers of black crust (stratification and thickness), and any possible by-products/side effects of the applied cleaning method(s) can also be achieved by these techniques.

The second arising question is: what are the desired surface texture, cohesion, and microstructure properties of the cleaned stone in terms of compatibility? In general, after cleaning the resulting surface should be smooth, with no extra cracks and fractures, meaning that the applied method does not act aggressively on the substrate. The cleaned stone surface's location and orientation on structure, degree of protection from rainwash and wind (microclimate), and environmental conditions (temperature, relative humidity, and pollution) will favor the development of the same decay pattern that existed on the surface before cleaning. Therefore, in the case of black crusts, the stone's surface after cleaning must hold such microstructural characteristics where the processes of sulfation and black depositions adsorption will take place at the lowest possible rate. Based on this assumption, it may be argued that a cleaned surface has to present low roughness and a low active surface area in order for it to be less susceptible to further decay. Although, this statement, at first sight, seems correct, it does not consider the possibility of authentic material loss in the attempt to achieve a cleaned surface with low roughness and thus of low susceptibility to further decay, a fact which is in contradiction with the requirement of authentic material preservation previously described. Therefore, taking this into consideration, the possibility of a conflict of interest between the loss of surface stone material and the stone's susceptibility to decay, the surface's microstructural parameters after cleaning should present similar threshold levels with the ones before cleaning. However, in the case of black crusts, the above conclusion can be varied.

Compared to the clear gypsum layer and the intact calcite stone substrate, the external layer of black depositions and gypsum found in black crusts typically present loose cohesion, friability, cracks, and fractures. Since the black depositions – gypsum external layer – are to be removed, the surface microstructure, texture, and cohesion after cleaning have to be under similar and/or slightly lower threshold levels as compared to those prior to cleaning.

Some research has been performed for the quantitative measurement of surface texture, cohesion, and microstructure before and after cleaning [30]. In recent years though, the advances of modern technology have led to the application of numerous methods for the 3D representation of a material's surface [9, 12, 31, 32] and the digital image processing of microscopy images [17, 33, 34]. However, these recent advances include the application of experimental techniques of different principles, which complicate the comparison of results and make the quantitative approach of surface microstructure challenging.

The third arising question concerns the aesthetics and the color that a surface after cleaning has to present. The aesthetics of a cleaned stone surface and historic buildings in general are still considered a rather ambiguous and subjective issue, although theoretical and statistical studies have attempted to describe aesthetics also taking into account public opinion [35, 36]. However, in conservation science, color parameters and their modification due to conservation interventions, like cleaning, are in most of the cases evaluated by colorimetry, employing in situ or in laboratory spectrophotometers/colorimeters and using the CIE $L^*a^*b^*$ color space 1976 [4, 10, 37, 38]. This common practice of the conservation scientific community led to the recently published European standard EN 15886:2010 [39]. According to EN 15886:2010 and the older ASTM D2244:1993 standard [40], in the CIE $L^*a^*b^*$ color space, ΔE is the total color difference which is calculated by Eq. 14.2, where $\Delta L^* = L^*_{after} - L^*_{reference}$ is the lightness difference (black to white), $\Delta a^* = a^*_{after} - a^*_{reference}$ is the red/green difference, and $\Delta b^* = b^*_{after} - b^*_{reference}$ is the yellow/blue difference. In the case of cleaning assessment, L^*_{after}, a^*_{after}, and b^*_{after} are the values measured after cleaning, where several approaches for the determination of the corresponding reference values can be found in literature.

In particular, a common approach is to assess the color modification after cleaning using ΔE as the color index, taking as reference points the values L^*, a^*, and b^* which are measured at the stone's surface before cleaning [4, 13], since it is not possible to know the initial color parameters of the stone façades of the monuments under investigation.

Other approaches consider measurements obtained on fresh-cut quarry samples [11] or on artificially aged quarry samples with a gypsum layer [41] as the L^*, a^*, and b^* reference values for the ΔE calculation. Other works record and compare the L^*, a^*, and b^* values of the cleaned surfaces to assess different treatments [16, 42].

In the case of cleaning applications with a Nd:YAG laser at 1064 nm, b^* values are considered important because they quantitatively describe the yellowing effect which is evident after this type of treatment [15, 41].

Considering the above, when black crusts are to be cleaned with ΔE used as the color modification index utilizing L^*, a^*, and b^* reference values from the blackish surface of the black crust prior to cleaning, increased ΔE values are expected since the goal of cleaning is the removal of black depositions in the external layer. After cleaning a black crust, an increase in the L^* parameter (black to white) is expected due to black depositions removal, where a^* and b^* parameters record the red/green and the yellow/blue coordinates, respectively, and their values depend on the actual color of the substrate.

It is obvious that ΔE values express the degree of black deposition removal and thus, in the case of whitish stone substrates, low ΔE values show inadequate removal of black depositions indicating insufficient cleaning. High ΔE values demonstrate successful removal of black depositions when whitish stone substrates are cleaned; however, they highlight that significant color modifications have taken place after cleaning, a fact which has to be prevented so as to preserve the monument's aesthetics. Therefore, medium ΔE values are desirable on a stone's surface after a successful cleaning, since they are indicative of an adequate removal of black depositions while preserving the monuments' aesthetics. Furthermore, it is worth mentioning that systematic recording of color parameters is very useful because it helps to produce cleaned surfaces of a homogeneous color. This is rather important in order to avoid undesirable color patchwork which negatively affects a monument's aesthetics, especially in the cases of adjacent surfaces that display different decay patterns, and have likely been cleaned by different methods.

In addition, systematic recording of color parameters is essential for the monitoring of stone façade soiling, providing data about the durability of a cleaned surface. In particular, the accumulation and adsorption of black depositions is directly connected with the pathology of monuments' stone façades in the framework of the application of periodical conservation interventions. Thus, the systematic recording of color parameters can be used as a soiling index, providing insight about a cleaned surface susceptibility to decay, as well as an approximation of when a new treatment should be applied.

The fourth arising question concerns the durability of a stone surface after cleaning, that is, how fast decay will develop. High durability or decreased susceptibility to decay, for a stone surface that presented black crust decay prior to cleaning, indicates the slow reactivation of the sulfation process as well as low adsorption of black depositions. As explained above, the parameters that describe surface texture, cohesion, and microstructure can indicate the durability of a cleaned stone surface. However, presently there is no clear correlation between the surface microstructural parameters and the durability of a cleaned stone surface. Furthermore, as already mentioned, color measurements can be used for the long-term assessment of soiling on stone surfaces after cleaning.

The nondestructive technique of colorimetry can be easily applied in situ on monuments, giving prompt and quantitative evidence on black deposition accumulation. In parallel, accelerated aging tests have been used to assess the durability of suggested cleaning interventions on black crusted stone surfaces. In particular, cleaned stone samples from the investigated monuments and quarry samples were artificially aged in a SO_2 atmosphere [43, 44]. Comparing the width of the developed gypsum layer between the monument and quarry samples, future susceptibility to sulfation was approximated for the cleaned stone surfaces. Besides these indirect approaches, the most objective way to evaluate the durability of stone surfaces after cleaning is the monitoring of their preservation state. This entails repeating the assessment process of the pilot cleaning interventions after several months, using the same experimental techniques and measuring the same parameters that were used in the first pilot cleaning assessment to acquire comparable data. This kind of approach for cleaning durability evaluation is rather difficult to be adopted, a fact that is reflected by the insufficient data found in literature. This weakness is mainly attributed to sampling restrictions that govern the cultural heritage field, as well as to public bodies and legal limitations regarding the time intervals among diagnostic studies, application of pilot cleaning and its assessment, and actual conservation works.

The above remark stresses the necessity of further advanced research and development of nondestructive techniques in the cultural heritage protection field, as well as the need to adopt the concepts of monitoring and periodical conservation interventions into legislation. In addition, there is no solid evidence as to when a reassessment of pilot cleaning interventions should take place. This is because decay development on a cleaned surface is controlled not only by endogenous stone parameters that are impacted by cleaning but also by exogenous parameters of micro- and macroclimate conditions, factors varying significantly from case study to case study. Thus, further research is required on this subject in order to substantiate the claim that artificial aging is an important technique. In parallel, there is a need for a methodological tool that can consider the characteristics of stone surfaces under investigation along with the spatial data of the monument and environment while considering their variation over time.

A geographical information system (GIS) can address all of these issues as several studies of building pathology representation and monument preservation have already demonstrated [45–48]. This is because of GIS' capabilities of recording, grouping, managing, and analyzing large volumes of spatially referenced data along with different attribute data sets which can include building materials data. Furthermore, GIS' potential for database elaboration, spatial analysis, and spatial and attribute data transformation and correlation can enable control and monitoring of the stone surfaces under examination.

Conclusively, the adopted cleaning assessment criteria for black crusts, the critical cleaning assessment parameters, and their pursuing goals, as well as the experimental techniques used in this work, are summarized in Fig. 14.1.

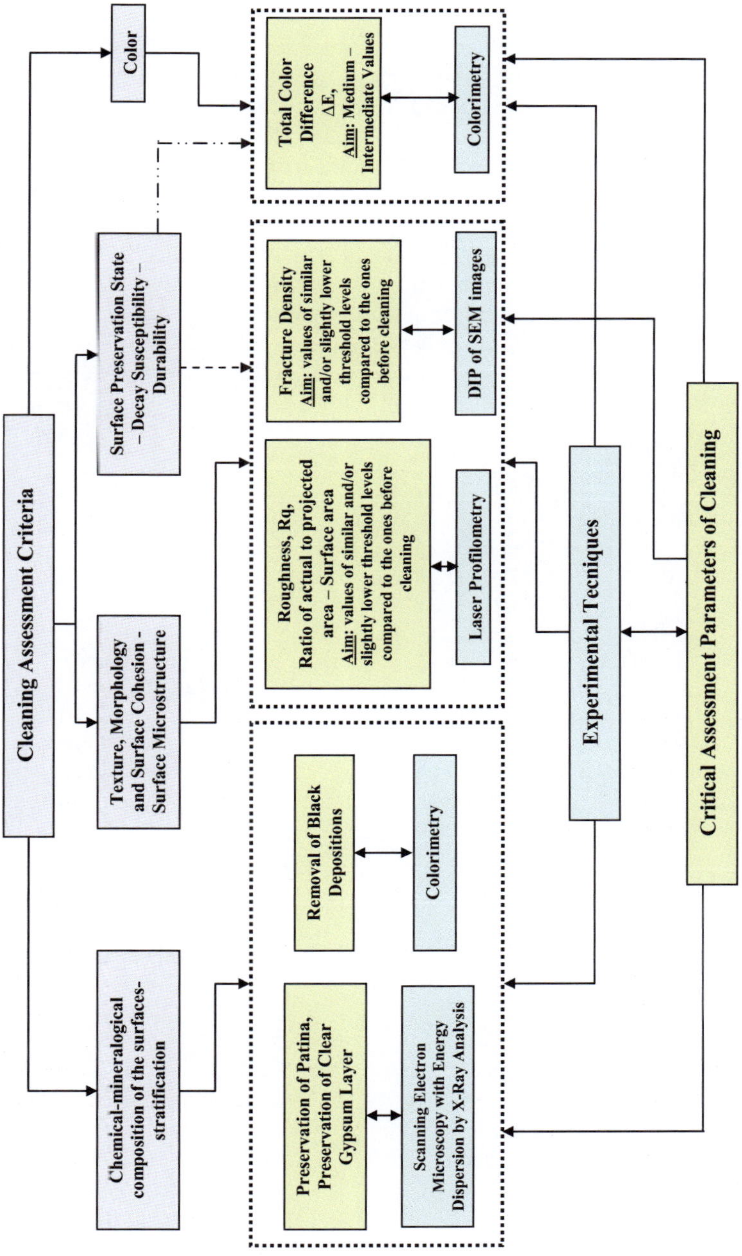

Fig. 14.1 Flowchart of the adopted cleaning assessment criteria, their critical parameters, and the experimental techniques used

14.3.2 Fuzzy Logic Modeling

The most popular fuzzy model suggested in literature is the method proposed in 1975 by E. Mamdani [49]. The way that classical mathematical modeling serves as a quantitative descriptor of physical and chemical phenomena is through systems of equations describing their mechanisms and interactions. When such a system is extremely complex to produce code or solve, the tools of artificial intelligence that resemble the way humans think and reach decisions through appropriate inference may replace the need of a detailed description for a system through methodologies covering machine learning or fuzzy reasoning. The latter is based on the theory of fuzzy sets [50]. This theory is different from the classical mathematical set theory where sets are essentially clusters of objects or entities (practically everything) in the sense that an object belongs to only one of different disjoint sets. Fuzzy reasoning is based on the concept that each object can belong to different disjoint sets through partial membership that is though belonging to a certain level to each one of the individual sets [51, 52]. The level of such membership is proportional to the resemblance of each object to the dominant characteristics of each set.

There are three different and discrete serial processes (steps) to make a fuzzy system produce a value inference or a decision based on a known set of input variables (serving as system arguments) appropriately scaled to be evaluated.

The first step is called fuzzification and is plainly the process of revealing the resemblance of the input object to the basic fuzzy sets used through the appropriate evaluation of the object's membership level to each fuzzy function [53]. In this work, the triangular membership functions are used. The second step is fuzzy inference where the contribution of each fuzzy set through the membership values evaluated in the previous step is convoluted through a set of fuzzy if-then-else rules to produce a fuzzy output [54]. In this work, the "min-max" inference technique has been used [55]. The last step is defuzzification where the outcome of the rule testing is appropriately weighed and descaled to produce the final system output that could be a scalar value or a decision [53]. The centroid defuzzification method [56] is used in this work. The process is appropriately presented in Fig. 14.2.

Furthermore, development of the model's knowledge base is completed by the following three steps: (1) selection of input and output parameters, (2) development of the fuzzy sets for all of the input and output parameters, and (3) development of the rule base.

A popular fuzzy model suggested in literature, and also used in this work, was proposed by Mamdani [57] and has the following formulation with respect to its fuzzy rules [53]:

$$\forall r \in R : if \underset{1 \le i \le n}{\wedge} \left(x_i \in A_i^r \right) \ then \ \underset{1 \le j \le m}{\wedge} \left(y_j \in B_j^r \right) \tag{14.3}$$

where n is the number of input variables, m stands for the number of output variables, $x_i, 1 '' i '' n$ are input variables, $A_i^r, 1 '' i '' n$ are fuzzy sets defined on the

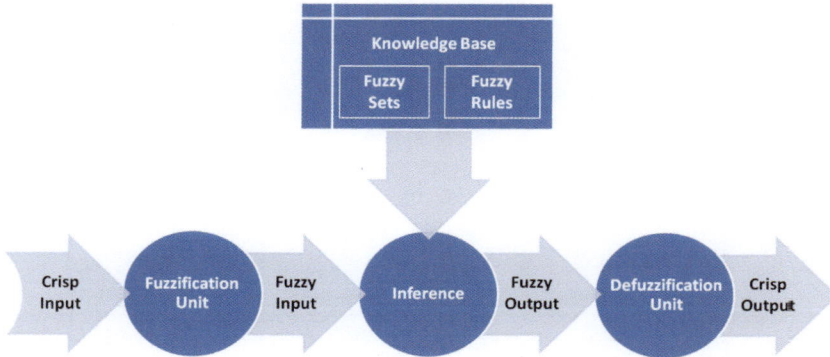

Fig. 14.2 The structure of a typical fuzzy logic model

respective universes of discourse, $y_j, 1'' j'' m$ are output variables, and $B_j^r, 1'' j'' m$ are fuzzy sets defined for the output variables.

In this work, the Mamdani type of fuzzy model was developed under Visual Studio using the C++ programming, and it is based on important influencing factors.

14.3.3 Architecture of the Decision-Making System

The architecture of the suggested decision-making system and the information flow within the system are presented in Fig. 14.3. There are two main parts of the suggested decision-making system. The first one includes the GIS-based graphical user interface for features and attributes retrieval, management, comparison, analysis, and correlation. Spatial classification of decay and pilot cleaning interventions on the representative investigation areas take place through the building of the thematic maps of decay and pilot cleaning interventions, respectively. Moreover, attribute databases consisting of the physical and chemical characteristics data are elaborated and linked to the attribute table of the corresponding GIS decay/pilot cleaning interventions' mapping project, thus resulting in relational databases (RDBs). Finally, the thematic map of "planning of conservation interventions" is the resulting output theme of both decay and pilot cleaning interventions' thematic maps after the application of the geo-processing analysis of the intersection operational tool.

This new thematic map includes spatial information and attributes before and after cleaning, accomplishing comparison and analysis of the recorded to space physicochemical characteristics of stone surfaces during different time periods. The second main part of the suggested decision-making system is the solver of the fuzzy logic model. The critical cleaning assessment parameters of the adopted criteria are used as input parameters of the model. These parameters also serve as the data

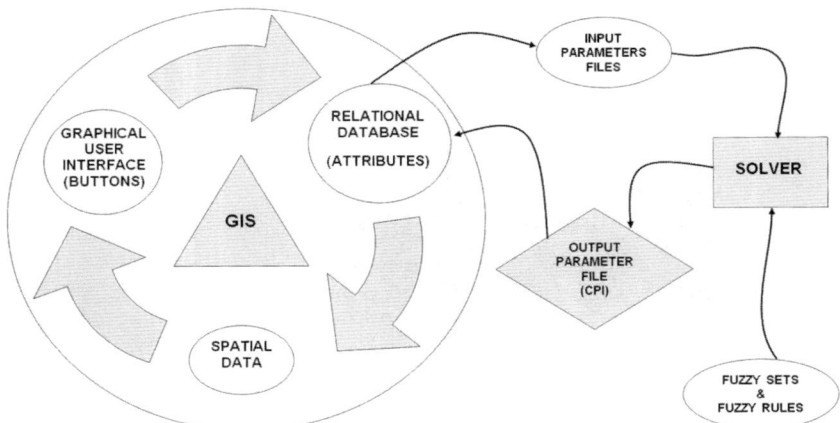

Fig. 14.3 Architecture of the integrated decision-making system for the assessment of cleaning performance

entries in the RDB of the thematic map of the pilot cleaning interventions. Thus, the input variables of the fuzzy model are (a) patina preservation index, PPI (%) (SEM-EDX results); (b) preservation index of gypsum layer, PIGy (%) (SEM-EDX results); (c) fracture density, FD (%) (DIP of SEM images results); (d) actual/projected area ratio, r (LP results); (e) roughness (μm) (measured parameter: Rq; LP results); and (f) total color difference, ΔE (colorimetry results). In the cases of the input parameters PPI and PIGy, input values are expressed in percentages in order to have comparable input data for the fuzzy model from different surfaces of black crusts. In particular, quantitative and qualitative observations deriving from SEM-EDX, regarding width, cohesion, continuity, and homogeneity of patina/gypsum layer, are expressed in percentage in comparison to the corresponding data before cleaning applications. Therefore, variances of the examined layers in width and cohesion, from different areas or buildings, can be classified in the designed fuzzy model. During the development of the fuzzy sets, the acceptance and nonacceptance threshold levels are configured for all input and output parameters, as shown in Table 14.1. The cleaning performance index (CPI) is defined as the output parameter, which classifies the cleaning interventions into nonaccepted, medium, accepted, and optimum/recommended ones. Additionally, 648 "IF-THEN" fuzzy rules were developed, using the logical operation "AND." The fuzzy rules relate successfully all the input conditions, giving specific and repeatable output results. An example of a fuzzy rule is shown: "IF the patina preservation index is high, AND the preservation index of gypsum layer is high, AND the fracture density is medium, AND the actual/projected area ratio is medium, AND the roughness is medium, AND the total color difference is medium, THEN cleaning performance index is high (that is recommended cleaning-Optimum)."

Table 14.1 Definition of fuzzy sets for all input and output parameters of the fuzzy model

Input parameters – output parameter	Definition of fuzzy sets			
Patina preservation index (%) (PPI)	Low (0–70%)		High (50–100%), accepted	
Preservation index of gypsum layer (%) (PGy)	Low (0–40%)	Medium (30–70%)	High (50–100%), accepted	–
Fracture density (%) (FD)	Low (0–10%)	Medium (8–25%), accepted	High (18–35%)	Extra high (30–60%)
Ratio of actual to projected area-surface area (r)	Low (1–1.5)	Medium (1.25–3), accepted	High (2.5–4)	–
Roughness, Rq (μm)	Low (0–10 μm)	Medium (5–20 μm), accepted	High (15–50 μm)	–
Total color difference (ΔE)	Low (0–6)	Medium (5–15), accepted	High (13–40)	–
Output parameter: CPI	Nonaccepted (0–4.5)	Medium (4–7.5)	Accepted (7–10)	Optimum – recommended (8–10)

The development of the fuzzy sets and the fuzzy rules (i.e., the knowledge base of the fuzzy logic model) was based on the knowledge and experience of human experts about stone decay and stone cleaning. Furthermore, the laboratory and field experimental results acquired by 30 different and real scenarios of cleaned black crusted marble surfaces, as well as the international bibliography of stone cleaning assessment, were taken into consideration. Therefore, an outcome was given (output parameter – CPI) to each possible combination of each input parameter set. The operation of the fuzzy logic model was based on the processes of fuzzification, inference, and defuzzification. It is worth mentioning though that data movement within the system is accomplished by two intermediate files. The first extracts data from the RDB, and the second loads and stores data into the RDB after the code execution (Fig. 14.3). In particular, the values of the critical cleaning assessment parameters (for every pilot cleaning) are extracted from the corresponding fields of the RDB of the cleaning interventions' thematic map to an intermediate file. This intermediate file feeds the solver (i.e., fuzzy model) with the critical cleaning assessment parameter values which are its input parameters. In succession, the executable code is retrieved along with the fuzzy sets and the fuzzy rules, whereas the results of the output variable (CPI) for every examined cleaning are stored into a new intermediate file. Data of this second intermediate file are retrieved by GIS, and the CPI values are stored in the RDB of the thematic map of the "planning of conservation interventions." All the abovementioned operations are provided to the user when the GIS graphical interface is employed along with the activation of two extra buttons specially created for the generation of the two intermediate files. In addition,

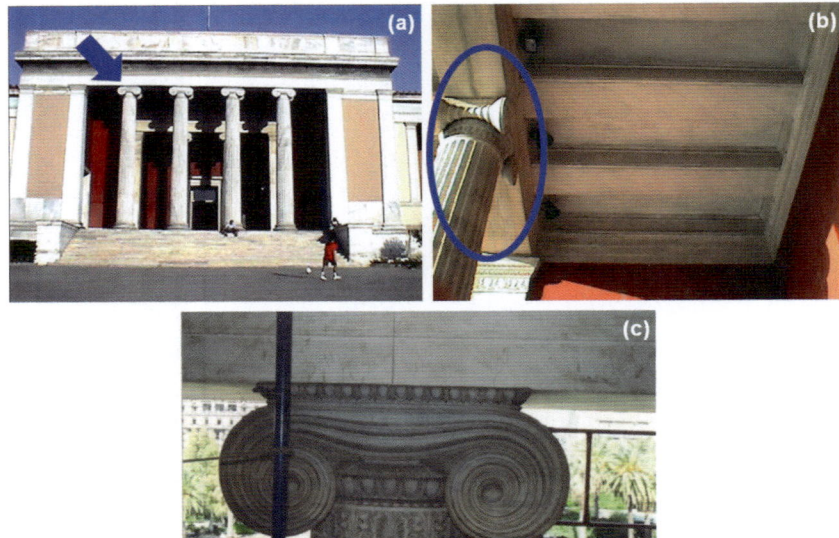

Fig. 14.4 The façade of the historic building of National Archaeological Museum (**a**); the representative investigation area of the capital (**b, c**)

cartographic attribution of CPI sets into the spatial entities of the thematic map of "planning of conservation interventions" can take place using the GIS capabilities. CPI is classified according to the threshold levels of the output parameter of the fuzzy model expressed as crisp sets.

14.3.4 Demonstration in Practice: The Case Study of the NAM Historic Building

Within the context of the above methodological approach, a representative surface of Pentelic marble at the NAM (National Archaeological Museum in Athens, Greece) was selected to act as pilot project for cleaning assessment. The capital under investigation is located on the first (from North) column of the propylon, has an eastern orientation, is totally protected from rainwash, and presented the characteristic decay pattern of black crust (Fig. 14.4). Figure 14.5 presents the investigation area after cleaning, while Table 14.2 lists the applied pilot cleaning interventions.

After SEM-EDX examination of the core samples from the anthemia relief surface, it was found that the black crust varied widely in regard to the width and the

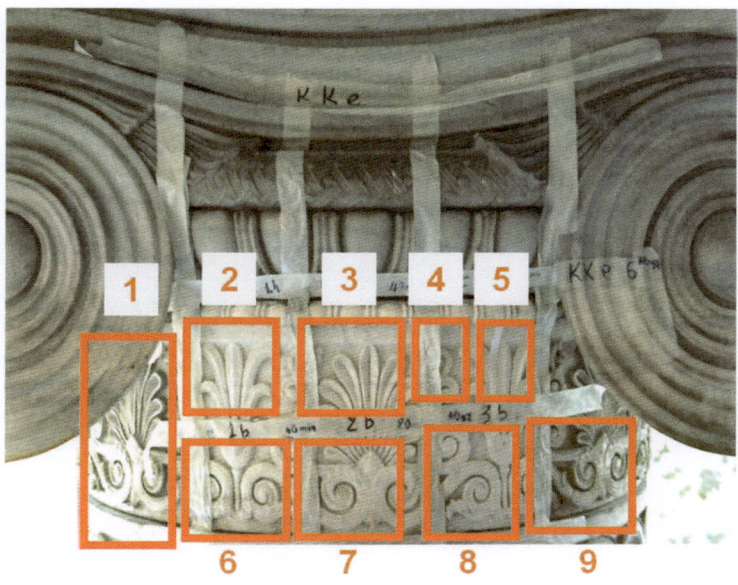

Fig. 14.5 The investigation area of the NAM, after the application of the pilot cleaning interventions

Table 14.2 Applied cleaning methods on the capital of NAM

Area code	Applied cleaning methods
1	Wet micro-blasting method, where spherical particles of $CaCO_3$ (diameter < 80 μm) were springing with a maximum function pressure of 0.5 bar; the proportion of water and spherical particles of $CaCO_3$ in the device's commixture barrel was 3:1
2	Poultice of ion-exchange resin with deionized water, applied for 60 min
3	Poultice of ion-exchange resin with 10% w/v $(NH_4)_2CO_3$ solution, applied for 40 min
4	Poultice of ion-exchange resin with 10% w/v $(NH_4)_2CO_3$ solution, applied for 10 min
5	Poultice of ion-exchange resin with deionized water, applied for 10 min
6	Poultice of ion-exchange resin with deionized water, applied for 30 min
7	Poultice of ion-exchange resin with 10% w/v $(NH_4)_2CO_3$ solution, applied for 20 min
8	Poultice AB57 (1 L deionized water, 30 g NH_4HCO_3, 50 g $NaHCO_3$, 25 g of disodium EDTA, 10 mL Desogen, 800 g sepiolite), double application for 5 and 15 min respectively
9	Poultice of ion-exchange resin with deionized water, applied for 20 min

cohesion of the gypsum layer as well as the presence and the location of barite within the crust. Representative SEM-EDX results of samples from the front-face of the anthemia relief are displayed in Fig. 14.6. Besides the differences regarding crust width, examination of any existing patina was also of high importance. In Fig. 14.6a, the white line between the macro- and microcrystalline gypsum is barite.

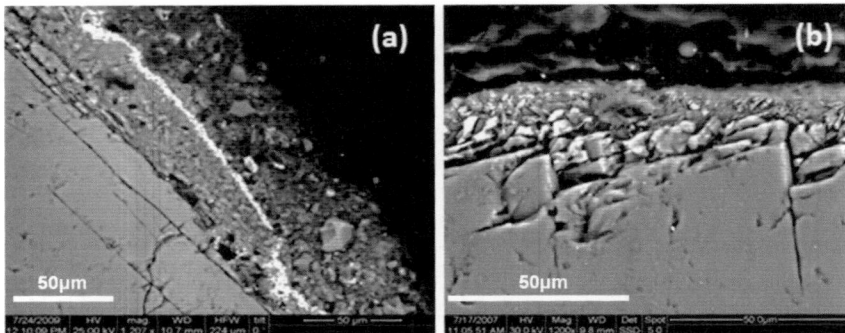

Fig. 14.6 Representative SEM images of the anthemia relief, before the pilot cleaning interventions; (**a**) front-face central part; (**b**) front-face right part

The thickness of this barite layer was of 10 μm in average. Barite is considered patina even though it is a nonuniform layer in relation to its width, continuity, cohesion, and location within the crust because it is the residue of lithopone pigment. Lithopone was used as a substrate for coloring decorations, and the presence of barite is of high historic significance since it is strong evidence that at the end of the nineteenth century the capitals of NAM's propylon were colored. In contrast, the examined decay sample from the front-face of the right part of the anthemia relief presented no barite (patina) as it is displayed in Fig. 14.6b. The other two cases of the front-face of the anthemia relief (central 2 and left part) presented a barite layer of 5 μm in average thickness on the surface of the gypsum layer or inside the gypsum layer.

Representative SEM results of the samples after pilot cleaning interventions are displayed in Fig. 14.7. In Tables 14.3 and 14.4, the classification of SEM-EDX results after cleaning is recorded, in order to rank the input values of both the parameters of PPI and PIGy, respectively, according to the characteristics of the decay reference samples. In particular, the decay samples of the central front-face areas 1 and 2 are the reference samples for the classification of the input values PPI and PIGy for the cleaning of areas 2, 3, 4, 5, 6, 7, and 8, whereas the left front-face sample is the reference decay sample for cleaning area 1 and the right front-face sample is the reference decay sample for cleaning area 9. The great variance of the black crust presented on the anthemia relief of the marble capital indicates the complexity of the cleaning assessment process in relation to the classification of actual input data, verifying the necessity of developing and using a decision support mechanism.

Table 14.5 presents the results of the used experimental techniques which also represent the values of the fuzzy system crisp inputs and the fuzzy model output CPI.

The pilot cleaning intervention which was recommended (CPI > 8) for the removal of the black crust was a poultice of ion-exchange resin with 10% w/v

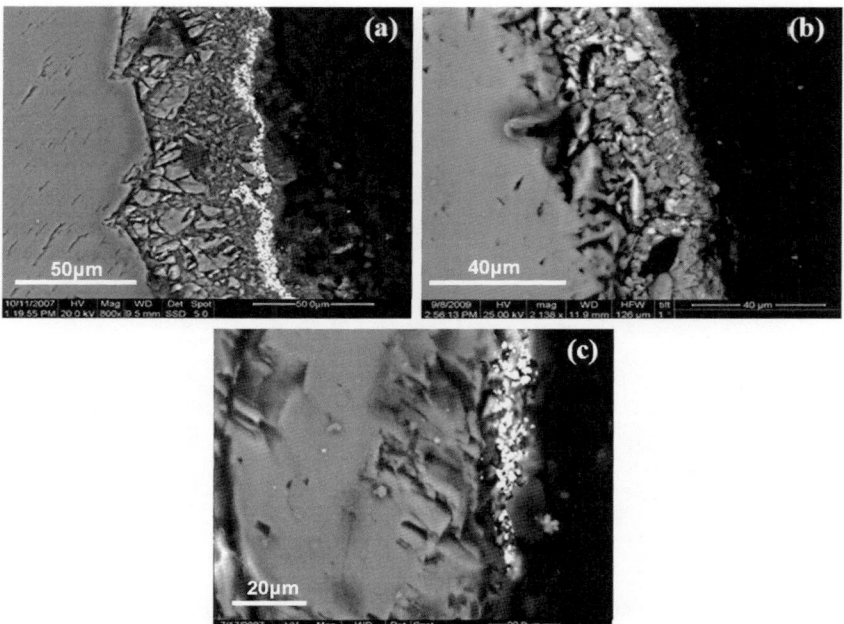

Fig. 14.7 Representative SEM images from the areas of the following pilot cleaning interventions: (**a**) ion-exchange resin with 10% w/v $(NH_4)_2CO_3$ solution, applied for 10 min; (**b**) ion-exchange resin poultice with deionized water, applied for 10 min; (**c**) wet micro-blasting method

$(NH_4)_2CO_3$ solution, applied for 10 min (code area 4). This cleaning intervention is classified as optimum, since it preserves the barite (patina) and the layer of microcrystalline gypsum (authentic material). In addition, the characteristics of surface microstructure (i.e., FD, r, Rq) belong to the corresponding medium sets, assuring that the accepted superficial material loss is in accordance with the medium reactivation of the sulfation process and medium adsorption of black particles that controls susceptibility to further decay. Furthermore, the aesthetic modifications that the surface underwent after cleaning were classified as accepted, since the value of total color difference belonged to the medium set. Application of poultice AB57, poultices of ion-exchange resin with deionized water for 60, 30, and 20 min (respective code areas 2, 6, 9), and poultices of ion-exchange resin with 10% w/v $(NH_4)_2CO_3$ solution, for 40 and 20 min (respective code areas 3, 7), were classified as not accepted cleanings (CPIs < 4.5). First, the removal of the desirable patina and microcrystalline gypsum has taken place. Secondly, even though in all of these cases of classified as not accepted, values of total color difference belonged to the medium set (except AB57 poultice where the recorded value is considered relatively high), the critical parameters of surface microstructure in most of these examined cases present relatively low values indicating unacceptable material loss.

Table 14.3 Recording of SEM-EDX results for the classification of the input values of patina preservation index (PPI)

Area code	Average width of barite layer (μm)	Barite location	PPI (%)
Central front-face 1	10	In between microcrystalline Gy and black depositions-Gy layer	–
Central front-face 2	5	On the surface of Gy layer	–
Left front-face	5	On the surface of Gy layer or inside of it	–
Right front-face	0	No barite	–
1	3	Barite areas on the top of Gy layer	65
2	0	No barite	1
3	0	No barite	1
4	5	On the surface of Gy layer or under the Gy layer	90
5	0	Barite grains on the top of Gy layer	25
6	0	No barite	1
7	0	No barite	1
8	0	No barite	1
9	0	No barite	1

Representative results for the front-face anthemia relief

Table 14.4 Recording of SEM-EDX results for the classification of the input values of preservation index of gypsum layer (PIGy)

Area code	Average total width of crust layer (μm)	Average width of Gy and black depositions layer (μm)	Average width of microcrystalline Gy layer (μm)	PIGy (%)
Central front-face 1	130	70	60	–
Central front-face 2	35	0	35	–
Left front-face	15	0	15	–
Right front-face	80	0	80	–
1	6	0	6	70
2	0	0	0	1
3	0	0	0	1
4	55	0	55	90
5	30	0	30	80
6	0	0	0	1
7	0	0	0	1
8	0	0	0	1
9	0	0	0	1

Representative results for the front-face anthemia relief

Table 14.5 Values of crisp inputs—results of the fuzzy model (CPI)

Area code	Values of crisp inputs—results of experimental techniques						Output
	PPI (%)	PIGy (%)	FD (%)	r	Rq (μm)	ΔE	CPI
1	65	70	36.6	2.593	30	6.48	6.65
2	1	1	26.4	1.581	11	6.57	3.36
3	1	1	20.3	1.896	15	6.13	3.36
4	90	90	16.1	1.914	17	7.21	8.14
5	25	80	23.1	1.542	11	9.56	6.62
6	1	1	11.3	1.562	9	4.75	3.36
7	1	1	27.9	2.345	22	4.91	3.36
8	1	1	11.2	1.740	10	15.3	3.36
9	1	1	36.5	1.549	10	8.74	3.36

Representative results for the front-face anthemia relief

Furthermore, in the case of the ion-exchange resin poultice with 10% w/v $(NH_4)_2CO_3$ solution applied for 20 min (code area 7), the surface's microstructural parameters (FD, r, Rq) exhibited high values, indicating high susceptibility to further decay of the exposed calcite surface.

The pilot cleaning intervention of the ion-exchange resin poultice with deionized water applied for 10 min (code area 5) was classified to have a medium performance (4.5 < CPI < 7). Here, although the layer of microcrystalline gypsum was preserved as well as the microstructure parameters and the total color difference showed relatively accepted values, the barite patina was not preserved at the required high percentages. It may be argued that this cleaning should be categorized as not accepted since the PPI is so low. However, the great variance that the barite layer shows in regard to width, location, and even its very existence must be taken into account. Finally, the pilot cleaning intervention of the wet micro-blasting method was also classified of medium performance (4.5 < CPI < 7). Despite the fact that patina and microcrystalline gypsum were preserved at relatively high percentages and the total color difference presented an acceptable medium value, the critical parameters of surface microstructure (FD, r, Rq) held high values. Therefore, this cleaned surface displays low durability, meaning that the reactivation of the sulfation process and the adsorption of black particles are expected to be high.

GIS thematic maps of decay and pilot cleaning interventions of the NAM capital were built and are presented in Figs. 14.8 and 14.9, respectively. Therefore, spatial classification of decay and pilot cleaning on the investigated marble surface was accomplished. After application of geo-processing procedure in both decay and pilot cleaning interventions' thematic maps, the resulting output theme is the conservation planning thematic map presented in Fig. 14.10. This output thematic map includes features of combined spatial info with attribute data from both input and overlay themes, that is decay and pilot cleaning interventions' themes, respectively. Data movement across the integrated decision-making system is accomplished by extracting the values of the critical cleaning assessment parameters from the RDB

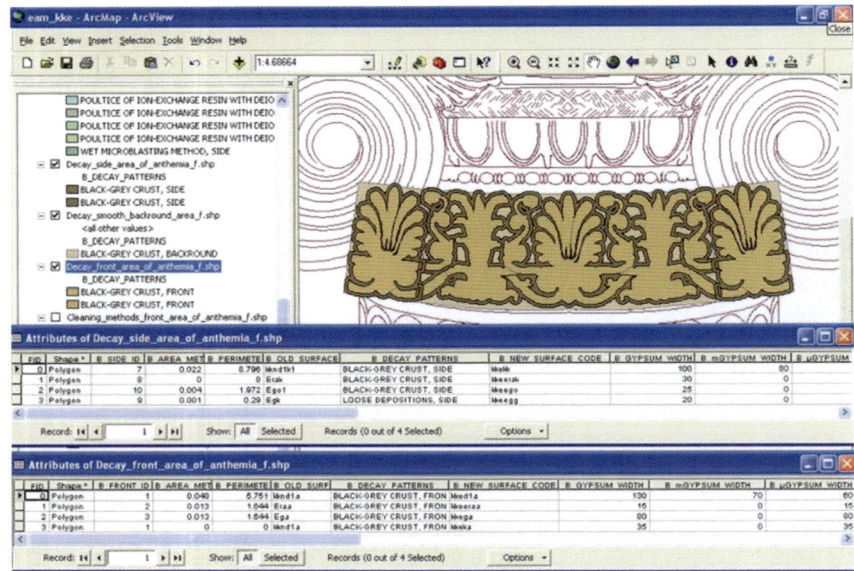

Fig. 14.8 GIS thematic map of decay for the capital surface with the corresponding RDBs

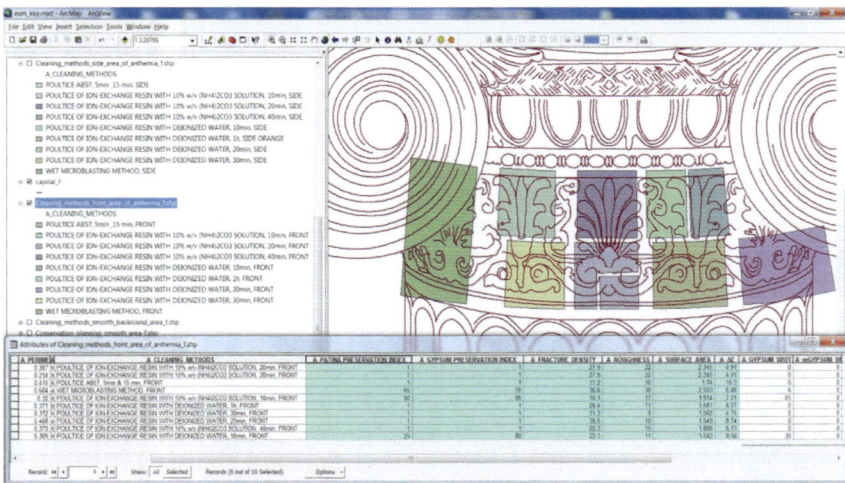

Fig. 14.9 GIS thematic map of pilot cleaning interventions for the capital surface with the corresponding RDB

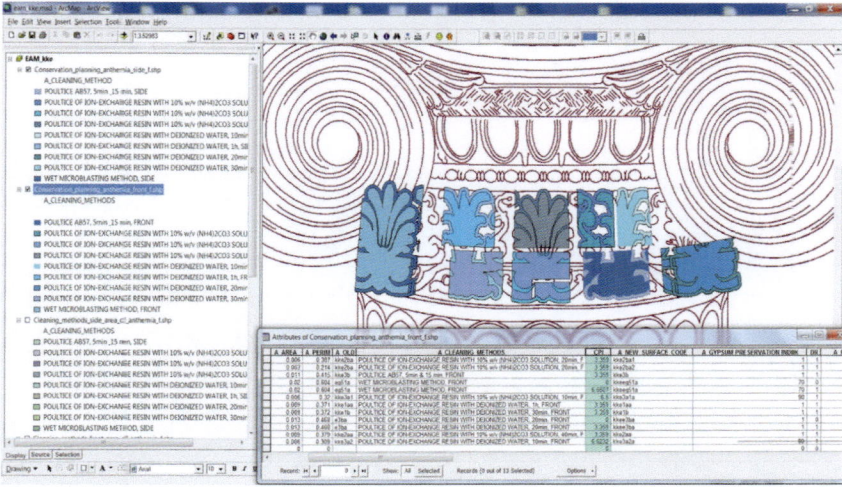

Fig. 14.10 GIS thematic map of "planning of conservation interventions" for the capital surface with the corresponding RDB

Table 14.6 Classification of CPI

Color	Set	Description
Red	0 < CPI < 4.5	Nonaccepted cleaning
Orange	4.5 < CPI < 7	Medium cleaning
Yellow	7 < CPI < 8	Accepted cleaning
Green	8 < CPI < 10	Optimum-recommended cleaning

of the pilot cleaning thematic map and by loading and storing the values of CPI into the RDB of the conservation planning thematic map after code execution.

Furthermore, CPI was classified according to the threshold levels of the output parameter of the fuzzy logic model expressed as crisp sets and in different colors as displayed in Table 14.6 and in Fig. 14.11.

Finally, the suggested methodology is presented in the flow chart of Fig. 14.12, summarizing the process used in this work.

14.4   Conclusion

An assessment methodology for stone cleaning is presented, while cleaning assessment criteria are established along with their critical parameters and the experimental techniques that can measure these criteria. The suggested methodology for the assessment of stone cleaning encompasses both endogenous and exogenous

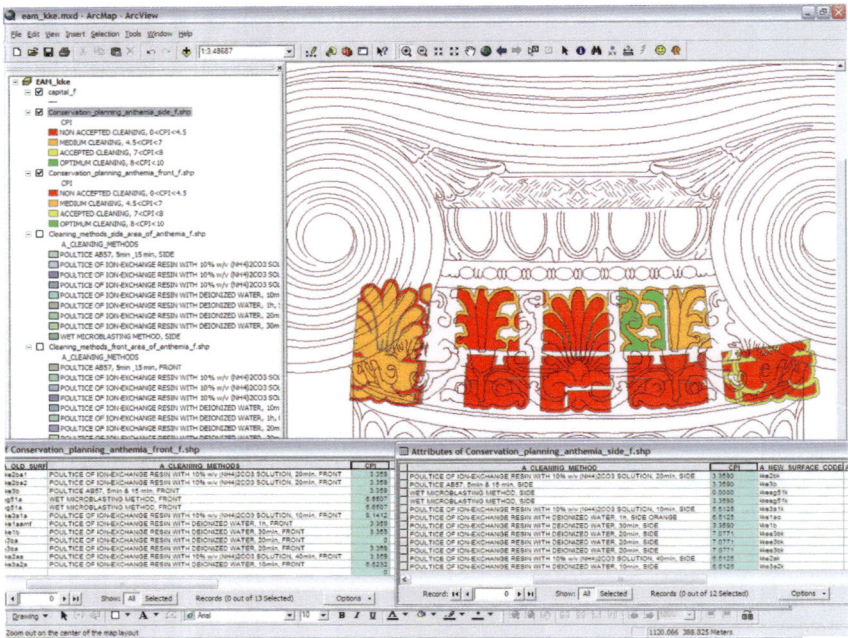

Fig. 14.11 Cartographic attribution of CPI sets into the spatial entities of the GIS thematic map of "planning of conservation interventions"

parameters that govern stone durability after cleaning, since it interrelates physical/ chemical characteristics of stone with spatial data relating to the monument and environment, taking into account their variation over time. This is accomplished by the development of an integrated decision-making system where a fuzzy logic model is incorporated into GIS thematic maps which depict decay patterns and applied pilot cleaning interventions. Cleaning performance of the stone surface under investigation can be constantly acquired under predefined assessment parameters and acceptance threshold levels. The suggested decision-making system has been demonstrated successfully in practice on 30 different field scenarios of cleaned sulfated marble surfaces, nine of which are presented in this work. It is concluded that the decision-making system for stone cleaning can also be used for other case studies for marble materials and compact limestones which present the decay pattern of black crusts, under the prerequisite that the input parameters will be measured by the specifications given in this work.

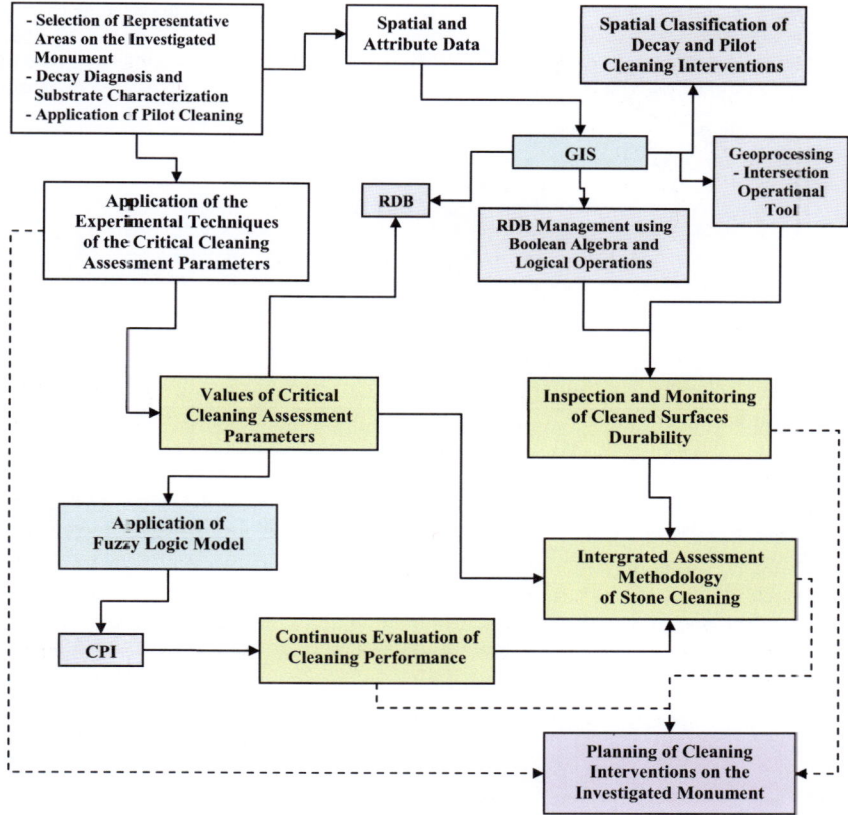

Fig. 14.12 Flowchart of the advanced methodology on the assessment of stone cleaning

References

1. Biscontin G, Zendri E, Bakolas A, Longega G, Driussi G. Moropoulou, A. Alcune considerazioni sullapulitura delle superfici architettoniche. Proc. of Symb. Scienza e Beni Culturali XI on Pulitura, Bressanone, Italy, 1995, pp. 625–631.
2. Fassina V. General criteria for the cleaning of stone: theoretical aspects andmethodology of application. In: Zezza F, editor. Stone Material in Monuments: Diagnosis and Conservation; Scuola Universitaria C.U.M. Conservazione dei Monumenti; Heraklion, Crete. Bari: Mario Adda Editore; 1993. p. 126–32.
3. Verges-Belmin V. Towards a definition of common evaluation criteria for the cleaning of porous building materials: a review. Sci Techn Cult Herit. 1996;5:69–83.
4. Moropoulou A, Delegou ET, Avdelidis NP, Koui M. Assessment of cleaning conservation interventions on architectural surfaces using an integrated methodology. In: Vandiver P, Goodway M, Druzik JR, Mass JL, editors. Materials issues in art and archaeology VI, vol. 712. Boston: Publishing Materials Research Society; 2002. p. 69–76.

5. Revez MJ, Rodrigues JD. Incompatibility risk assessment procedure for the cleaning of built heritage. J Cult Herit. 2016;18:219–28.
6. Senesi GS, Carrara I, Nicolodelli G, DMBP M, De Pascale O. Laser cleaning and laser-induced breakdown spectroscopy applied in removing and characterizing black crusts from limestones of Castello Svevo, Bari, Italy: a case study. Microchem J. 2016;124:296–305.
7. Pozo-Antonio JS, Rivas T, López AJ, Fiorucci MP, Ramil A. Effectiveness of granite cleaning procedures in cultural heritage: a review. Sci Total Environ. 2016;571:1017–28.
8. Gioventù E, Lorenzi PF, Villa F, Sorlini C, Rizzi M, Cagnini A, Griffo A, Cappitelli F. Comparing the bioremoval of black crusts on colored artistic lithotypes of the Cathedral of Florence with chemical and laser treatment. Int Biodeterior Biodegrad. 2011;65:832–9.
9. Iglesias-Campos MÁ, Prada Pérez JL, Fortes SG. Spot analysis to determine technical parameters of microblasting cleaning for building materials maintenance. Constr Build Mater. 2017;132:21–32.
10. Appolonia L, Bertone A, Brunetto A, Vaudan D. The St Orso priory the comparison and testing of cleaning methods. J Cult Herit. 2000;(14):105–10.
11. Gaspar P, Hubbard C, McPhail D, Cummings AA. Topographical assessment and comparison of conservation cleaning treatments. J Cult Herit. 2003;4:294–302.
12. Pouli P, Zafiropoulos V, Balas C, Doganis Y, Galanos A. Laser cleaning of inorganic encrustation on excavated objects: evaluation of cleaning result by means of multi-spectral imaging. J Cult Herit. 2003;4:338–42.
13. Pozo-Antonio JS, Ramil A, Rivas T, López AJ, Fiorucci MP. Effectiveness of chemical, mechanical and laser cleaning methods of sulphated black crusts developed on granite. Constr Build Mater. 2016;112:682–90.
14. Guidetti V, Uminski M. Ion exchange resins for historic marble desulphatation and restoration. In: 9th International Congress on Deterioration and Conservation of Stone, Venice, 2000, pp. 327–333.
15. Verges-Belmin V, Dignard C. Laser yellowing: myth or reality. J Cult Herit. 2003;4:238–44.
16. Siedel H, Hubrich K, Kusch HG, Wiedemann G, Neumeister K, Sobott R. Results of Laser cleaning on encrusted oolithic limestone of angel sculptures from the Cologne cathedral. In: 9th International Congress on Deterioration and Conservation of Stone, Venice, 2000, pp. 583–590.
17. Mossotti VG, Eldeeb AR, Fries TL, Coombs MJ, Naude VN, Soderberg L, Wheeler GS. The effect of selected cleaning techniques on Berkshire Lee marble; A scientific study at Philadelphia City Hall, U.S. Geological Survey, Prof. Paper 1635, Virginia 2002. http://geo-pubs.wr.usgs.gov/prof-paper/pp1635/. Last accessed date: October 2017.
18. Delegou ET, Krokida M, Avdelidis NP, Moropoulou A. Assessment of cleaning interventions on marble surfaces using pulsed thermography. In: Paipetis AS, Matikas TE, Aggelis DG, Van Hemelrijck D, editors. Emerging technologies in non-destructive testing, vol. 2012. London: CRC Press. p. 31–6. ISBN: 978-0-415-62131-1.
19. Delegou ET, Doulamis A, Moropoulou A. Decision making system on the assessment of cleaning interventions using combined fuzzy C-means and neural networks. Herit. Protect. From Document. to Interven. Proceeding of EU-CHIC International Conference on Cultural Heritage, Split, Croatia, May 29–June 1, 2012, pp. 191–4.
20. Stout KJ, Blunt LA. Application of 3-D topography to bio-engineering. J Int Mach Tools Manuf. 1995;35:219–29.
21. Amoroso G, Fassina V. Stone decay and conservation. New York: Elsevier; 1983.
22. Rodrigues JD. Defining, mapping and assessing deterioration patterns in stone conservation projects. J Cult Herit. 2015;16:267–75.
23. Larson JH, Madden C, Surtherland I. Ince Blundell: preservation of an important collection of classical sculpture. J Cult Herit. 2000;1:79–87.
24. Verges-Belmin V. Pseudomorphism of gypsum after calcite, a new textural feature accounting for the marble sulfation mechanism. Atm Environ. 1994;28(2):295–304.

25. Skoulikidis T, Papakonstantinou-Ziotis P. Mechanism of sulfation by atmospheric SO_2 of lime-stones and marbles of the ancient monuments and statues: I. Observations in situ (Acropolis) and measurements in the laboratory. Brit Cor J. 1981;16:63–9.

26. Skoulikidis T, Charalambous D. Mechanism of sulfation by atmospheric SO_2 of limestones and marbles of the ancient monuments and statues, II. Hypothesis concerning the rate determining steps in the process of sulphation, and its experimental confirmation. Brit Cor J. 1981;16 70–7.

27. Skoulikidis T, Papakonstantinou E, Galanos A, Doganis Y. Conservation of the west frieze. Study for the restoration of the Parthenon. 1995;3c:3–15.

28. Maravelaki-Kalaiztaki P. Black crusts and patinas on Pentelic marble from the Parthenon and Erechtheum (acropolis, Athens): characterization and origin. Anal Chim Acta. 2005;532:187–98.

29. Fassina V. New findings on past treatments carried out on stone and marble monuments surfaces. Sci Total Environ. 1995;167:185–203.

30. Young M, Uquhart D. Abrasive cleaning of sandstone buildings and monuments: an experimental investigation. In: Stone cleaning and the nature, soiling and decay mechanisms of stone: proceedings of the International conference held in Edinburgh, UK, 14–16 April 1992, pp.126–138.

31. Avdelidis NP, Delegou ET, Almond DP, Moropoulou A. Surface roughness evaluation of marble by 3D laser profilometry and pulsed thermography. NDT & E Inter. 2004;37(7):571–5.

32. Lee JM, Steen WM. In-process surface monitoring for laser cleaning processes using a chromatic modulation technique. Int J Adv Manuf Technol. 2001;17:281–7.

33. Kapsalas P Maravelaki-Kalaitzaki P, Zervakis M, Delegou ET, Moropoulou A. Optical inspection for quantification of decay on stone surfaces. NDT & E Inter. 2007;40:2–11.

34. Moropoulou A, Delegou ET, Vlahakis V, Karaviti E. Digital processing of SEM images for the assessment of evaluation indexes of cleaning interventions on Pentelic marble surfaces. Mater Charact. 2007;58(11):1063–9.

35. Andrew C.Towards an aesthetic theory of building soiling. In: Stone Cleaning and the Nature, Soiling and Decay Mechanisms of Stone. Proceedings of the International Conference. 14–16 April 1992, London: Donhead, Edinburgh, UK, pp. 63–81.

36. Brimblecombe P, Grossi CM. Aesthetic thresholds and blackening of stone buildings. SciTotal Environ. 2005;349(1):175–89.

37. Fort R, Mingarro F, Lopez de Azcona MC, Rodriguez Blanco J. Chromatic parameters as performance indicators for stone cleaning techniques. Color Res Applic. 1999;25:442–6.

38. Grossi CM, Esbert RM, Diaz Pache F, Alonso FJ. Soiling in building stones in urban environments. BuildEnviron. 2003;38:147–59.

39. EN 15886: Conservation of cultural property—Test methods—Color measurement of surfaces, 2010.

40. ASTM D2244: Standard practice for calculation of color tolerances and color differences from instrumentally measured color coordinates, 1993.

41. Klein S, Fekrsanati F, Hildenhagen J, Dickmann K, Uphoff H, Marakis Y, Zafiropoulos V. Discolouration of marble during laser cleaning by Nd:YAG laser wavelengths. Appl Surf Sci. 2001;171:242–51.

42. Delegou ET, Avdelidis NP, Karaviti E, Moropoulou A. NDT&E techniques and SEM-EDS for the assessment of cleaning intervention on pentelic marble surfaces. J X-Ray Spectrom. 2008;37:435–43.

43. Kouzeli K. Black crust removal methods in use. Their effects on pentelic marble surfaces. 7th Int Congr Deterior Conserv Stone Lisbon. 1992;3:1147–55.

44. Moropoulou A, Kefalonitou S. Efficiency and counter effects of cleaning treatment on limestone surfaces—investigation of Corfu venetian fortress. Build Environ. 2002;37:1181–91

45. Inkpen RJ, Fontana D, Collier P. Mapping decay: integrating scales of weathering within GIS. Earth Surf Proc Landf. 2001;26:885–900.

46. Salonia P, buildings NAH, decay t. Data recording, analyzing and transferring in an ITC environment. Int Arc of the Photogram Rem Sens Spatial Inf Sc. 2002;XXXIV(5/W12):302–6.

47. Inkpen R, Duane B, Burdett J, Yates T. Assessing stone degradation using an integrated database and geographical information system (GIS). Environ Geol. 2008;56:789–801.
48. Delegou ET, Tsilimantou E, Oikonomopoulou E, Sayas J, Ioannidis C, Moropoulou A. Mapping of building materials and conservation interventions using GIS: the case of Sarantapicho acropolis and Erimokastro acropolis in Rhodes. Inter J Herit Digit Era. 2013;2(4):631–53.
49. Mamdani EH, Assilian S. An experiment in linguistic synthesis with a fuzzy logic controller. Int J Man-Mach Stud. 1975;7(1):1–13.
50. Vakalis D, Sarimveis H, Kiranoudis CT, Alexandridis A, Bafas GV. A GIS based operational system for wild-land fire crisis management I. Mathematical modeling and simulation. Appl Math Model. 2004;28:389–410.
51. Klir JG, Yuan B. Fuzzy sets and fuzzy logic: theory and applications. Englewood Cliffs: Prentice Hall; 1995.
52. Kazaras K, Konstandinidou M, Nivolianitou Z. Enhancing road tunnel risk assessment with a fuzzy system based on the CREAM methodology. Chem Engin Trans. 2013;31:349–54.
53. Keramitsoglou I, Cartalis C, Kiranoudis CT. Automatic identification of oil spills on satellite images. Environ Model Softw. 2006;21(5):640–52.
54. Konstandinidou M, Nivolianitou Z, Kiranoudis C, Markatos NC. A fuzzy modeling application of CREAM methodology for human reliability analysis. Reliab Engin System Safet. 2006;91(6):706–16.
55. Zadeh LA. Outline of a new approach to the analysis of complex systems and decision processes. IEEE Trans Syst Man Cybern. 1973;3:28–44.
56. Driankov D, Hellendoorn H, Reinfrank M. An introduction to Fuzzy Control. Berlin: Springer; 1993.
57. Mamdani EH. Application of fuzzy algorithms for simple dynamic plants. Proc Electr Engin IEE. 1974;121(12):1585–8.

Index

A

Acrylates, 151
Acrylic resins, 30, 169
Adam statue, 28
Adequate water repellency, 93
Ag NPs, 241, 242
Alberese stone, 154, 155
Alcoholic colloidal solutions, 132
Alkoxysilanes, 164, 169
  advantages, 187
  carbonate stones, 190
  consolidants, 191
  cracking tendency, 200
  DBTL, 189
  hydrophobicity, 188
  limestones, 191
  MTMOS, 189
  sol-gel reactions, 188, 189
  stone consolidants, 187, 188
  TEOS, 187, 188
Alpha®SI30, 8
Amine (3-aminopropyl)triethoxysilane
  (ATS), 167
Amorphous calcium carbonate (ACC), 159
Amylamine, 80, 92
Anodization, 127
Antifungal protection, 126, 137–142
Antimicrobial nanoparticles, 280–288, 293
  Ag
    A. niger and *S. parvullus*, 285
    biocidal effect, 285
    biocidal treatment, 286, 287
    bioreceptivity, 288
    Cladosporium sp., 285
    COMAS, 288
    EPS, 287
    GMAg-Si, 288
    phototrophs, 288
    ROS, 280
  Cu, 280, 289–291
    gram-positive and negative bacteria, 279
  TiO$_2$
    Ag NPs, 284
    A. niger, 281
    biocidal effect, 281, 293
    biocidal treatment, 282, 283
    fungal colonization, 293
    nanocomposite, 284
    phototrophs, 281
    ROS, 280
    toxicity, 279
  Zn, 291, 292
  ZnO, 280
Artificial aging tests, 187
Atomic layer deposition (ALD), 127
Avrami–Erofeev model, 159, 160

B

B-72/Epotek composite system, 34
Bio-calcarenite, 161
Biocalcareous, 154, 164
Biocidal effect, 237, 238, 270, 271
Biocoatings
  hydrophobicity/water repellency, 226
  lichen, 247
  moisture, 227
  photochemical reactions, 226
  testings, 227
  toxicity, 226

© Springer International Publishing AG 2018
M. Hosseini, I. Karapanagiotis (eds.), *Advanced Materials for the Conservation of Stone*, https://doi.org/10.1007/978-3-319-72260-3

Printed in Great Britain
by Amazon